Ocean Renewable Energy Systems (ORES); Wave–Structure Interaction Analysis and Design Methods

Ocean Renewable Energy Systems (ORES); Wave–Structure Interaction Analysis and Design Methods

Guest Editors

Constantine Michailides
Wei Shi

MDPI

Basel • Beijing • Wuhan • Barcelona • Belgrade • Novi Sad • Cluj • Manchester

Guest Editors
Constantine Michailides
Department of Civil Engineering
International Hellenic University (IHU)
Thessaloniki
Greece

Wei Shi
State Key Laboratory of Coastal and Offshore Engineering
Dalian University of Technology
Dalian
China

Editorial Office
MDPI AG
Grosspeteranlage 5
4052 Basel, Switzerland

This is a reprint of the Special Issue, published open access by the journal *Journal of Marine Science and Engineering* (ISSN 2077-1312), freely accessible at: https://www.mdpi.com/journal/jmse/special_issues/Ocean_Renewable_Energy_Systems.

For citation purposes, cite each article independently as indicated on the article page online and as indicated below:

Lastname, A.A.; Lastname, B.B. Article Title. *Journal Name* **Year**, *Volume Number*, Page Range.

ISBN 978-3-7258-6516-1 (Hbk)
ISBN 978-3-7258-6517-8 (PDF)
https://doi.org/10.3390/books978-3-7258-6517-8

Contents

About the Editors

Constantine Michailides

Constantine Michailides is Associate Professor of Marine Works at Dept. of Civil Engineering, International Hellenic University. His research interests focus on numerical analysis, experimental testing and structural/field monitoring of marine structures, wave–structure interaction, offshore wind turbines (fixed bottom and floating) and wave energy converters. He has been involved in more than 20 national and E.U. research projects such as 'FORCYS', 'HydrOTEC', 'ARCWIND', and 'Marina Platform'. He has published one book and over 120 refereed journal and conference papers. He was referee for more than 30 scientific journals and conferences, and he is on the Editorial Board of six international journals. He is the corresponding author of the 'Chapter 6: Pressures: Marine Infrastructure' of the development of the third World Ocean Assessment (WOA III) for the United Nations. Constantine listed among the World's Top 2% Scientists for the hears 2022, 2023 and 2024 (Stanford/Elsevier, 2024).

Wei Shi

Wei Shi is a professor of Deepwater Engineering Research Centre at Dalian University of Technology (DUT). He received PhD from POSTECH in Korea and worked as an assistant professor at Konkuk University and a postdoc at NTNU in Norway. His research interests include numerical modeling of offshore renewable energy systems, wind turbine drivetrain dynamics, and ice–structure interaction for offshore wind turbines. Meanwhile, he has good experience in dynamic analysis of mechanical systems.

His main research interests include the dynamics of offshore wind turbines under complex metocean conditions. He is the PI of many projects including NSFC, Key projects from MOST, and other national and provincial projects. He published more than 70 SCI-index papers among of them 50 are first-author or corresponding authors. He has 28 Chinese patents and 8 US patents. Meanwhile, he achieved several first prizes and second prizes from several different academic societies.

Preface

A large number of OWT farms are already in operation while WECs arrays are in the pre-commercial phase. At the same time, possible combined ORES or co-located farms and arrays, combining the two technologies are being researched. Successful steps forward, towards safety, sustainability and resilience, are required from both technologies for different reasons but with the same target.

ORES are operating in a very challenging environment where dynamic excitation loads from different sources and with different characteristics need to be considered. The development of numerical analysis models and design methods of ORES as single systems or in OWTs farm and WECs array configurations is a very demanding task. New methods for addressing the wave–structure interaction effects are continuously being developed. Design methods covering the whole life-cycle range of ORES (conception, design, installation, operation, maintenance, dismantling) are continuously redeveloped and reassessed. Novel techniques (e.g. artificial intelligence and smart sensors) are being tested and integrated with various numerical models and design methods.

Constantine Michailides and Wei Shi
Guest Editors

Journal of
Marine Science and Engineering

MDPI

Article

Analysis of Dynamic Characteristics of an Ultra-Large Semi-Submersible Floating Wind Turbine

Zhixin Zhao [1,2], Xin Li [1,2,*], Wenhua Wang [1,2] and Wei Shi [1,3]

1 State Key Laboratory of Coastal and Offshore Engineering, Dalian university of Technology, Dalian 116024, Liaoning, China; zhixinzhao@mail.dlut.edu.cn (Z.Z.); wang19891018cool@163.com (W.W.); weishi@dlut.edu.cn (W.S.)
2 Institute of Earthquake Engineering, Faculty of Infrastructure Engineering, Dalian University of Technology, Dalian 116024, Liaoning, China
3 Deepwater Engineering Research Center, Offshore Renewable Energy Research Center, Dalian University of Technology, Dalian 116024, Liaoning, China
* Correspondence: lixin@dlut.edu.cn; Tel.: +86-411-84707784

Received: 10 April 2019; Accepted: 28 May 2019; Published: 1 June 2019

Abstract: An initial design of the platform for the moderate water depth (100 m) is performed by upscaling of an existing 5 MW braceless semi-submersible platform design to support the DTU (Danish University of Science and Technology) 10 MW wind turbine. To investigate the dynamic characteristics of the ultra-large semi-submersible floating offshore wind turbine (FOWT), an aero-hydro-servo-elastic numerical modeling is applied to carry out the fully coupled time-domain simulation analysis. The motion responses of the ultra-large semi-submersible FOWT are presented and discussed for selected environmental conditions. Based on the quasi-static and dynamic analysis methods, the influence of the dynamic effects of the mooring lines on the platform motion responses and mooring line tension responses are discussed. Subsequently, the difference in the motion responses and structural dynamics of the DTU 10 MW and NREL (National Renewable Energy Laboratory) 5 MW FOWT is studied due to the difference in turbine properties. The simulation results reveal that the excitation of the low-frequency wind loads on the surge and pitch motions, the tower-base fore-aft bending moments and the mooring line tension response becomes more prominent when the size of the wind turbine increases, but the excitation action of the 3P effect on the structural dynamics of the 5 MW FOWT is more obvious than those of the 10 MW FOWT.

Keywords: semi-submersible floating offshore wind turbine; fully coupled numerical model; dynamic characteristic; power spectra; dynamic effects of mooring lines

1. Introduction

Over the past few decades, offshore wind energy has emerged as an attractive alternative to conventional power generation and established itself as a major source of environmentally-friendly, inexhaustible, economical, and importantly renewable energy [1]. China's offshore wind energy resources can reach 750 GW [2]. The development of the offshore wind energy is characterized by more consistent wind, less space constraints and visual impacts on the residents, which has become the main area of the development of China's wind energy industry.

In shallow waters (<50 m), bottom-fixed substructures, such as gravity-based structures, mono-piles, multi-piles, high-piles, jackets, and buckets are usually applied to support the offshore wind turbines. However, many promising sites are at deeper water depths (>50 m), where bottom-fixed substructures can no longer be used [3]. A floating structure can be used as an alternative from an economic perspective [4], such as spar-buoys [5], semi-submersible [6], barges [7], and tension leg platforms (TLP) [8], which have the advantages of reduced noise during installation, no required scour protection, and cheap anchor installation for the mooring system. However, the floating offshore wind turbine (FOWT) is a rigid and flexible hybrid multi-body system with the complex interaction (aerodynamic-hydrodynamic coupling) of wind turbine (including blades, hub, and nacelle)-tower-support platform-mooring system, which makes the analysis of the dynamic characteristics of the FOWT system extremely complex and challenging [9].

At present, considerable research on the dynamic characteristics of 5 MW FOWT has been conducted by domestic scholars. Jonkman et al. [10] developed a HydroDyn module coupled with the original aero-servo-elastic FAST code for land-based wind turbines and analyzed various TLP and barge designs for the NREL (National Renewable Energy Laboratory) 5 MW wind turbines. A TLP support platform for the 5 MW wind turbine was the focus of a study by Matha [11], whereby its performance was compared to that of other floating platform concepts in load computations, and its major instabilities were identified. Li et al. [12] investigated the wind-wave-induced global motion responses and mooring tension responses of the 5 MW OC4-DeepCwind semi-submersible FOWT, and a quadratic damping model was developed to gain a better consideration of the viscous performance. Ma et al. [13] analyzed the motion responses and structural dynamics of the 5 MW OC3-Hywind Spar-type FOWT in different sea states and investigated the coupling effects between various motion modes of the FOWT. Furthermore, the coupled wind-wave-induced analyses of the 5 MW OC3-Hywind FOWT of different hydrodynamic models based on the Morison formula, Pressure integration method, and Panel method was conducted by Madjid [14], then, the effect of damping and inertia forces of the mooring lines on the mooring line tension responses was discussed. Guo et al. [15] also analyzed the influence of the dynamic behaviors (including the structural inertial effect and fluid drag) of the mooring lines on the integrated motion responses of the 5 MW Spar-type FOWT based on the improved three-dimensional dynamic catenary theory.

The recent trend in the offshore wind industry is going to larger wind turbines, farther offshore and deeper water due to lack of offshore wind resources in shallow water area [15]. An ultra-large wind turbine can absorb more wind power, potentially reduce the installation and construction costs, take advantage of economies of scale, and thus capitalize on the stronger and more consistent wind resources farther offshore [16]. The DTU (Danish University of Science and Technology) 10 MW wind turbine [17] jointly proposed by the DTU and Vestas Wind Technologies corporation has become an extensively accepted reference wind turbine. Owing to the ultra-large wind turbine requirements of larger support structures, Xue [18], Tian [19], and Islam [20] formulated a preliminary design of a floater to support the 10 MW wind turbine by upscaling the existing 5 MW OC3-Hywind, MIT-TLP, and WindFloat concepts. This design satisfied the basic requirements of hydrostatic stability and dynamic properties of the 10 MW FOWTs. Xu et al. [21] also considered the DTU 10 MW wind turbine as the research object and used FAST to analyze the dynamic characteristics of the OC3-Hywind Spar-type FOWT under the joint action of the wind-wave-induced loads. However, the ultra-large FOWT was faced with more complicated marine environmental conditions given the increased turbine size. The aeroelastic effects of the giant flexible blades, the large and strong nonlinear motions of the support platform, the strong nonlinear coupling effects between the mooring system and support platform are still to be extensively and thoroughly investigated.

In this study, the DTU 10 MW wind turbine is selected as the research object, and the columns of the braceless semi-submersible wind turbines are connected by pontoons rather than braces, which can reduce design complexity and the construction cost of additional heave plate and avoid the stress concentration effect at the joint [22]. Therefore, an initial design is performed by upscaling of an existing 5 MW braceless semi-submersible platform design proposed by Luan et al. [23] to support the DTU 10 MW wind turbines in the moderate water depths (100 m). Aero-hydro-servo-elastic numerical modeling is used to conduct fully coupled time-domain simulation analyses to investigate the dynamic characteristics of the ultra-large semi-submersible FOWT for the selected environmental conditions. Furthermore, the impact of additional dynamic behaviors of the mooring line on the integrated motion responses of the FOWT is studied based on the mooring dynamic analysis modeling, including the hydrodynamic damping and inertia forces of the mooring lines, and compared with the quasi-static modeling. Additionally, the difference in the motion responses and structural dynamics of the NREL 5 MW and DTU 10 MW FOWTs is studied due to the complexity of the dynamic characteristics and coupling mechanism of the ultra-large FOWTs, which can provide a reference for the design and engineering practices of the ultra-large FOWTs.

2. Theoretical Background

2.1. Coupled Motion Equation of the FOWT

Based on the theory of aerodynamics, hydrodynamics and multi-body dynamics, the coupled motion equation of the FOWTs under the joint action of the wind-wave induced loads can be expressed by [24]

$$(M + A_\infty)\ddot{x}(t) + \int_{-\infty}^{\infty} h(t-\tau)\dot{x}(\tau)d\tau + (K_m(x) + K_h)x(t) = F_{exc}(x, \dot{x}, t), \quad (1)$$

where M is the mass matrix of the floating system, A_∞ is the added mass matrix at infinite frequencies, K_h is the hydrostatic restoring stiffness matrix, K_m is the nonlinear restoring stiffness matrix from the mooring system, $\ddot{x}$, $\dot{x}$, and x, are the acceleration, velocity, and displacement of the platform, respectively, $h(t-\tau)$ is the retardation function which represents the fluid memory effect, and F_{exc} is the excitation force which includes the aerodynamic force, hydrostatic force, radiation force, diffraction force, and nonlinear viscous drag force.

2.2. Hydrodynamic Loads

Four parts of the hydrodynamic properties of the floating platform are calculated in this model. (1) The linear hydrostatic restoring force, (2) the added mass and radiation damping from the linear wave radiation including the free-surface memory effects, (3) the incident-wave excitation from the linear diffraction in regular or irregular seas, and (4) the nonlinear viscous drag from the incident-wave kinematics and the platform motions. Therefore, the total force $F_i^{Platform}$ acting on the floating platform is expressed by [25]

$$F_i^{Platform} = F_i^{Hydrostatic} + F_i^{Radiation} + F_i^{Diffraction} + F_i^{Drag}, \quad (2)$$

where $F_i^{Hydrostatic}$ is the hydrostatic force, $F_i^{Radiation}$ is the radiation force, $F_i^{Diffraction}$ is the diffraction force, F_i^{Drag} is the nonlinear viscous drag force.

The radiation force includes contributions from the added mass and radiation damping, which are proportional to the acceleration and velocity of the floating platform motions, respectively. The radiation force can be given in time domain by

$$F_i^{Radiation} = -A_\infty \ddot{x}_j(t) - \int_0^t h_{ij}(t-\tau)\dot{x}_j(t)d\tau, \quad (3)$$

where $h_{ij}(\tau)$ is the retardation function which can be described by

$$h_{ij} = \frac{2}{\pi}\int_0^\infty B_{ij}(\omega)\cos(\omega t)d\omega, \quad (4)$$

in which B_{ij} is the linear radiation damping matrix.

The diffraction force refers to the total excitation loads acting on the floating platform from the incident waves. In considering stochastic sea states defined by an appropriate wave spectrum, the wave elevation $\zeta(t)$ and diffraction force $F_{\mathrm{i}}^{\mathrm{Diffraction}}$ can be determined by [26,27]

$$\zeta(t) = \frac{1}{2\pi}\int_{-\infty}^{\infty} W(\omega)\sqrt{2\pi S_{\zeta}^{\text{2-sided}}(\omega)}e^{\mathrm{j}\omega t}\mathrm{d}\omega, \tag{5}$$

$$F_{\mathrm{i}}^{\mathrm{Diffraction}} = \frac{1}{2\pi}\int_{-\infty}^{\infty} W(\omega)\sqrt{2\pi S_{\zeta}^{\text{2-sided}}(\omega)}X_{\mathrm{i}}(\omega,\beta)e^{\mathrm{j}\omega t}\mathrm{d}\omega, \tag{6}$$

where $S_{\zeta}^{\text{2-sided}}(\omega)$ is the two-sided power spectral density of the wave elevation per unit time, ω is the circular frequency of the wave, $W(\omega)$ is the Fourier transform of a realization of a white Gaussian noise(WGN) time-series process with zero mean and unit variance, $X_{\mathrm{i}}(\omega,\beta)$ is the ith component of the frequency- and direction-dependent complex incident-wave-excitation force on the support platform per unit wave amplitude.

The nonlinear viscous drag load [28] can be calculated according to the Morison equation, which is proportional to the square of relative velocity between the fluid and the floating platform and can be expressed by

$$F_{\mathrm{i}}^{\mathrm{Drag}} = \frac{1}{2}C_{\mathrm{D}}\rho_{\mathrm{w}}A\left[u_{\mathrm{i}}(\mathrm{t}) - \dot{x}_{\mathrm{i}}(\mathrm{t})\right]\left|u_{\mathrm{i}}(\mathrm{t}) - \dot{x}_{\mathrm{i}}(\mathrm{t})\right|, \tag{7}$$

where $u_{\mathrm{i}}(\mathrm{t})$ is the wave particle velocity estimated at the instantaneous position of the center of gravity (COG), C_{D} is the viscous drag coefficient in the specific direction with a projection area A.

The hydrostatic force $F_{\mathrm{i}}^{\mathrm{Hydrostatic}}$ includes the buoyancy force from Archimedes' principle, and the linear hydrostatic force and moment resulting from the effects of the water-plane area and the center of buoyancy (COB) as the floating platform is displaced, which can be described by [25]

$$F_{\mathrm{i}}^{\mathrm{Hydrostatic}} = \rho_{\mathrm{w}}gV_0v_{\mathrm{i}3} - C_{\mathrm{ij}}^{\mathrm{Hydrostatic}}x_{\mathrm{j}}(t), \tag{8}$$

where ρ_{w} is the density of the seawater, V_0 is the displaced volume of fluid when the floating platform is in its undisplaced position, $v_{\mathrm{i}3}$ is $(i,3)$ component of the Kronecker-Delta function, and non-zero only when the degrees of freedom (DOFs) $i = 3$, $C_{\mathrm{ij}}^{\mathrm{Hydrostatic}}$ is the (i,j) component of the linear hydrostatic-restoring matrix from the effects of water-plane area and the COB.

2.3. Aerodynamic Loads

The aerodynamic loads on the turbine are calculated based on the blade element momentum (BEM) theory [29], which assumes that the blades can be divided into many small elements, and the forces of each element are superimposed along the span of the blades in order to calculate the total forces and moments exerted on the turbine. According to the blade element theory, the aerodynamic loads on a single blade element can be described by

$$\mathrm{d}T = \mathrm{d}L\cos\varphi + \mathrm{d}D\sin\varphi = \frac{1}{2}\rho_{\mathrm{a}}V_{rel}{}^2(C_{\mathrm{L}}\cos\theta + C_{\mathrm{D}}\sin\theta)c\mathrm{d}r, \tag{9}$$

$$\mathrm{d}Q = \mathrm{d}L\sin\varphi - \mathrm{d}D\cos\varphi = \frac{1}{2}\rho_{\mathrm{a}}V_{rel}{}^2(C_{\mathrm{L}}\sin\theta - C_{\mathrm{D}}\cos\theta)c\mathrm{d}r, \tag{10}$$

where $\mathrm{d}T$ and $\mathrm{d}Q$ is the thrust and moment acting on a single blade element, respectively, C_{L} and C_{D} is the lift and drag coefficients of the airfoil, respectively, ρ_{a} is the air density, c is the chord of the airfoil, V_{rel} is the relative speed, θ is the inflow angle, and r is the distance of the airfoil section from the blade root.

The aerodynamic effects of the tip losses, the hub losses, and the skewed wake are incorporated. The unsteady aerodynamics is enabled based on the semi-empirical Beddoes–Leishman dynamic stall model [30].

2.4. Mooring Line Tension

The mooring line tension can be calculated using a multi-segmented quasi-static method or dynamic analysis method. The multi-segmented quasi-static method [31] is based on the classic catenary theory, which allows the evaluation of the mooring system by considering the average mooring line loads, the seabed friction of each mooring line, and the nonlinear geometric restoring forces for the catenaries and taut mooring systems. However, the individual line bending stiffness, the inertia and the hydrodynamic damping forces of the mooring lines are neglected. As shown in Figure 1, a single mooring line with undercover length in a local coordinate system is demonstrated. Then the Newton–Raphson iteration scheme is used to solve the mooring line tension in which the force analysis formulation of the mooring line is given by [25]

$$\begin{aligned} X_F(H_F, V_F) = L - \tfrac{V_F}{\omega_M} + \tfrac{H_F}{\omega_M}\ln\left[\tfrac{V_F}{H_F} + \sqrt{1+\left(\tfrac{V_F}{H_F}\right)^2}\right] + \\ \tfrac{H_F L}{EA} + \tfrac{C_B\omega_M}{2EA}\left[-\left(L-\tfrac{V_F}{\omega_M}\right)^2 + \left(L - \tfrac{V_F}{\omega_M} - \tfrac{H_F}{C_B\omega_M}\right)\cdot MAX\left(L - \tfrac{V_F}{\omega_M} - \tfrac{H_F}{C_B\omega_M}, 0\right)\right] \end{aligned} \tag{11}$$

$$Z_F(H_F, V_F) = \frac{H_F}{\omega_M}\left[\sqrt{1+\left(\frac{V_F}{H_F}\right)^2} - \sqrt{1+\left(\frac{V_F-\omega_M L}{H_F}\right)^2}\right] + \frac{1}{EA}\left(V_F L - \frac{\omega_M L^2}{2}\right), \tag{12}$$

where H_F and V_F are the horizontal and vertical components of the effective tension in the mooring line at the fairlead, respectively, H_A and V_A are the horizontal and vertical components of the effective tension in the mooring line at the anchor, respectively, $x(s)$ and $z(s)$ are the horizontal and vertical coordinates of the reference points of the mooring line, respectively, ω_M is the apparent weight of a line in fluid per unit length of line, L is the total unstretched length of a mooring line, A is the cross-sectional area of the mooring line, C_B is the coefficient of the static-friction drag between the seabed and a mooring line, and EA is the extensional stiffness of a mooring line.

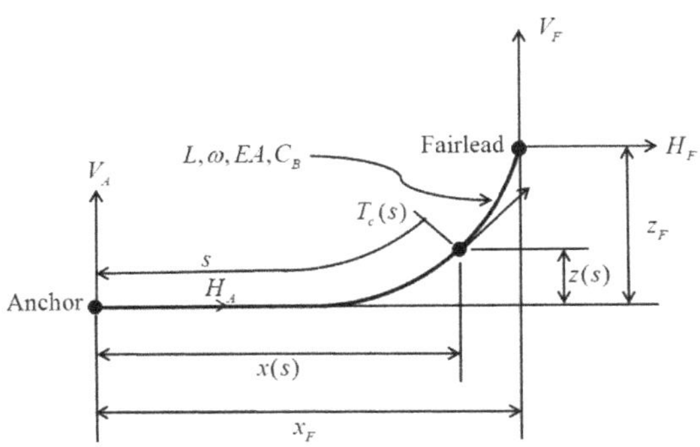

Figure 1. A single mooring line with undercover length in a local coordinate system.

The dynamic analysis of the mooring line that captures mooring stiffness and damping forces in the axial direction, weight, and buoyancy effects, seabed contact force (without friction), and hydrodynamic loads from the mooring motion using the Morison equation. In this paper, the dynamic analysis of the mooring line is based on a lumped-mass modeling approach [32,33] to discretize the cable dynamics over the length of the mooring line. As shown in Figure 2, the mooring line is broken up into N evenly-sized line segments connecting $N+1$ node points. The indexing starts at the anchor (or lower end), with the anchor node given a value of zero, and the mooring line segment between nodes 0 and 1 given an index of 1/2. The right-handed inertial reference frame with the z-axis being measured positive up from the water-plane is defined in this model, and each node's position is defined by a vector r, which contains the x, y, and z coordinates of the node position. Each segment $i + 1/2$ of a mooring line has identical properties of unstretched length, diameter, density, and Young's modulus. This allows mooring lines with different sets of properties and supports features such as point masses and line interconnections. The detailed calculation of the dynamic analysis model of the mooring line can refer to [34].

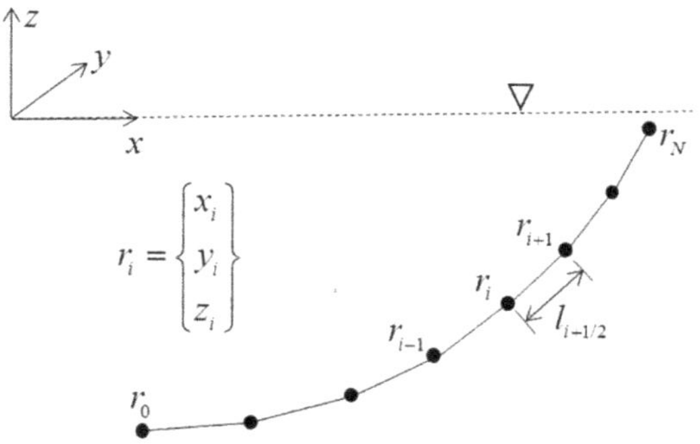

Figure 2. Mooring line discretization and indexing.

In this paper, the dynamic analysis method of the mooring line tension is used in the fully coupled analysis of the 5 MW and 10 MW FOWT. The multi-segmented quasi-static method is only used when analyzing the dynamic effects of the mooring line (including hydrodynamic damping and inertia forces of the mooring lines) on the motion responses of the 10 MW FOWT.

3. Baseline Model Description and Numerical Tool

3.1. Design of the FOWT

Taking DTU 10 MW wind turbine as the research object, the specifications of the DTU 10 MW wind turbine is as shown in Table 1. However, owing to the platform height above still water level (SWL), the tower of the DTU 10 MW wind turbine was shortened by 12.3 m. The overall properties of the new tower are listed in Table 2.

The braceless semi-submersible platform can offer the advantage of reducing the number of welding connections, improving the floater reliability as it is less sensitive to fatigue and corrosion hot-spot, and also simplifies the fabrication process [35]. Therefore, referring to the size of the NTNU (Norwegian University of Science and Technology)braceless semi-submersible platform [23] for the NREL 5 MW wind turbine, a scaled-up model of the braceless semi-submersible platform to hold the DTU 10 MW wind turbine was completed based on the Froude scaling rather than redesigning the floating platform completely to improve the efficiency of the design process [36]. The main properties of the platform are given in Table 3 and the side and top views of the braceless semi-submersible platform are demonstrated in Figure 3. The mooring system is arranged as shown in Figure 4, and properties of the mooring system are shown in Table 4. In addition, the layout of the FOWT is presented in Figure 5.

Table 1. Specifications of the wind turbine [20].

Turbine Type	DTU 10 MW	NREL 5 MW
Wind regime	IEC Class 1A	IEC Class 1B
Rotor orientation, configuration	Clockwise, upwind, three blades	Clockwise, upwind, three blades
Control	Variable speed, collective pitch	Variable speed, collective pitch
Cut-in, rated, cut-out wind speed (m/s)	4.0, 11.4, 25	3.0, 11.4, 25
Rated thrust (kN)	1500	750
Rotor, hub diameter (m)	178.3, 5.6	126, 3
Hub height (m)	119	90
Drive train	Medium speed, multiple-stage gearbox	High speed, multiple-stage gearbox
Rated generator speed (rpm)	480.0	1173.7
Gearbox ratio	50:1	97:1
Rated tip speed (m/s)	90	80
Hub overhang(m), shaft tilt (°), precone (°)	7.07, 5, −2.5	5.0, 5, −2.5
Blade prebend (m)	3.332	0.000
Rotor mass, nacelle mass(t)	229.0, 446.0	110, 240.0

Table 2. Properties of the tower of the floating offshore wind turbine (FOWT).

Turbine Type	DTU 10 MW	NREL 5 MW
Elevation of tower top above SWL (m)	115.63	87.6
Elevation of tower base above SWL (m)	12.30	10
Overall tower length (m)	103.33	77.6
Tower top diameter and thickness (m)	5.50, 0.02	3.87, 0.019
Tower base diameter and thickness (m)	8.00, 0.036	6.5, 0.027
Tower steel density (kg·m^{-3})	8500	8500
Overall (integrated) tower mass (t)	527.362	249.65
Center of gravity above SWL on tower centerline (m)	56.00	43.35

Table 3. Properties of the platform of the FOWT.

Turbine Type	DTU 10 MW	NREL 5 MW
d_c (m)	8	6.5
d_s (m)	8	6.5
h_{cu} (m)	12.30	10
h_{su}(m)	24.60	20
h_b (m)	29.52	24
d_{pw} (m)	11.07	9
d_{ph} (m)	7.38	6
d_{cs} (m)	50.43	41
d_{cse} (m)	55.965	45.5
Depth of water (m)	100	100
COG of the platform during operation phase (m)	(0, 0, −30.092)	(0, 0, −24.36)
Displacement (m^3)	19257.13	10555
Platform steel density (kg·m^{-3})	7850	7850
Platform mass, including ballast water (t)	17942.21	9789
Platform steel mass (t)	3259	1804
COG of the platform steel mass	(0, 0, −19.089)	(0, 0, −12.31)

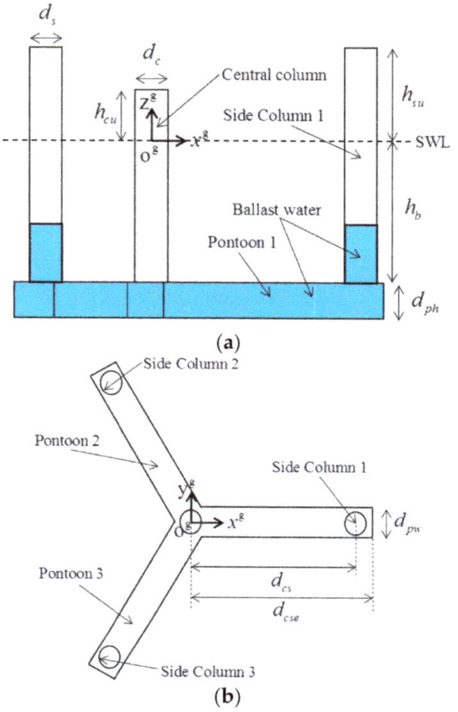

Figure 3. Side (**a**) and top (**b**) views of the braceless semi-submersible platform [23].

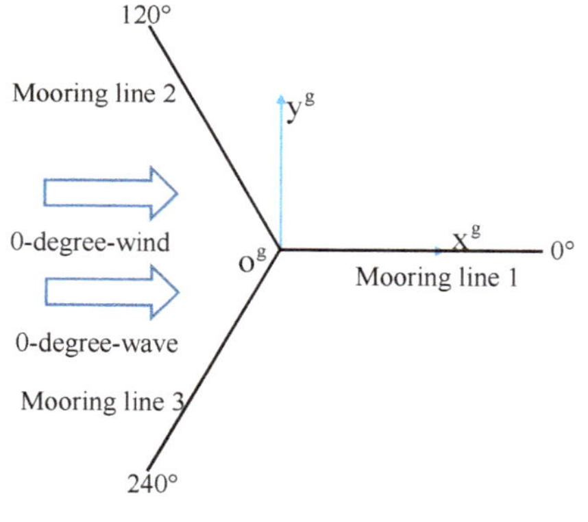

Figure 4. Arrangement of the mooring system.

Table 4. Properties of the mooring system.

Number of the Mooring Line	3
Mooring line type	Spiral rope
Transversal drag coefficient	1.2
Longitudinal drag coefficient	0.02
Transversal added-mass coefficient	0
Longitudinal added-mass coefficient	1
Length of each line (m)	666.5
Mass of per length (kg·m^{-1})	466
Diameter of the mooring line (m)	0.153
Equivalent axial stiffness (N)	2.5×10^9
Fairlead distance from platform center (m)	54.3

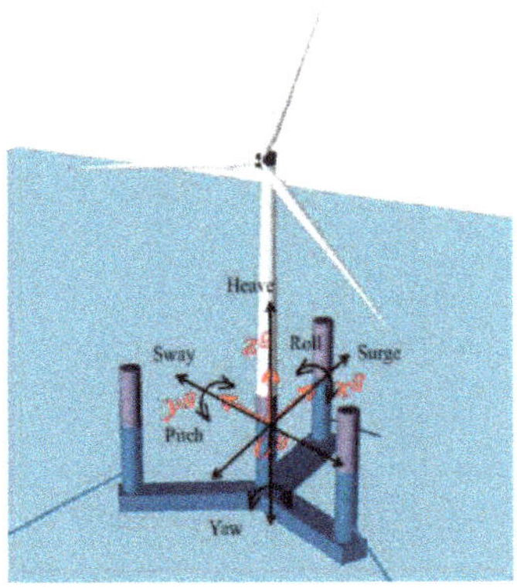

Figure 5. Layout of the FOWT.

3.2. Introduction of Numerical Tool

Using FAST, fully coupled aero-hydro-servo-elastic time domain simulations for the ultra-large FOWT can be conducted. The modules used in FAST are shown in Figure 6. The aerodynamic loads of the DTU 10 MW wind turbine were calculated by AeroDyn v14 [29], while the influence of the tower on the local airflow was not considered. Based on a combination of the linear potential flow theory and Morison equation, HydroDyn was used to calculate the hydrodynamic loads. According to a linearized model proposed by Frank Lemmer [37], a single-input-single-output (SISO) Proportional-Integral (PI) controller was chosen for the ServoDyn. The modules of MAP++ for the quasi-static method [38] and the modules of MoorDyn [39] for the dynamic method were respectively utilized to calculate the mooring line tensions. The structural dynamic response of the integrated FOWT, including the rotor, drivetrain, nacelle, tower, and platform were simulated in ElastoDyn.

During the analysis, two additional modules were introduced. One is ANSYS-AQWA [40], which was used to calculate the hydrodynamic coefficients of the braceless semi-submersible platform in the frequency domain, including the added mass, potential damping, and first-order wave exciting forces. The panel mesh model of the braceless semi-submersible platform below the SWL is shown in Figure 7. The results of hydrodynamic coefficients were converted to the WAMIT [41] format to perform fully coupled time-domain simulations in FAST. The other is TurbSim [42], which was used to generate the 3D turbulent wind velocity fields which were input into AeroDyn.

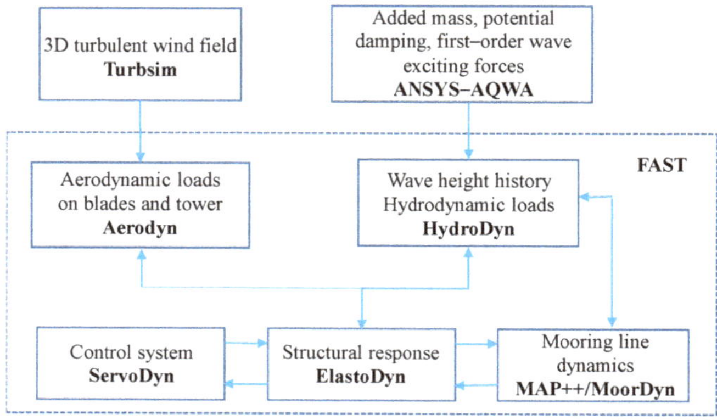

Figure 6. Overview of FAST used in the fully coupled analysis.

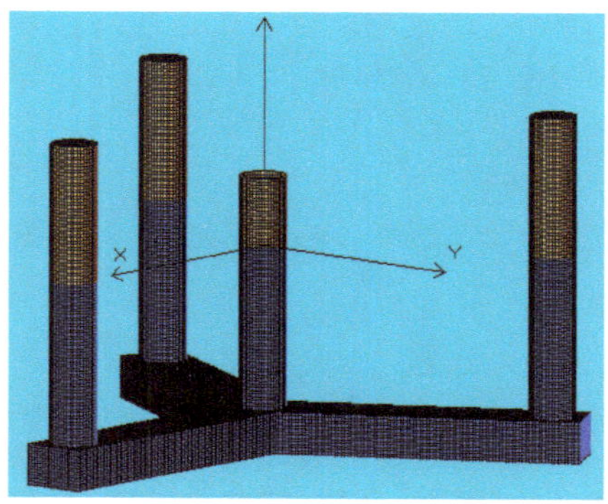

Figure 7. Panel mesh model of the braceless semi-submersible platform.

At each time step, the dynamic equilibrium equations of the rotor, drivetrain, nacelle, platform, and mooring system, with the rotor's rotational speed and blade pitch angle regulated by the ServoDyn, were solved in Elastodyn and Map++/MoorDyn. Then, the platform motions were transferred to HydroDyn to update the hydrodynamic loads. Meanwhile, the displacements, velocities, and accelerations of the blade and tower elements were transferred to the AeroDyn to update the aerodynamic loads. The updated loads will be applied to the dynamic equilibrium equations at the next time step [3].

4. Simulation Results and Discussion

4.1. Free Decay Test

Free decay simulation tests were conducted in FAST by applying a specific initial tower and platform displacement without the existence of wind, wave, or current. Time histories of decay tests were recorded, such as those shown in Figure 8 for the surge decay histories of the 5 and 10 MW FOWTs. Based on the Fourier transforms of the numerical time histories, the natural frequencies of the 5 and 10 MW FOWTs are obtained, as listed in Table 5. It can be observed that the rotor frequency (1P) is much larger than the natural frequencies of the platform and much smaller than the natural frequencies of tower, and the blade passing frequency (3P) is larger than the natural frequencies of the tower. Therefore, resonance will not occur during the operation of the FOWTs.

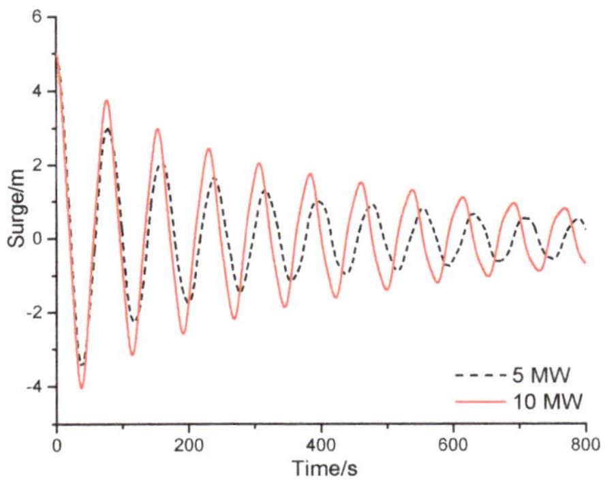

Figure 8. Time histories of surge decay test.

Table 5. Natural frequencies of the NREL (National Renewable Energy Laboratory) 5 MW and DTU (Danish University of Science and Technology) 10 MW FOWTs.

Turbine Type	DTU 10 MW	NREL 5 MW
Surge/Sway (Hz)	0.013	0.015
Heave (Hz)	0.036	0.039
Roll/Pitch (Hz)	0.039	0.042
Yaw (Hz)	0.012	0.012
1st tower side-side (Hz)	0.368	0.436
1st tower fore-aft (Hz)	0.379	0.442
Minimum rotor speed (rpm)	6.0	6.9
Maximum rotor speed (rpm)	9.6	12.1

4.2. *Dynamic Response of the FOWT*

4.2.1. Environmental Conditions

A series of representative load cases (LCs) were selected to investigate the dynamic characteristics of the ultra-large semi-submersible FOWT, the dynamic effects of mooring lines and the difference in dynamic characteristics of 5 and 10 MW FOWTs, as shown in Table 6. In the LC4 and LC7 cases, a turbulent wind field over a 200 × 200 m vertical plane centered on the rotor was generated in TurbSim using IEC Kaimal spectra based on NTM (normal turbulence model) conditions and turbulence category A. However, steady wind was selected for other load cases. The JONSWAP spectrum was used to generate the irregular wave time series. As can be observed in Figure 4, the wind and waves were correlated and directionally aligned in the positive X-direction.

Table 6. Definition of load cases.

Load Cases (LCs)	Wind Speed (m/s)	H_s (m)	T_p (s)	Turbulence Intensity	Turbine Status
LC1	11.4	-	-	-	operating
LC2	-	2.5	10.2	-	operating
LC3	11.4	2.5	10.2	-	operating
LC4	11.4	2.5	10.2	0.15	operating
LC5	18	2.5	10.2	-	operating
LC6	18	4.1	10.5	-	operating
LC7	18	4.1	10.5	0.13	operating

4.2.2. Statistics of Motion Simulation Results for the 10 MW FOWT

Based on the environmental conditions in Table 6, the fully coupled time domain simulation of the 10 MW FOWT was conducted in different cases. Each simulation lasted 4200 s and corresponded to an hour dynamic analyses because the first 600 s were filtered to eliminate the startup transient effects. This study only compared and analyzed the surge, heave, and pitch motion responses, which were critical for the braceless semi-submersible FOWT. Table 7 summarizes the statistical values of the platform motion responses. First, the steady-state responses of the integrated 10 MW turbine-platform-mooring system in all the simulations are found to be in a reasonable range, such as the mean pitch angle is significantly less than the design value of 10° [43]. Therefore, the initial design of the braceless semi-submersible platform for the DTU 10 MW wind turbine by upscaling of the existing 5 MW platform design is feasible. This provides a reference for the design of support platform of the ultra-large wind turbines. However, more detailed adjustment of size and layout, strength checks for fatigue and ultimate limit states, hydrodynamic performance and economic analysis of the braceless semi-submersible platform need to be further implemented to hold the ultra-large wind turbines.

Table 7. Statistical results of the platform motions in different cases.

Mode	Statistic	LC1	LC2	LC3	LC4	LC5	LC6	LC7
Surge (m)	Maximum	7.77	0.45	8.19	10.77	3.97	4.35	5.43
	Minimum	7.72	−0.43	7.31	1.45	3.12	2.72	1.93
	Mean	7.75	−0.01	7.76	6.09	3.53	3.53	3.52
	STD	0.01	0.15	0.15	1.87	0.14	0.25	0.64
Heave (m)	Maximum	1.14	1.61	1.43	1.52	1.55	1.84	1.91
	Minimum	1.06	0.95	0.77	0.82	0.92	0.63	0.60
	Mean	1.10	1.28	1.10	1.15	1.23	1.23	1.23
	STD	0.03	0.11	0.11	0.12	0.11	0.18	0.21
Pitch (°)	Maximum	4.20	0.07	4.40	5.47	2.14	2.30	4.74
	Minimum	4.20	−0.36	3.99	1.01	1.74	1.63	−0.56
	Mean	4.20	−0.14	4.20	3.38	1.94	1.94	2.04
	STD	0.00	0.08	0.08	0.81	0.08	0.13	0.93

Additionally, the load effects of wind and wave were analyzed in detail by comparing the statistical values of the system motions of the 10 MW FOWT subjected to the single wind loads, the single wave loads, and the joint action of the wind-wave induced loads. The wind loads determine the mean value of the system motions of the 10 MW FOWT. The standard deviation of the system motions of the 10 MW FOWT are governed by the wave loads, which increase significantly when the wave height increases. The extreme value of the system motions and the amplitude of its change increased significantly in turbulent wind cases. At the rated wind speed case, platform responses are characterized by the large mean value of surge and pitch motions owing to the maximum aerodynamic loads. However, the mean value of the heave motion is slightly increased. This can be attributed to the 5° angle between the rotation shaft of the rotor of the DTU 10 MW wind turbine and the horizontal plane. Additionally, the low-frequency aerodynamic force in the rotor is a downward force in the vertical direction and affected the mean value of the heave motion.

4.2.3. Comparative Analysis of the Motion Responses for the 10 MW FOWT

Analysis of Wind and Wave Load Effects of the 10 MW FOWT

The time histories of the system motions for different combinations of loads are illustrated in Figure 9. In the single wind loads case (LC1), the surge motion is balanced at 7.75 m along the surge direction after the integrated system of the FOWT reached the quasi-equilibrium state. In the single wave loads case (LC2), the integrated system slightly vibrates at the initial position of the surge motion. Under the joint action of the wind-wave induced loads (LC3), the integrated system vibrates around an offset value of 7.76 m. Therefore, the wind loads determine the mean value of the surge motion, and the wave loads mainly cause the oscillation of the surge motion, which determines the amplitude of the system's surge motion. Similarly, the same trend is observed in heave and pitch motions.

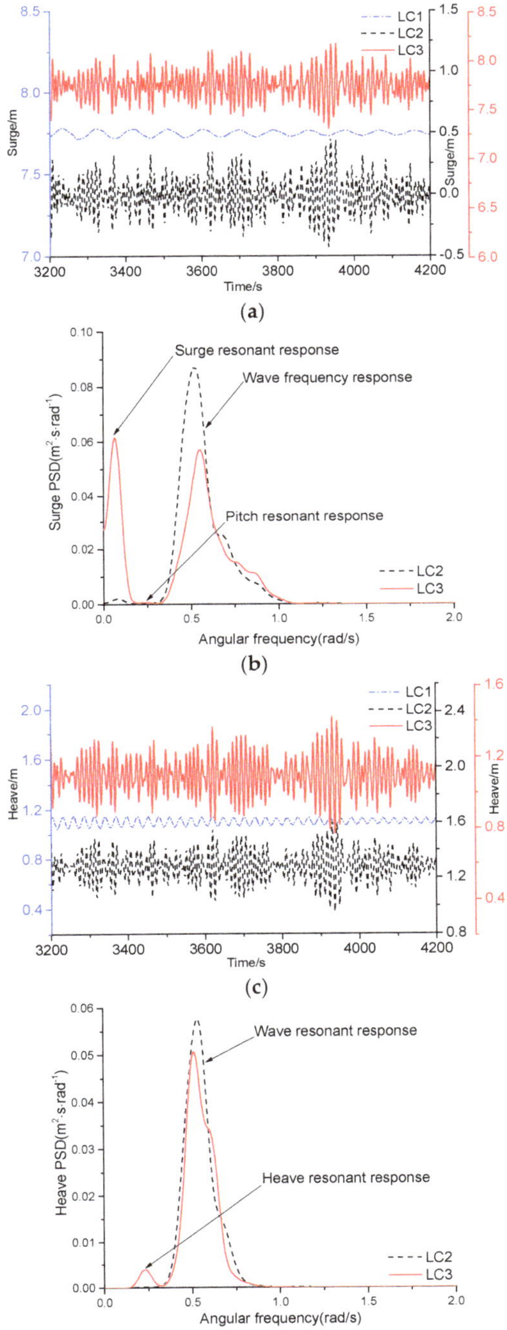

Figure 9. *Cont.*

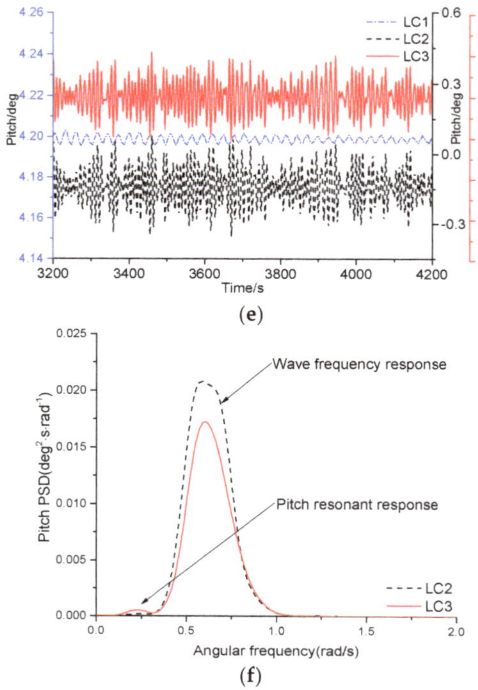

Figure 9. Power spectra and time histories of the platform motions for different load cases. (**a**) Time histories of surge motion; (**b**) power spectra of surge motion; (**c**) time histories of heave motion; (**d**) power spectra of heave motion; (**e**) time histories of pitch motion; (**f**) power spectra of pitch motion.

In Figure 9, the power spectra of surge, heave and pitch motions are compared in two cases: (a) with wave loads only, and (b) with joint wind-wave induced loads to observe and compare the load effects of the wind and wave. Accordingly, the resonant response could occur if the excitation frequencies are close to the natural frequencies of the system motions. Under the joint action of the wind-wave induced loads, it can be found from Figure 9 that there is a large resonance response at the surge natural frequency of 0.082 rad/s, the heave natural frequency of 0.226 rad/s and the pitch natural frequency of 0.245 rad/s. In the single wind loads case, the surge, heave, and pitch motions are mainly dominated by the wave frequency response and the wave frequency response for the wave only case is significantly higher than those values for the wind-wave induced loads case, which is caused by the positive aerodynamic damping induced by the wind loads. Furthermore, it can be observed that the natural frequencies of the system motions differ considerably from the wave peak spectral frequency so that the response energies excited by the wind-wave induced loads do not coincide in the frequency range. This indicates that the wind and wave loads affect the system motions of the FOWT separately in a decoupled way.

Comparison of Motion Responses of 10 MW FOWT for the Rated and Over-Rated Wind Speeds

The power spectra of the system motions of FOWT in the over-rated wind speed case (LC5) and in the rated wind speed case (LC3) are compared in Figure 10. It can be observed that the surge and pitch resonant responses are significantly increased in LC5 with the blade-pitch controller, while the heave resonant response and the wave-frequency response of the system motions of FOWT have hardly changed. This resonant phenomenon is caused by negative damping from the blade-pitch controller in the over-rated wind speed range, according to [44,45].

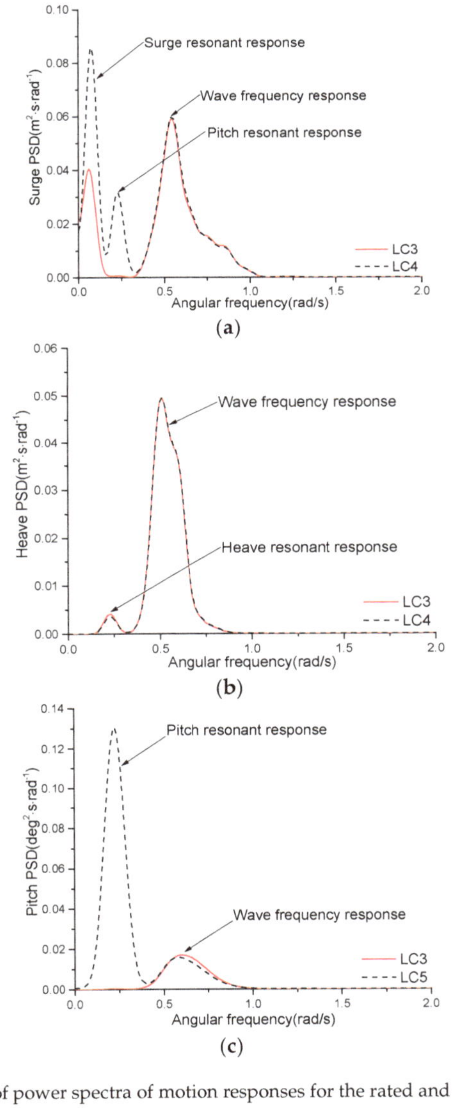

Figure 10. Comparison of power spectra of motion responses for the rated and over-rated wind speeds. (**a**) Power spectra of surge motion; (**b**) power spectra of heave motion; (**c**) power spectra of pitch motion.

Comparison of Motion Responses of the 10 MW FOWT for the Steady and Turbulent Wind

The time histories of surge motion under the steady and turbulent wind cases are compared in Figure 11a,c. The system motions are highly oscillating in time owing to the turbulent effects, which is consistent with the statistical results in Table 7. The maximum and standard deviation of the system motions increased significantly. This is owing to the turbulent wind which acted on the rotor with a time-varying longitudinal force and moments.

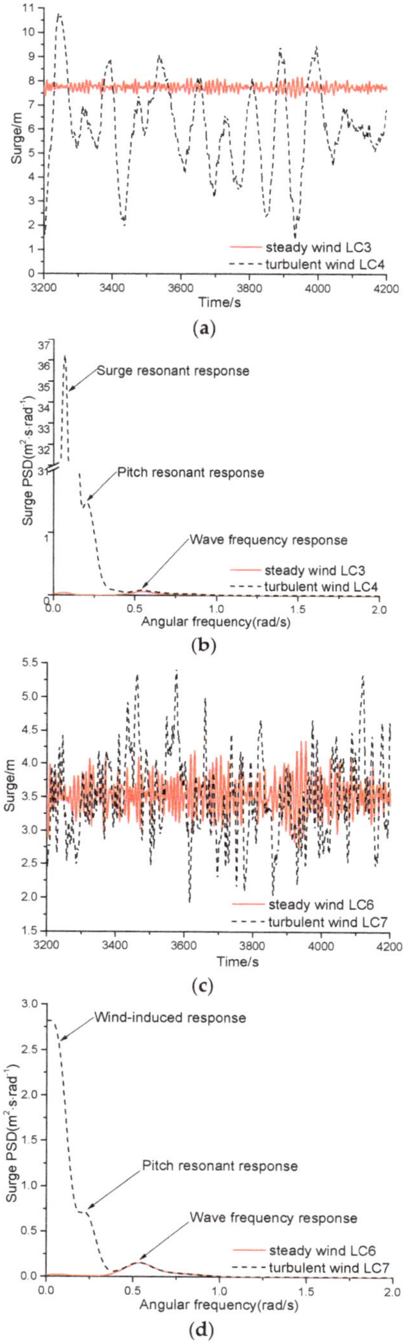

Figure 11. Comparison of time histories and power spectra of surge motion for the steady and turbulent wind cases. (**a**) Time histories of surge motion; (**b**) power spectra of surge motion; (**c**) time histories of surge motion; (**d**) power spectra of surge motion.

The power spectra of surge motion between the turbulent and steady wind cases are compared in Figure 11b,d. In the rated wind-speed case, Figure 11b shows that the turbulent wind loads excite the natural frequencies of the rigid body's surge and pitch motions, and the surge motion of the 10 MW FOWT is mainly dominated by the low-frequency surge and pitch resonant responses. This can be clearly explained as the turbulent wind has energy in the low-frequency region. However, there is no obvious difference in the wave frequency response of the surge motion between the turbulent and steady wind cases. Furthermore, the surge motion is also governed by the low-frequency component above the rated wind speed case of the turbulent wind as indicated in Figure 11d. However, the amplitude of the low-frequency resonant responses is reduced compared to those of the rated wind speed case.

4.3. *Dynamic Analysis of Mooring Lines for the 10 MW FOWT*

4.3.1. Dynamic Analysis of Mooring Lines Based on the Dynamic Analysis Modeling

The dynamic analysis of the mooring line of the 10 MW FOWT based on a lumped-mass method is carried out by using the FAST-MoorDyn module in the rated wind speed case (LC3). As shown in Figure 12a, there is a significant difference between the tension responses of the downstream and upstream mooring lines. This is observed since the Mooring line 1 (ML1) is the only mooring line acting downstream. Consequently, the tension and fluctuating components of ML1 attained smaller values than those of the Mooring line 2 (ML2) and the Mooring line 3 (ML3).

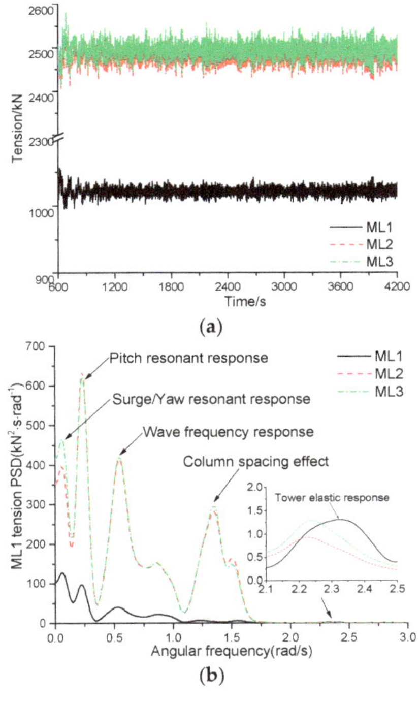

Figure 12. Time histories and power spectra of mooring line tension response in the rated wind speed (LC3) case. (**a**) Time histories of mooring line tension; (**b**) power spectra of mooring line tension.

Figure 12b illustrates the power spectra of mooring line tension response in the LC3 case. The mooring system of the FOWT is mainly induced by the low-frequency, wave-frequency and high-frequency components in the marine environment. The low-frequency components are related to slowly-varying platform motions primarily induced by low-frequency wind loads, such as surge, yaw and pitch. At the same time, an obvious peak occurs at the surge natural frequency of 0.082 rad/s and the pitch natural frequency of 0.245 rad/s, respectively. This indicates that there is a coupling effect between the platform motions and the mooring line tension response. The wave-frequency component is mainly caused by the action of the wave loads, and an obvious peak around the wave peak spectral frequency of 0.616 rad/s exists. The high-frequency components are mainly induced by the elastic response of the tower, which is excited by the harmonics of the wave loads [46].

4.3.2. Analysis of Dynamic Effects of Mooring Lines Based on Quasi-Static and Dynamic Models

In the process of the fully coupled analysis of the 10 MW FOWT, the mooring system was simulated using the dynamic mooring module (MoorDyn) and the multi-segmented quasi-static mooring module (MAP++) to study the dynamic effects of the mooring lines. The statistical results of the system motions and tension responses for the LC3 case are listed in Table 8. It can be clearly observed that the dynamic effects of the mooring lines do not significantly affect the mean values of motions, or the tension responses of the 10 MW FOWT. However, the system motions and the mooring line tension responses based on the dynamic (MoorDyn) mooring model yield larger standard deviations than those elicited by the quasi-static (MAP++) model owing to dynamic amplification effects.

Table 8. Comparison of dynamic analysis of mooring lines for the LC3 case.

	Surge (m)		**Heave (m)**		**Pitch (°)**		**Mooring Line Tension (kN)**	
	Mean	**STD**	**Mean**	**STD**	**Mean**	**STD**	**Mean**	**STD**
Quasi-static	7.67	0.14	1.06	0.10	4.20	0.08	1025.58	2.52
Dynamic	7.76	0.15	1.10	0.11	4.20	0.08	1021.15	5.16

The power spectra of platform motion responses based on the dynamic (MoorDyn) and quasi-static (MAP++) mooring models were compared, as shown in Figure 13. The peak values of low-frequency resonant responses based on the quasi-static mooring model is significantly higher than that of dynamic mooring model, especially for the surge and heave motions. However, there is no significant change in the wave-frequency responses. Therefore, the dynamic effects of the mooring lines significantly affects the low-frequency resonant responses of the platform motions, while the quasi-static mooring model ignoring the dynamic effects of the mooring lines significantly overestimates the amplitude of the low-frequency resonant responses.

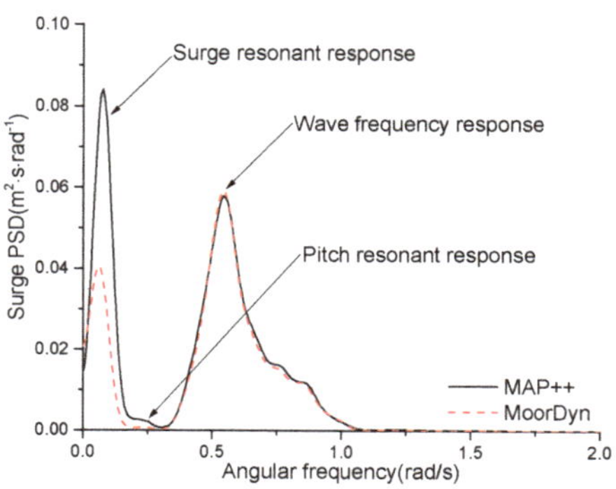

(a)

Figure 13. *Cont.*

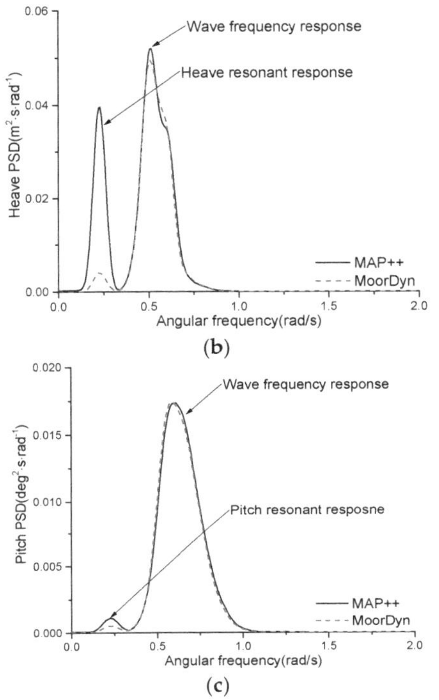

Figure 13. Comparison of motion responses based on the dynamic and quasi-static models. (**a**) Power spectra of surge motion; (**b**) power spectra of heave motion; (**c**) power spectra of pitch motion.

Since the dynamic effects of the mooring lines have significant influences on the standard deviation of the mooring line tension responses and the resonant response of the platform motions, the dynamic effects of the mooring lines must be fully taken into account in the design of the mooring system of the ultra-large FOWT so as to accurately check the fatigue and ultimate limit states of the mooring lines, and to optimize the mooring system.

5. Comparison and Analysis of the Dynamic Characteristics of the 10 MW and 5 MW FOWT

The design data of a 5 MW braceless semi-submersible FOWT was listed in detail in [23], and the same mooring system was used to investigate the difference in dynamic characteristics of the NREL 5 MW and DTU 10 MW FOWT under the joint action of wind-wave induced loads (LC3 and LC6) due to the different in turbine properties. Then, a comparative analysis associated with the dynamic characteristics of the 5 MW and 10 MW FOWT was conducted by FAST, including the main natural frequencies, motion responses and structural dynamics.

5.1. Comparison of Natural Frequencies for the 5 and 10 MW FOWT

Free decay motion simulations were first conducted to study the resonant properties of the coupled system of the FOWTs. Therefore, time histories of the surge decay test and main natural frequencies of the coupled system of the 5 and 10 MW FOWTs were compared, as shown in Figure 8 and Table 5. The surge motion of the 5 MW FOWT decays faster than that of the 10 MW FOWT, but the natural frequencies of the six DOFs of the rigid body motions of the 5 and 10 MW support platforms are nearly the same. Furthermore, the first-order natural frequency of the tower of the NREL 5 MW and DTU 10 MW FOWTs are significantly different.

5.2. Comparison of Motion Responses for the 5 and 10 MW FOWT

The simulation results of the platform motions of the 5 MW FOWT are listed in Table 9, which are compared with those of the 10 MW FOWT in the LC3 and LC6 cases (Table 7). The fluctuation of the motion responses of the 5 MW FOWT is very obvious for the rated and over-rated wind speed cases, but the mean value of the surge motion of the 10 MW FOWT is 29.1% and 19.3% higher than that of the 5 MW FOWT, respectively, which indicates that the 10 MW FOWT needs the mooring system to provide a larger surge restoring force to reduce the mean value of the surge motion. In addition, the 10 MW FOWT can offer large restoring forces and moments for the heave and pitch motions owing to its large structural weight and buoyancy, but it also suffers from the greater wind and wave loads. Accordingly, the mean values of the heave and pitch motions of the 10 MW FOWT are significantly higher than those of the 5 MW FOWT for the rated and over-rated speed cases.

Table 9. Statistics of platform motions of the 5 MW FOWT.

Mode	Load Case	Maximum	Minimum	Mean	STD
Surge (m)	LC3	6.65	5.40	6.01	0.21
	LC6	4.06	1.86	2.96	0.36
Heave (m)	LC3	0.78	0.00	0.34	0.15
	LC6	1.27	−0.40	0.45	0.27
Pitch (°)	LC3	3.24	2.69	2.97	0.09
	LC6	1.54	0.54	1.04	0.17

Figure 14 shows a comparison of the power spectra of surge and pitch motion responses for the NREL 5 MW and DTU 10 MW FOWTs in the LC3 and LC6 cases. The amplitude of the surge and pitch resonant response of the 10 MW FOWT excited by the low-frequency wind loads is higher than that of the 5 MW FOWT for the LC3 and LC6 cases, but the wave-frequency response of the surge and pitch motion excited by the wave loads is obviously smaller. Therefore, it can be predicted that the excitation of the low-frequency wind loads on the motion responses of the FOWTs becomes more prominent from the relative proportions of low-frequency and wave-frequency kinetic energy of the 5 and 10 MW FOWTs when the size of the wind turbine increases.

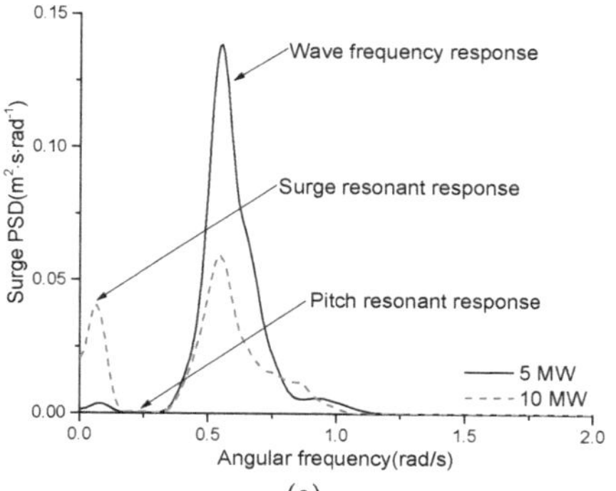

(a)

Figure 14. *Cont.*

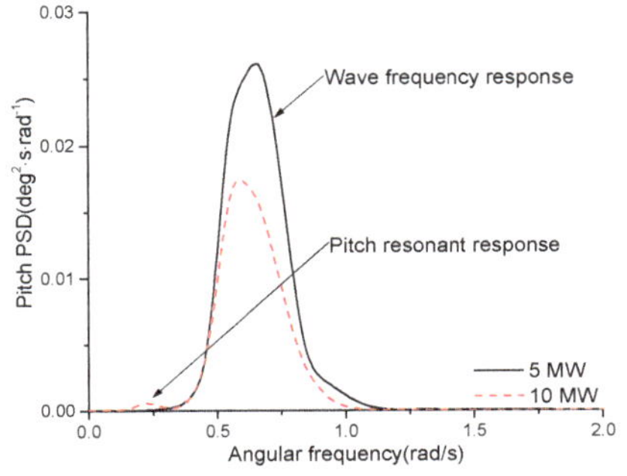

(b)

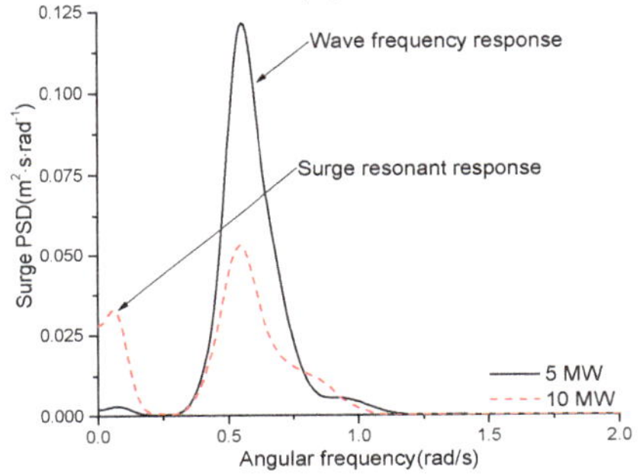

(c)

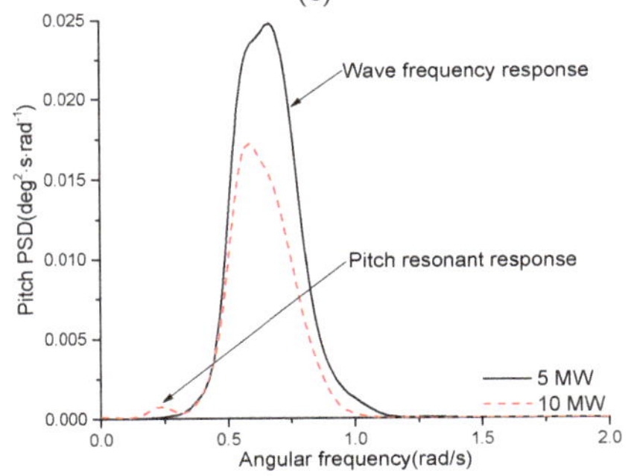

(d)

Figure 14. Comparison of power spectra of platform motion responses. (**a**) Power spectra of surge motion for the LC3 case; (**b**) power spectra of pitch motion for the LC3 case; (**c**) power spectra of surge motion for the over-rated wind speed (LC6) case; (**d**) power spectra of pitch motion for the LC6 case.

5.3. Comparison of Structural Dynamics for the 5 and 10 MW FOWT

As shown in Table 10, a comparison of the structural dynamics of the DTU 10 MW and NREL 5 MW FOWT in the LC3 and LC6 cases was conducted for the out-of-plane blade-root bending moments (RootMyc1), the rotor thrust, the fore-aft force and bending moments in the tower base (TwrBsFxt and TwrBsMyt) and the ML1 tension. Structural loads diagram of the FOWT system is presented in Figure 15. The mean and maximum values of the structural loads of the 10 MW FOWT are more than two times than those of the 5 MW FOWT for the LC3 and LC6 cases. In addition, the fluctuation of the structure loads of the 10 MW FOWT is very significant, especially for the LC6 case with a relatively high significant wave height. This provides a future reference for the design, application, and the reliability analysis for the ultra-large FOWT.

Table 10. Statistics of structural loads of the 5 and 10 MW FOWT.

Issue	Load Case	Turbine	Maximum	Minimum	Mean	STD
RootMyc1 (kN·m)	LC3	5 MW	11,030.00	8957.00	10,120.37	569.76
		10 MW	31,700.00	26,910.00	29,577.44	1113.86
	LC6	5 MW	6459.00	2655.00	4643.67	973.99
		10 MW	15,980.00	9524.00	12,847.27	1935.73
Thrust (kN)	LC3	5 MW	908.00	826.50	868.47	13.04
		10 MW	1984.00	1776.00	1878.32	33.15
	LC6	5 MW	548.50	378.40	467.93	26.70
		10 MW	1211.00	785.90	997.13	71.49
TwrBsFxt (kN)	LC3	5 MW	1175.00	827.00	1009.70	54.94
		10 MW	2715.00	1990.00	2348.44	120.63
	LC6	5 MW	724.40	163.60	454.64	92.06
		10 MW	1731.00	518.70	1111.01	199.66
TwrBsMyt (kN·m)	LC3	5 MW	87,030.00	63,910.00	76,202.47	3709.71
		10 MW	259,600.00	191,000.00	224,712.38	11,288.86
	LC6	5 MW	55,010.00	15,760.00	35,860.25	6316.27
		10 MW	163,800.00	49,500.00	103,991.09	19,004.32
ML1 tension (kN)	LC3	5 MW	418.20	378.60	396.81	3.28
		10 MW	1042.00	1003.00	1021.15	5.16
	LC6	5 MW	582.70	388.00	492.56	6.61
		10 MW	1431.00	1312.00	1368.16	9.17

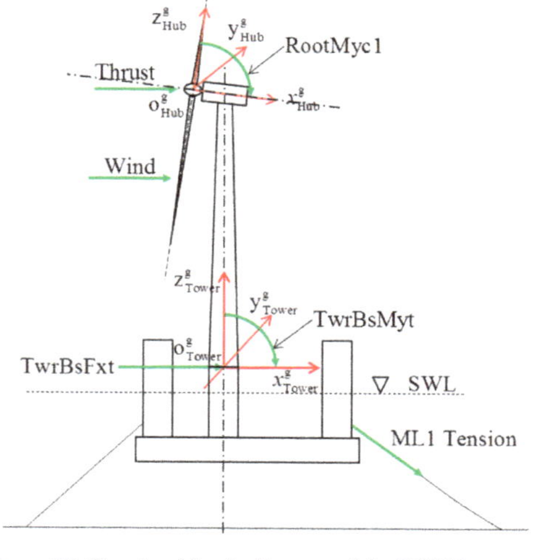

Figure 15. Structural loads diagram of the FOWT system.

Figure 16 depicts a comparison of the power spectra of RootMyc1 response for NREL 5 MW and DTU 10 MW FOWTs in the LC3 and LC6 cases. It is readily apparent that the RootMyc1 response of the 5 and 10 MW FOWTs are mainly stimulated by the 1P effect for the LC3 and LC6 cases, and the stimulating of the 1P effect on the RootMyc1 response of the 10 MW FOWT is more obvious than that of the 5 MW FOWT for the rate wind speed case. However, the excitation of the 1P effects on the RootMyc1 response of 10 MW FOWT above the rated wind speed case is diminished. In addition, the excitation of the wave loads on the RootMyc1 response of the 10 MW FOWT is prominent for the LC3 and LC6 cases, but the excitation of the 3P effect on the RootMyc1 response of the 5 MW FOWT is significant.

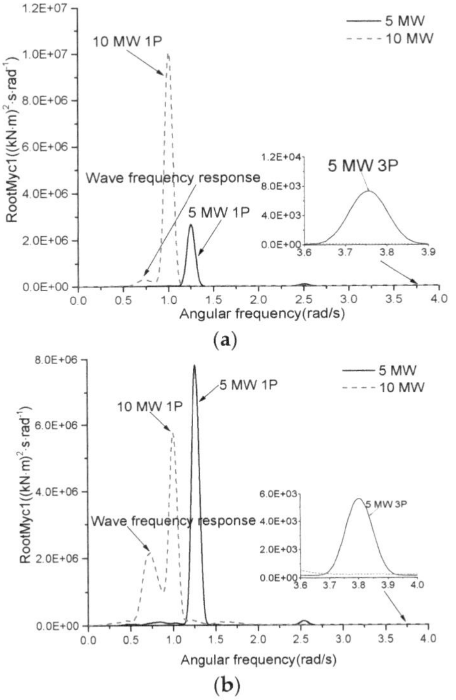

Figure 16. Comparison of the power spectra of RootMyc1 response. (**a**) Power spectra of RootMyc1 response for the LC3 case; (**b**) power spectra of RootMyc1 response for the LC6 case.

Power spectrum analyses of the TwrBsMyt response for NREL 5 MW and DTU 10 MW FOWTs were also performed in the LC3 and LC6 cases, as depicted in Figure 17. It demonstrates that the TwrBsMyt responses of the 5 and 10 MW FOWTs are mainly dominated by the wave loads for the LC3 and LC6 cases, and the amplitude of the TwrBsMyt responses of the 10 MW FOWT excited by the wave loads is significantly higher than those of the 5 MW FOWT. Furthermore, the excitation of the 3P effect on the TwrBsMyt response of the 5 MW FOWT is very prominent. However, the excitation of the low-frequency wind loads and the tower elastic response on the TwrBsMyt response of the 10 MW FOWT is significant, especially for the over-rated speed case.

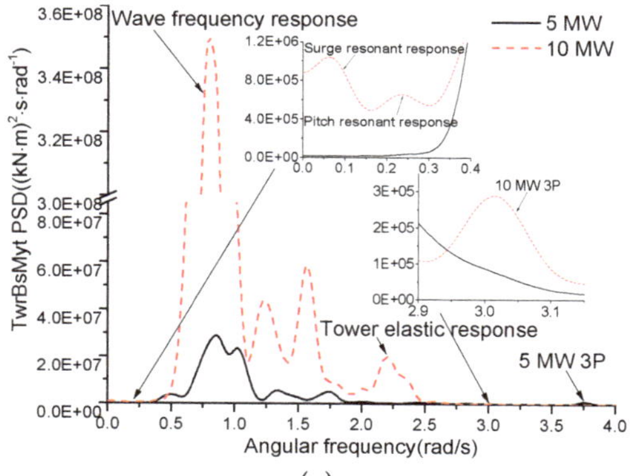

(a)

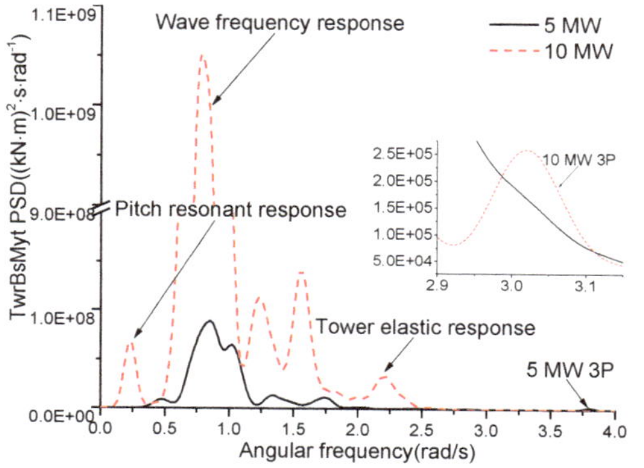

(b)

Figure 17. Comparison of the power spectra of TwrBsMyt response. (**a**) Power spectra of TwrBsMyt response for the LC3 case; (**b**) power spectra of TwrBsMyt response for the LC6 case.

As shown in Figure 18, the downstream mooring line (ML1) was selected as the research object and the comparison of the power spectra of ML1 tension response for NREL 5 MW and DTU 10 MW FOWTs is demonstrated in the LC3 and LC6 cases. The most significant difference of ML1 tension response is in the low-frequency and wave-frequency components. As the low-frequency resonant response of the 10 MW FOWT induced by the wind loads is more significant, this results in a significantly higher low-frequency component of the ML1 tension response for the 10 MW FOWT than that of the 5 MW FOWT. By contrast, the excitation effect of the wave loads on the ML1 tension response for the 5 MW FOWT is more significant, especially for the over-rated speed case with a relatively high significant wave height. In addition, the 3P effect has a significant impact on the ML1 tension response of the 5 MW FOWT, but the elastic response of the tower has a more obvious stimulation effect on the ML1 tension response of the 10 MW FOWT.

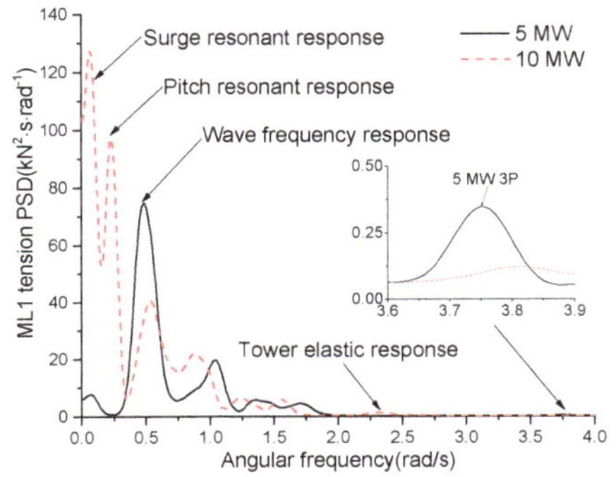

(**a**)

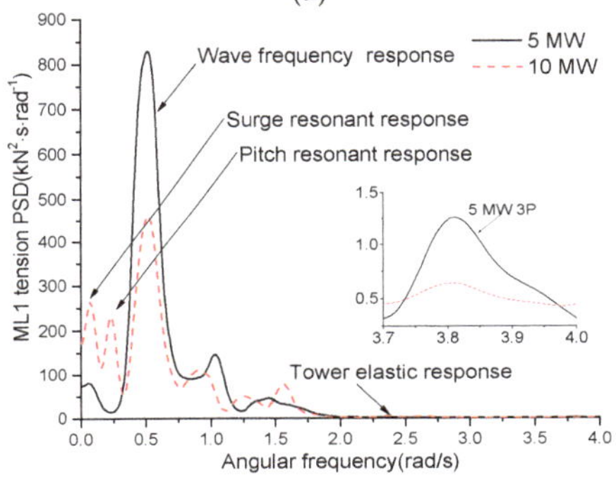

(**b**)

Figure 18. Comparison of the power spectra of Mooring line 1 (ML1) tension response. (**a**) Power spectra of ML1 tension response for the LC3 case; (**b**) power spectra of ML1 tension response for the LC6 case.

6. Conclusions

In this study, an initial design was performed by upscaling of an existing 5 MW braceless semi-submersible platform design to support the DTU 10 MW wind turbines and the fully coupled time domain simulation was conducted to investigate the dynamic characteristics of an ultra-large semi-submersible FOWT under typical sea states. The influence of the dynamic effects of the mooring line on the motions of the integrated motion responses of the FOWT was highlighted. Meanwhile, the motion responses and structural dynamics of the DTU 10 MW and NREL 5 MW braceless semi-submersible FOWTs were compared and studied. The conclusions drawn from the study can be summarized as follows:

(1) The aerodynamic and wave loads induced motion responses differently and independently. Specifically, in the case of turbulent wind, the low-frequency excitation of the wind loads on the system motions should be more prominent so that the fluctuating components of the system motions increase significantly. Additionally, the surge and pitch resonant responses are significantly increased owing to the negative damping from the blade-pitch controller in the over-rated wind case.

(2) The dynamic effects of the mooring line significantly reduced the low-frequency resonant responses of the system motions but had minor effects on the mean values of the integrated motion responses and the mooring tension responses. Accordingly, the integrated motion responses and the mooring tension responses yielded large standard deviations owing to the

dynamic amplification effects. This could provide a valuable reference for the safety evaluation and the mooring line design of the ultra-large FOWT.

(3) With the increase in the turbine size, the integrated motion responses and structural dynamics were significantly increased, but the integrated motion responses were in a reasonable range. This showed the initial design of the support platform for the DTU 10 MW wind turbine by upscaling of the existing 5 MW platform design is feasible. Correspondingly, the dynamic behaviors between the motion responses and structural dynamics of the 5 and 10 MW FOWTs were significantly different. It can be seen that the low-frequency excitations of the wind loads on the surge and pitch motions, the tower-base fore-aft bending moments and the mooring line tension responses of the 10 MW FOWT were more prominent than those of the 5 MW FOWT, but the 3P effect on the structural dynamics of the 5 MW FOWT was significant.

In general, the studies conducted herein provided prominence to the feasibility of the application of a braceless semi-submersible platform to hold the DTU 10 MW reference wind turbine for specific environmental conditions. This provides a reference for the design and application of the ultra-large FOWT in the future. However, if the braceless semi-submersible platform is used to support the ultra-large wind turbines in practice, more design requirements such as the adjustment of size and layout, detailed strength checks for fatigue and ultimate limit states, hydrodynamic performance and economic analysis, and a long-term extreme analysis should be conducted in further study.

Author Contributions: Z.Z. and X.L. formulated the problem addressed in this study; Z.Z., X.L., W.W., and W.S. formed the basis for the integrated FOWT system design and carried out all the modeling and numerical studies; all the authors contributed to the first draft and final version of the paper.

Acknowledgments: The authors would like to acknowledge the support from the National Natural Science Foundation of China (Grant No. 51879040, 51709039).

Conflicts of Interest: The authors declare no conflict of interest.

References

1. Liu, J.; Thomas, E.; Manuel, L.; Griffith, D.; Ruehl, K.; Barone, M. Integrated system design for a large wind turbine supported on a moored semi-submersible platform. *J. Mar. Sci. Eng.* **2018**, *6*, 9. [CrossRef]
2. Luo, C.X. Latest Developments in the World's Wind Power Industry. *Sino-Glob. Energy* **2012**, *17*, 32–39.
3. Han, Y.; Le, C.; Ding, H.; Cheng, Z.; Zhang, P. Stability and dynamic response analysis of a submerged tension leg platform for offshore wind turbines. *Ocean Eng.* **2017**, *129*, 68–82. [CrossRef]
4. Snyder, B.; Kaiser, M.J. Ecological and economic cost-benefit analysis of offshore wind energy. *Renew. Energy* **2009**, *34*, 1567–1578. [CrossRef]
5. Jeon, S.H.; Cho, Y.U.; Seo, M.W.; Cho, J.R.; Jeong, W.B. Dynamic response of floating substructure of spar-type offshore wind turbine with catenary mooring cables. *Ocean Eng.* **2013**, *72*, 356–364. [CrossRef]
6. Roddier, D.; Cermelli, C.; Aubault, A.; Weinstein, A. WindFloat: A floating foundation for offshore wind turbines. *J. Renew. Sustain. Energy* **2010**, *2*, 033104. [CrossRef]
7. Philippe, M.; Babarit, A.; Ferrant, P. Modes of response of an offshore wind turbine with directional wind and waves. *Renew. Energy* **2013**, *49*, 151–155. [CrossRef]
8. Bachynski, E.E.; Moan, T. Design considerations for tension leg platform wind turbines. *Mar. Struct.* **2012**, *29*, 89–114. [CrossRef]
9. Butterfield, S.; Musial, W.; Jonkman, J.M.; Sclavounos, P. *Engineering Challenges for Floating Offshore Wind Turbines*; National Renewable Energy Laboratory (NREL): Golden, CO, USA, 2007.
10. Jonkman, J.M.; Matha, D. Dynamics of offshore floating wind turbines—analysis of three concepts. *Wind Energy* **2011**, *14*, 557–569. [CrossRef]
11. Matha, D. *Model Development and Loads Analysis of an Offshore Wind Turbine on a Tension Leg Platform with a Comparison to Other Floating Turbine Concepts: April 2009*; National Renewable Energy Laboratory (NREL): Golden, CO, USA, 2010.

12. Li, L.; Hu, Z.; Wang, J. Dynamic responses of a semi-type offshore floating wind turbine. In Proceedings of the 33rd International Conference on Ocean, Offshore and Arctic Engineering, San Francisco, CA, USA, 8–13 June 2014.
13. Ma, Y. Research on Dynamic Analysis for a Spar Type Offshore Floating Wind Turbine. Master's Thesis, Shanghai Jiao Tong University, Shanghai, China, 2014.
14. Karimirad, M. Modeling aspects of a floating wind turbine for coupled wave–wind-induced dynamic analyses. *Renew. Energy* **2013**, *53*, 299–305. [CrossRef]
15. Sun, X.; Huang, D.; Wu, G. The current state of offshore wind energy technology development. *Energy* **2012**, *41*, 298–312. [CrossRef]
16. Harrison, R.; Hau, E.; Snel, H. *Large Wind Turbines: Design and Economic*; Wiley: Chichester, UK, 2000.
17. Bak, C.; Bitsche, R.; Yde, A. Light Rotor: The 10-MW reference wind turbine. In Proceedings of the EWEA 2012-European Wind Energy Conference & Exhibition, Copenhagen, Denmark, 16–19 April 2012.
18. Xue, W.F. Design, Numerical Modelling and Analysis of a Spar Floater Supporting the DTU 10 MW Wind Turbine. Master's Thesis, Norwegian University of Science and Technology, Trondheim, Norway, 2016.
19. Tian, X.S. Design, Numerical Modelling and Analysis of a TLP Floater Supporting the DTU 10 MW Wind Turbine. Master's Thesis, Norwegian University of Science and Technology, Trondheim, Norway, 2016.
20. Islam, M.T. Design, Numerical Modelling and Analysis of a Semi-Submersible Floater Supporting the DTU 10 MW Wind Turbine. Master's Thesis, Norwegian University of Science and Technology, Trondheim, Norway, 2016.
21. Xu, Y.; Hu, Z.; Liu, G. Kinetic characteristics research of the 10 MW-level offshore floating wind turbine. *Ocean Eng.* **2018**, *36*, 44–51.
22. Luan, C.Y.; Gao, Z.; Moan, T. Design and analysis of a braceless steel 5-mw semi-submersible wind turbine. In Proceedings of the 35th International Conference on Ocean, Offshore and Arctic Engineering, Busan, Korea, 19–24 June 2016.
23. Luan, C. Design and Analysis for a Steel Braceless Semi-Submersible Hull for Supporting a 5-MW Horizontal Axis Wind Turbine. Ph.D. Thesis, Norwegian University of Science and Technology, Trondheim, Norway, 2018.
24. Cheng, Z.; Madsen, H.A.; Gao, Z.; Moan, T. A fully coupled method for numerical modeling and dynamic analysis of floating vertical axis wind turbines. *Renew. Energy* **2017**, *107*, 604–619. [CrossRef]
25. Jonkman, J.M. *Dynamics Modeling and Loads Analysis of an Offshore Floating Wind Turbine*; National Renewable Energy Laboratory (NREL): Golden, CO, USA, 2007.
26. Jonkman, J.M.; Robertson, A.; Hayman, G.J. *HydroDyn User's Guide and Theory Manual*; National Renewable Energy Laboratory: Golden, CO, USA, 2014.
27. Faltinsen, O. *Sea Loads on Ships and Offshore Structures*; Cambridge University Press: Cambridge, UK, 1993.
28. Zhao, Y.; Yang, J.; He, Y.; Gu, M. Dynamic response analysis of a multi-column tension-leg-type floating wind turbine under combined wind and wave loading. *J. Shanghai Jiao Tong Univ.* **2016**, *21*, 103–111. [CrossRef]
29. Moriarty, P.J.; Hansen, A.C. *AeroDyn Theory Manual*; National Renewable Energy Laboratory: Golden, CO, USA, 2005.
30. Hansen, M.O.L.; Sørensen, J.N.; Voutsinas, S. State of the art in wind turbine aerodynamics and aeroelasticity. *Prog. Aerosp. Sci.* **2006**, *42*, 285–330. [CrossRef]
31. Jonkman, J.M. Dynamics of offshore floating wind turbines-model development and verification. *Wind Energy* **2009**, *12*, 459–492. [CrossRef]
32. Giuseppe, R.T.; Alberto, M.A.; Luigia, R. Dynamic modelling of spar bouy wind turbine. In Proceedings of the 33rd International Conference on Ocean, Offshore and Arctic Engineering, Trondheim, Norway, 25–30 June 2017.
33. Masciola, M.; Robertson, A.; Jonkman, J.; Coulling, A.; Goupee, A. Assessment of the importance of mooring dynamics on the global response of the DeepCwind floating semisubmersible offshore wind turbine. In Proceedings of the Twenty-Third International Offshore and Polar Engineering Conference, Anchorage, AK, USA, 30 June–5 July 2013.
34. Hall, M.; Goupee, A. Validation of a lumped-mass mooring line model with DeepCwind semisubmersible model test data. *Ocean Eng.* **2015**, *104*, 590–603. [CrossRef]
35. Vittori, F.E. Design and Analysis of Semi-Submersible Floating Wind Turbines with Focus on Structural Response Reduction. Master's Thesis, Norwegian University of Science and Technology, Trondheim, Norway, 2015.

36. Leimeister, M. Rational Upscaling and Modelling of a Semi-Submersible Floating Offshore Wind Turbine. Master's Thesis, Norwegian University of Science and Technology, Trondheim, Norway, 2016.
37. Lemmer, F.; Raach, S.; Schlipf, D. Prospects of Linear Model Predictive Control on a 10 MW Floating Wind Turbine. In Proceedings of the 34th International Conference on Ocean, Offshore and Arctic Engineering, Houston, TX, USA, 13–19 November 2015.
38. Masciola, M.; Jonkman, J.; Robertson, A. Implementation of a multisegmented, quasi-static cable model. In Proceedings of the 23th International Offshore and Polar Engineering Conference, Anchorage, AK, USA, 30 June–5 July 2013.
39. Hall, M. *MoorDyn—Users Guide*; Department of Mechanical Engineering, University of Maine: Orono, ME, USA, 2015.
40. ANSYS A.W., Inc. *AQWA Manual Release 15.0*; ANSYS A.W., Inc.: Canonsburg, PA, USA, 2013.
41. M. WAMIT Inc. *WAMIT v7.0 Manual*; M. WAMIT Inc.: Chestnut Hill, MA, USA, 2013.
42. Jonkman, B.J. *TurbSim User's Guide: Version 1.50*; National Renewable Energy Laboratory (NREL): Golden, CO, USA, 2009.
43. Wei, T. The Study of Offshore Floating Wind Turbine on a Tension Leg Platform. Master's Thesis, Jiangsu University of Science and Technology, Jiangsu, China, 2014.
44. Jonkman, J.M. Influence of control on the pitch damping of a floating wind turbine. In Proceedings of the 46th AIAA Aerospace Sciences Meeting and Exhibit, Reno, NV, USA, 7–10 January 2008.
45. Nielsen, F.G.; Hanson, T.D.; Skaare, B. Integrated dynamic analysis of floating offshore wind turbines. In Proceedings of the 25th International Conference on Offshore Mechanics and Arctic Engineering, Hamburg, Germany, 4–9 June 2006.
46. Karimirad, M.; Michailides, C. V-shaped semisubmersible offshore wind turbine: An alternative concept for offshore wind technology. *Renew. Energy* **2015**, *83*, 126–143. [CrossRef]

Journal of
Marine Science and Engineering

MDPI

Article

Motion Control of Pentapod Offshore Wind Turbines under Earthquakes by Tuned Mass Damper

Wenhua Wang [1,2], Xin Li [1,2,*], Zuxing Pan [3] and Zhixin Zhao [1,2]

1 State Key Laboratory of Coastal and Offshore Engineering, Dalian university of Technology, Dalian 116024, China
2 Institute of Earthquake Engineering, Faculty of Infrastructure Engineering, Dalian University of Technology, Dalian 116024, China
3 Marine Department, Renewable Energy Engineering Institute, Power China Huadong Engineering Corporation Limited, Hangzhou 311122, China
* Correspondence: lixin@dlut.edu.cn; Tel.: +86-411-8470-7784

Received: 9 June 2019; Accepted: 6 July 2019; Published: 16 July 2019

Abstract: The dynamic characteristics of a bottom-fixed offshore wind turbine (OWT) under earthquakes are analyzed by developing an integrated analysis model of the OWT. Further, the influence of the interactions between the rotor and support system on the structural responses of the OWT subjected to an earthquake is discussed. Moreover, a passive control method using a tuned mass damper (TMD) is applied to the OWT to control the responses under earthquakes. The effects of the mass ratio, location and tuned frequency of the TMD on controlling structural responses of the OWT under different recorded seismic waves are studied.

Keywords: offshore wind turbine; earthquakes; motion control; integrated analysis model; TMD

1. Introduction

Earthquakes are potential risk factors for offshore wind farms in the coastal regions of China, such as the Bohai bay. High capacity offshore wind turbines (OWT) will be constructed along the coast of China in the next five years, where they may be subjected to high intensity earthquakes. Hence, earthquake loads would be one of the dominant loads to consider for OWT designs in such regions.

Bazeos et al. [1], Witcher [2], Hänler et al. [3], Zhao and Maisser [4] adopted the suggested load combinations in the OWT design standards of GL and DNV to perform the seismic analysis of OWTs. Bazeos et al. [1] considered the pseudostatic aerodynamic loads suggested by Riziotis and Madsen [5] and seismic loads in the stability analysis of the tower of wind turbines (WTs). Witcher [2] applied the corrected aeroelastic interaction model of the WT to perform the analysis under combined seismic and wind loads, and the states of the WTs under operation, emergency shutdown, and parked cases were modelled. Hänler et al. [3], Zhao and Maisser [4] modelled the soil-foundation interaction in the analysis and proved the importance of the higher mode shapes in earthquake analyses.

Moreover, the dynamic model tests of the OWT under combined seismic, wind, and wave loads were also performed by Prowell et al. [6,7], Zheng et al. [8], and Wang et al. [9,10]. Prowell et al. [6,7] investigated the responses of an OWT under seismic and wind loads, and proved that aerodynamic damping can reduce the global response to some degree in the FA direction. Zheng et al. [8] experimentally researched the interaction of seismic and wave loads, and recommended that proper combinations of these loads should be applied in the seismic analysis of OWTs to obtain realistic structural responses. Wang et al. [9,10] performed a series of model tests of OWTs under combined seismic, wind, and wave loads, discovered the aerodynamic and hydrodynamic damping effects, and concluded that the aerodynamic and hydrodynamic loads may have comparable effects on the responses when the peak ground acceleration (PGA) of the seismic excitation is small.

On the other hand, Penzien et al. [11] recommended that proper plastic response should be considered in the seismic analysis of offshore structures. Therefore, Kim et al. [12], Nuta et al. [13], and Sadowski et al. [14] investigated the inelastic response of OWTs under seismic loads. Kim [11] performed the seismic fragility analysis of OWTs considering the non-linear effects of soil-pile interaction and suggested that the applied ground motion should be calculated for each soil layer to obtain equitable fragility curves of the structure. Nuta et al. [13] investigated the probability of damage in tubular steel towers at varying seismic hazard levels, defined the fragility curves of the towers by considering different damage states in the analysis, and proved that large safety factors must be considered in the design against the phenomenon of overloading under seismic loads. Sadowski et al. [14] researched the responses of the tower under seismic loads with respect to geometric imperfections and discussed the influence of the imperfections on the capacity of the structure.

In the research on OWTs under seismic loads, the scholars gradually realised that proper structural control strategies should be applied to the WTs to reduce the responses under earthquakes. Some researchers have applied passive control methods using tuned mass dampers (TMDs), multiple tuned mass dampers (MTMDs), and tuned liquid column dampers (TLCD) on the OWTs to reduce the structural responses under the operating aerodynamic and hydrodynamic loads. In the studies of Stewart and Lacker [15], an optimal TMD was mounted on the OWT to reduce the responses of fixed and floating OWTs; however, the control effect of TMD was found to be insignificant on the monopile OWT due to the discrepancies between the dominated frequency of the response under wave load and the TMD's tuning frequency. Dinh and Basu [16] investigated the control effects of MTMD on a spar-type floating OWT and proved that the damping of TMD is insignificant to the reduction of structural responses, and the TMD should be tuned around the dominated frequencies of the structural responses to achieve significant control effects. Colwell and Basu [17] conducted numerical analyses and discovered that the fatigue life of the tower of OWT could be increased by the implantation of TLCD on the structure.

In this paper, the structural responses of a Pentapod OWT under different seismic waves are analyzed based on an integrated seismic analysis model in FAST to research the dynamic characteristics of the OWT under earthquake loads. Subsequently, a TMD is mounted on the OWT with the intention of reducing the structural responses under earthquakes. Finally, the influence of the TMD parameters on the structural responses control of the OWT under different seismic waves are discussed.

2. Theories for the Seismic Analysis of OWT

2.1. Seismic Analysis of an Integrated OWT Model

Generally, the equation of motion for structures subjected to earthquake loads can be written as follows:

$$[M]\{\ddot{u}(t)\} + [C]\{\dot{u}(t)\} + [K]\{u(t)\} = -[M]\{\ddot{u}_g\}, \tag{1}$$

where $[M]$, $[C]$, and $[K]$ are the structural mass, damping, and stiffness matrices, respectively; $\{\ddot{u}\ (t)\}$, $\{\dot{u}\ (t)\}$, and $\{u(t)\}$ are the vectors of structural acceleration, velocity, and displacement, respectively; and $\{\ddot{u}_g\}$ is the vector of input ground acceleration.

For the seismic analysis of OWTs, the rotor and nacelle are commonly simplified as concentrated masses in Equation (1), hence the flexibility of the rotor and interaction between the rotor and support system are neglected. To compute response of OWTs more accurately, an integrated coupling analysis model suitable for computing responses of OWTs subjected to earthquakes is suggested. Based on the combined modal and multibody dynamics formulation, the model of rotor nacelle assembly (RNA) is established by twelve degrees of freedom (DOFs). The flexibility of the drive-shaft system is modelled by three DOFs; nine DOFs are used to model the motion of the rotor system in the flapwise and edgewise directions, respectively [18]. Hence, the interactions of the rotor, servo, and support system are taken into account in the updated seismic analysis of OWTs.

Significantly different from the wind and wave loads, earthquakes are typical wide-band stochastic processes. A seismic module is added to the coupled model of the OWT under wind and wave loads in FAST, as shown in Figure 1. FAST is a time-domain numerical tool to capture the coupled aero-hydro-servo-elastics response of OWTs. According to Figure 1, it can be seen that the FAST is mainly consist of AroDyn, HydroDyn, ServoDyn, ElastoDyn and SubDyn modules for the coupled analysis of bottom fixed OWTs. AeroDyn and HyoDyn is the aero- and hydro-dynamics module, respectively. The mechanical control strategies of the OWT shall be implemented in the ServoDyn module, such as the variable speed control, blade pitch control and the emergency shutdown. The responses of the blades and the tower shall be analyzed based on the ElastoDyn module. Meanwhile, the finite element model of the substructure of the bottom fixed OWT is established in the SubDyn module by the linear beam element.

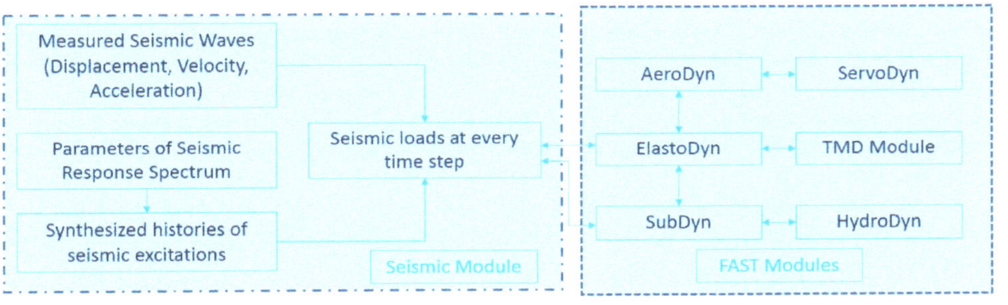

Figure 1. Recompiled integrated seismic analysis model of offshore wind turbine (OWT) in FAST.

For the recompiled FAST with the seismic module, the earthquakes can be the user-provided input or the synthetically generated time histories based on the suggested earthquake response spectrum. The user input seismic waves can be supplied in terms of acceleration, velocity or displacement. For the synthetic time histories, the artificial seismic wave is generated based on the user specified parameters. The earthquakes can be applied in any combination of three directions specified in the global coordinate system as shown in Figure 2, such as the two horizontal and one vertical direction.

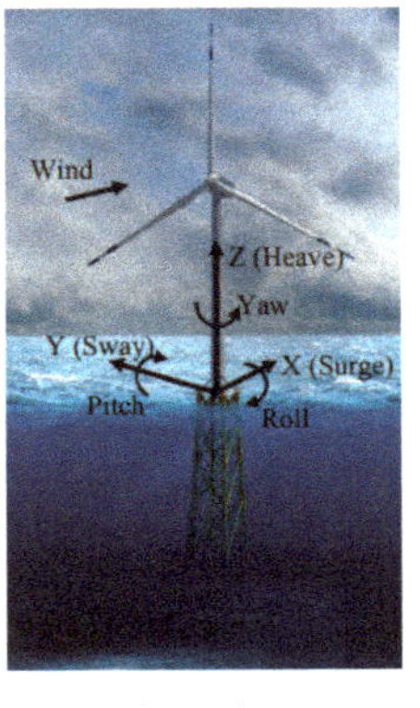

Figure 2. The global coordinate system in FAST.

The procedure for the coupled analysis of OWTs can be summarized as follows.

In procedure 1, the seismic module receives the parameters of the numerical model of the OWT, such as the mass of the rotor blades, tower, and substructure, from the ElastoDyn and SubDyn modules of FAST.

In procedure 2, the seismic module reads the histories of the recorded seismic waves or synthesised seismic waves according to the parameters in the input files.

In procedure 3, the seismic module calculates the relevant seismic loads and delivers it to the ElastoDyn module of FAST at every time step when the simulation time reaches the threshold time of earthquake occurrence.

2.2. Integrated Model of OWT for Seismic Analysis

The reference OWT structure is redesigned by combined the NREL 5-MW baseline WT [19] and a practical Pentapod OWT substructure constructed in the eastern coastal regions in China as shown in Figure 3a. The rating power of the practical OWT is identical with the NREL 5 MW baseline WT. The sectional geometries of the tower such as both the diameter of the tower of the practical and baseline wind turbine are also same. So the Pentapod substructure of the practical OWT can be directly applied to the baseline wind turbine, as shown in Figure 3a. While the height of the substructure is redesigned in order to satisfy the hub height of the baseline wind turbine, and the dimensions of the redesigned Pentapod is shown in Figure 3b,c. Meanwhile, the dynamic characteristics and ultimate capacities of the reference OWT is checked based on the DNV GL standards [20] in order to ensure the safety of the structure.

Further, the integrated coupling analysis model of the reference OWT is established in FAST with the updated Seismic module. The schematic diagram of the fully coupled analysis model under the seismic loads is shown in Figure 4.

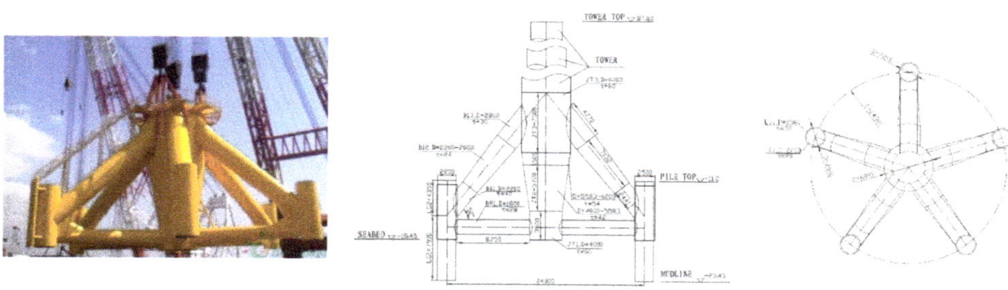

(**a**) Practical Pentapod substructure (**b**) Vertical view (**c**) Plan view

Figure 3. Dimensions of the reference OWT.

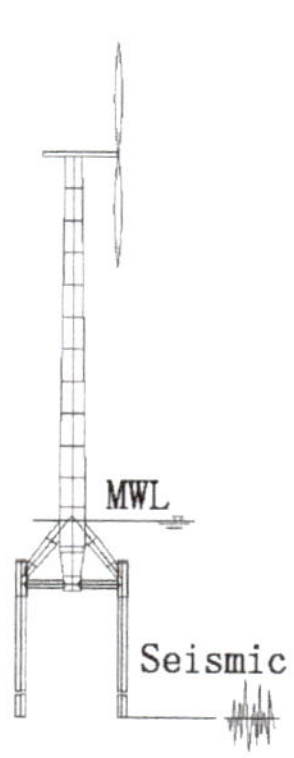

Figure 4. Schematic of the integrated analysis model of the reference OWT under earthquakes.

2.3. Load Cases for Seismic Analysis of OWT

Presently, the semi-integrated analysis method is widely adopted to design and check OWT on the base of the linear combination of seismic, wind and wave loads recommended by the DNVGL [21] and IEC [22] standards, indicating that the interactions of seismic, wind and wave loads are neglected. However, an integrated model of OWT under earthquakes is established to carry out sophisticated analysis in the paper. Then, the dynamic response of a parked OWT with multi-year mean water level under seismic excitations is studied in order to reveal the dynamic characteristics of the structure under the seismic cases, as shown in Figure 4.

As listed in Table 1, the dynamic characteristics and structural responses of the OWT in the standstill condition under the recorded seismic waves are investigated. Based on the geological conditions, requirements of the standards [22], and the seismic fortification intensity of the offshore wind farm, the parameters of the recorded seismic excitations are determined and applied in the integrated analysis of OWTs. Figure 5a shows the acceleration time histories of the selected seismic excitations, such as the El Centro, Taft, Northridge, and Chichi waves. From the figure, it can be observed that the Chichi wave has the largest peak ground acceleration (PGA), which is 0.37 g. The Fourier amplitudes of the selected seismic waves are shown in Figure 5b. The Figure demonstrates that the seismic excitations comprise of abundant frequency components in the range of 0.1–10 Hz, which includes the lower natural frequencies of general offshore structures.

The seismic excitations listed in Table 1 are applied in the F–A direction which is consistent with the X axis in the global coordinate system as shown in Figure 2.

Table 1. Load cases for the seismic analysis of offshore wind turbine (OWT).

Seismic Waves	PGA (g)	State	Winds and Waves	Seismic Direction
El Centro	0.21	Parked	-	In F-A direction
Taft	0.16	Parked	-	In F-A direction
Northridge	0.13	Parked	-	In F-A direction
Chichi	0.37	Parked	-	In F-A direction

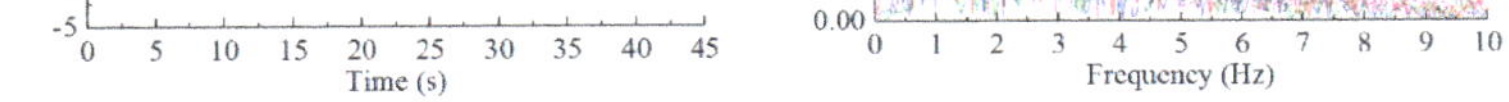

(**a**) Time histories (**b**) Fourier amplitudes

Figure 5. Measured seismic excitations.

3. Structural Responses of the OWT under Seismic Loads

3.1. Dynamic Characteristics of OWT

The fully coupled analysis model of the reference OWT is created in FAST based on the parameters of the NREL 5-MW baseline WT and the geometries of the Pentapod substructure. An equivalent pile model is used to model the pile-soil interaction in the fully coupled analysis model. The length of the equivalent pile is assigned as 18.26 m, which is 8.3 times the pile diameter, based on the similarity of the fundamental frequency. Dynamic characteristics of the fully coupled model are analyzed by the free-decay testing method.

The free-decay testing method is an extensively used method to calculate the natural frequencies of structures. Based on the free-decay testing method, an initial displacement or rotation angle of the structure shall be designated, then the natural frequencies of the structure can be determined based on the Fourier amplitudes or power spectral density functions of the decayed histories of the structural displacement or acceleration.

During the study, an initial displacement is designated at the blade tip, tower top and tower base, respectively. The natural frequencies of the blades and the integrated structure are derived based on the decayed accelerations of the blades and the tower. Figure 6 shows the Fourier amplitudes of the decayed histories of the blade tip and tower accelerations, and the corresponding analysis results are listed in Table 2.

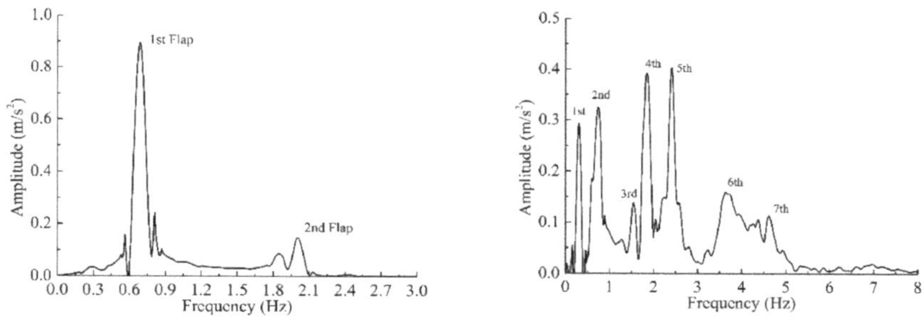

(**a**) Fourier amplitudes of the blade tip acceleration (**b**) Fourier amplitudes of the tower acceleration

Figure 6. Free decay test of the reference OWT.

Table 2. Natural frequencies of the integrated analysis model of the reference OWT in FAST.

Order of the Frequency	Values (Hz)	Note
1st OWT	0.305	In F-A direction
1st Flap	0.690	First blade collective flap mode
2nd OWT	0.742	
3rd OWT	1.544	In F-A direction
4rd OWT	1.824	
2nd Flap	2.029	Second blade collective flap mode
5th OWT	2.521	
6th OWT	3.640	In F-A direction
7th OWT	4.620	

3.2. Acceleration Responses

In this section, comparisons of tower accelerations under different types of seismic excitations are carried out. The recorded El Centro and Taft seismic waves were applied to the parked OWT along the FA direction, which is perpendicular to the rotating plane of the OWT. Moreover, the threshold time for the activation of the seismic loads was considered in the analysis to account for the stochastic nature of the seismic waves. The threshold value was 150 s for the seismic analysis of a parked OWT, and the total simulation time was 400 s.

Variations in the dynamic amplification factors (DAFs) with the height of the tower under different seismic excitations are recorded, as shown in Figure 4, to investigate the influence of the seismic excitation type. The DAF under earthquake loads can be expressed as Equation (2).

$$\text{DAF} = \frac{\text{PSA}}{\text{PGA}}, \tag{2}$$

where PSA is the peak acceleration response at a location, e.g., the maximum or minimum values of the acceleration time histories at the nacelle; PGA is the peak ground acceleration, e.g., the PGA of El Centro seismic wave = 0.21 g, as shown in Figure 5a.

Compared to the other seismic waves, the Northridge seismic wave stimulates the maximum DAFs along the height of the tower, as shown in Figure 7, and the maximum DAF under the Northridge wave is about 6.36, as listed in Table 3, at an elevation of 60 m on the tower. Statistics of the acceleration time histories at the tower top and the maximum tower accelerations corresponding to the maximum DAFs under different seismic waves are listed in Table 3. It can be observed that the Chichi seismic wave stimulates the maximum tower acceleration due to its high PGA. On the other hand, the PGA of the Northridge seismic wave is comparatively smaller than that of the El Centro and Chichi seismic waves; however, it stimulates significant accelerations at the tower top, comparable to the responses under the El Centro and Chichi seismic waves.

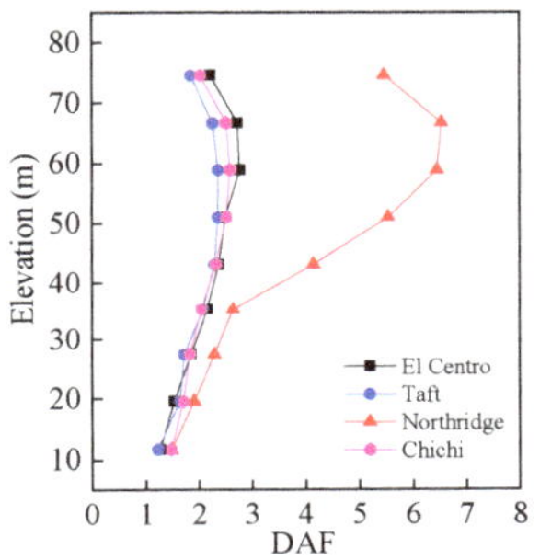

Figure 7. Dynamic amplification factors (DAFs) along the tower height under seismic excitations.

Table 3. Maximum values of the nacelle and tower acceleration time histories.

Maximum Values	Seismic Excitations			
	El Centro	Taft	Northridge	Chichi
Peak ground acceleration (PGAs) (m/s^2)	2.06	1.57	1.27	3.63
Nacelle acceleration (NAAs) (m/s^2)	1.58	0.97	0.75	2.87
DAF of NAA	0.77	0.62	0.59	0.79
Acceleration at the tower top (TTAs) (m/s^2)	4.66	2.82	6.84	7.28
DAF of TTA	2.26	1.80	5.39	2.01
Maximum acceleration along the height of the tower (MTAs) (m/s^2)	5.82	3.62	8.08	9.29
DAF of MTA	2.83	2.31	6.36	2.56

The dynamic characteristics of the accelerations at the tower top and the maximum tower accelerations are illustrated in Figure 8. The following results can be obtained:

(1) Other than the first two natural frequencies of OWT, the higher-order frequency components dominate the tower accelerations under earthquakes, especially for the Northridge seismic wave.
(2) For the nacelle accelerations of the OWT under seismic excitations, the responses are mainly dominated by the first two natural frequencies of the OWT.
(3) Under the El Centro and Taft seismic waves, the fifth natural frequency of the OWT dominates the tower accelerations, as shown in Figure 8a,b.
(4) Figure 8c demonstrates that the higher-order frequency components are stimulated under the Northridge seismic wave, which influence the tower accelerations significantly.
(5) Under the Chichi seismic wave, the influence of the first three natural frequencies of the OWT on the tower accelerations is non-negligible, as shown in Figure 8d.

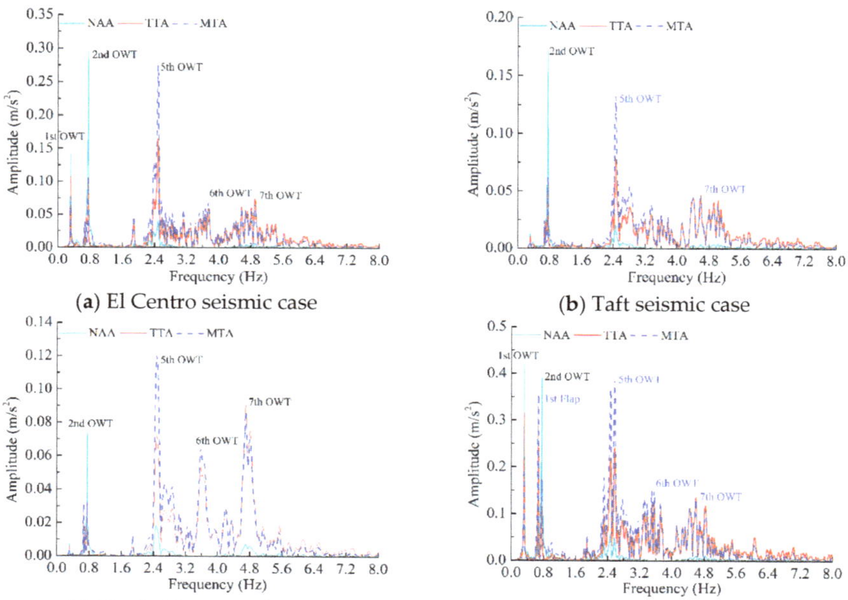

(a) El Centro seismic case (b) Taft seismic case (c) Northridge seismic case (d) Chichi seismic case

Figure 8. Fourier amplitudes of nacelle and tower accelerations under seismic excitations.

From these comparisons, the dominant frequencies of tower accelerations are found to vary significantly due to the aforementioned differences between the seismic waves.

3.3. Mudline Bending Moment

Figure 9a,b show the mudline bending moments under different seismic excitations. Compared to the other three seismic excitations, it can be seen that the Chichi seismic wave stimulates the largest bending moments, which can be attributed to the significant PGA of the Chichi seismic wave. The dynamic characteristics of bending moments in the frequency domain are shown in Figure 9c,d. Different results can be found by comparing them with the dynamic characteristics of the tower accelerations. The second mode shape of the OWT dominates the bending moments, unlike the responses of the tower accelerations shown in Figure 7. Moreover, for the Chichi seismic wave, the influence of the blade's first flap mode of the rotor system cannot be neglected, as shown in Figure 9d.

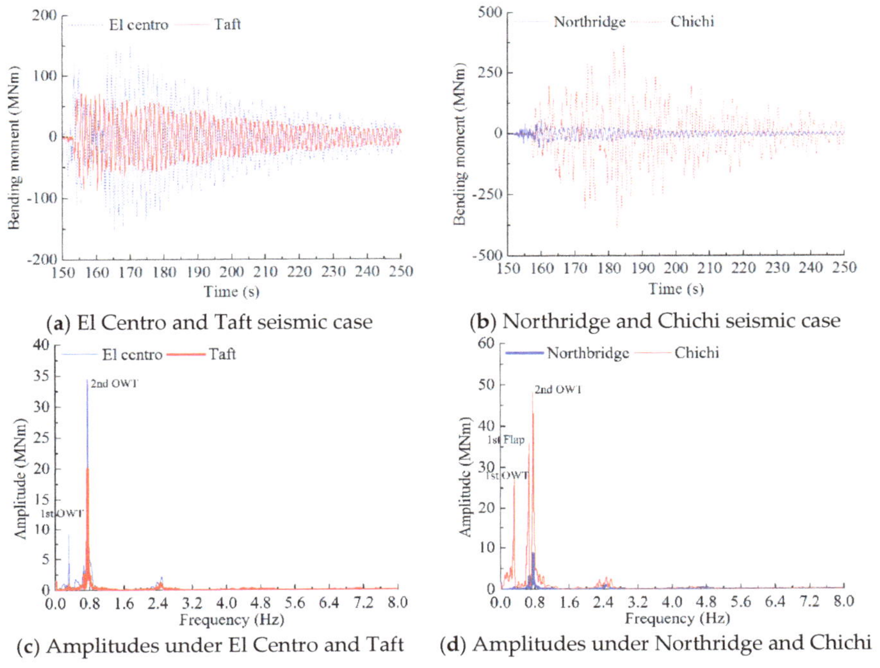

Figure 9. Mudline bending moments under seismic excitations.

4. Vibration Control of OWT using TMD

4.1. Governing Equation of Motion of OWT with TMD

During the study, a TMD is added to the tower of the OWT based on the released TMD module [23] in FAST. So the passive control method of TMD is applied to control the response of OWT under the seismic cases, as shown in Figure 1. The coupled governing equation of motion of OWT with TMD under earthquakes can be written as:

$$[M]\{\ddot{u}(t)\} + [C]\{\dot{u}(t)\} + [K]\{u(t)\} - \{E\}\left(C_{\mathrm{TMD}}\dot{u}_{\mathrm{TMD}} + K_{\mathrm{TMD}}u_{\mathrm{TMD}}\right) = -[M]\{\ddot{u}_g\} \tag{3}$$

$$M_{\mathrm{TMD}}\ddot{u}_{\mathrm{TMD}}(t) + C_{\mathrm{TMD}}\dot{u}_{\mathrm{TMD}}(t) + K_{\mathrm{TMD}}u_{\mathrm{TMD}}(t) = -M_{\mathrm{TMD}}\ddot{u}_{\mathrm{OWT}}(t), \tag{4}$$

where M_{TMD}, C_{TMD}, and K_{TMD} are the mass, damping, and stiffness of the TMD, respectively; $\{\ddot{u}_{\mathrm{TMD}}(t)\}$, $\{\dot{u}_{\mathrm{TMD}}(t)\}$, and $\{u_{\mathrm{TMD}}$ (t)$\}$ are the parameters of TMD's acceleration, velocity, and displacement, respectively; $\{E\}$ is a unit vector that represents the location of the TMD on the OWT; and $\ddot{u}_{\mathrm{OWT}}(t)$ is the corresponding velocity of the OWT with respect to the location of the TMD.

The selection of parameters of TMD will determine control effect on the OWT responses under earthquakes. The optimal design standards of TMD suggested by Zhou [24] and Connor [25] can be expressed as Equations (5) and (6).

$$f_{\mathrm{opt}} = \frac{(1 + 0.5\mu_{\mathrm{m}})^{\frac{1}{2}}}{(1 + \mu_{\mathrm{m}})} \tag{5}$$

$$\zeta_{\mathrm{opt}} = 0.5\sqrt{\mu_m}, \tag{6}$$

where f_{opt} is the optimized frequency ratio of TMD, $f_{opt} = f_{TMD}/f_{OWT}$; f_{TMD} is the tuning frequency of TMD; f_{OWT} is the selected natural frequency of OWT; μ_m is the mass ratio, $\mu_m = M_{TMD}/M_{OWT}$; M_{OWT} is the mass of OWT; and ζ_{opt} is the optimized damping ratio of TMD.

Then, the parameters of TMD can be calculated by μ_m, f_{opt} and ζ_{opt} based on Equations (7)–(9).

$$M_{TMD} = \mu_m \cdot M_{OWT} \tag{7}$$

$$f_{TMD} = f_{opt} f_{OWT};\ K_{TMD} = M_{TMD}(2\pi f_{TMD})^2 \tag{8}$$

$$C_{TMD} = 4\pi\zeta_{opt} f_{TMD} M_{TMD}. \tag{9}$$

4.2. Design of TMDs for OWT under Earthquakes

The parameters of the tuning frequency, locations and mass are essential for designing TMD. According to the dynamic characteristics of the Pentapod OWT under earthquakes, the first two natural frequencies of the OWT which are the dominant frequencies of the tower accelerations and mudline bending moments, are designated as the tuning frequencies of TMDs. Meanwhile, the nacelle and the tower base were selected as the alternative locations for TMDs, as shown in Figure 10. Further, the mass ratios such as 1%, 2% and 3% were selected to determine the optimized parameters of TMD in Equations (5) and (6). The upper limit of the TMD's mass ratio was selected as 3%, which is a recommended value for the motion control of offshore structures under earthquakes by Zhou [24]. Table 4 lists the calculated parameters of the TMD based on the optimal design formulas and the suggested mass ratios. As listed in Table 4, the tuning frequency of TMD is slightly different from the selected natural frequency of OWT due to the influence of f_{opt}, such as the tuning frequency of TMD01 is 0.303Hz, and the corresponding natural frequency of OWT is 0.305 Hz.

Table 4. Optimized parameters of tuned mass dampers (TMD) for the vibration control of OWT under seismic excitations.

TMD No.	Optimised Parameters			Parameters of TMD				TMD Location
	μ_m (%)	f_{opt}	ζ_{opt}	M_{TMD} (kg)	f_{TMD} (Hz)	K_{TMD} (N/m)	C_{TMD} (N/ms^{-1})	
TMD 01	1.0	0.993	0.05	14,897	0.303	52,148	2787	Nacelle
TMD 02	2.0	0.985	0.07	29,795	0.300	102,769	7826	
TMD 03	3.0	0.978	0.09	44,692	0.298	151,923	14,272	
TMD 04	1.0	0.993	0.05	14,897	0.303	52,148	2787	Tower base
TMD 05	2.0	0.985	0.07	29,795	0.300	102,769	7826	
TMD 06	3.0	0.978	0.09	44,692	0.298	151,923	14,272	
TMD 07	1.0	0.993	0.05	14,897	0.736	311,315	6810	Nacelle
TMD 08	2.0	0.985	0.07	29,795	0.734	613,518	19,120	
TMD 09	3.0	0.978	0.09	44,692	0.726	906,962	34,871	
TMD 10	1.0	0.993	0.05	14,897	0.736	311,315	6810	Tower base
TMD 11	2.0	0.985	0.07	29,795	0.734	613,518	19,120	
TMD 12	3.0	0.978	0.09	44,692	0.726	906,962	34,871	

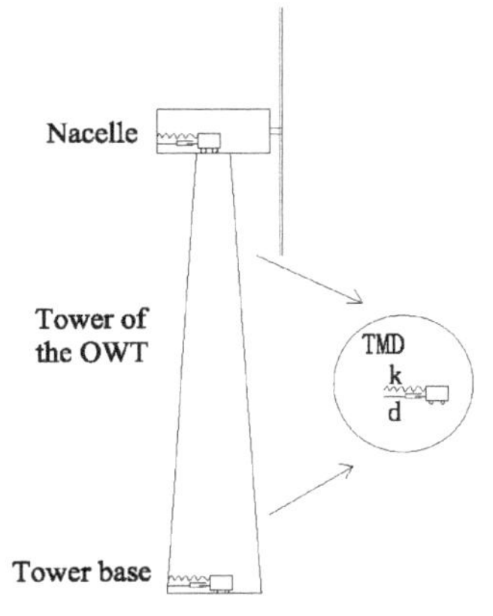

Figure 10. Schematic of the alternative locations for the tuned mass dampers (TMD) mounted on the OWT.

Table 5 lists the load cases used for investigating the influence of TMD parameters on the reduction in the structural responses of the OWT under seismic excitations. To investigate the influence of PGA on the control effects of TMD, El Centro seismic waves with different PGAs are also considered as cases T1–T48, as listed in Table 5.

Table 5. Load cases for the vibration control of OWT under earthquake load.

Case No.	Seismic excitation	State of OWT	TMD No.
T1–T12	El Centro (0.1 g)		TMD01–12
T13–T24	El Centro (0.15 g)		TMD01–12
T25–T36	El Centro (0.2 g)		TMD01–12
T37–T48	El Centro (0.4 g)	Parked	TMD01–12
T49–T60	Taft		TMD01–12
T61–T72	Northridge		TMD01–12
T73–T84	Chichi		TMD01–12

4.3. *Influence of Seismic Waves with Varying PGAs on Vibration Control*

4.3.1. Control Effect of TMD on the NAA

According to the dynamic characteristics of nacelle acceleration (NAA) under the El centro seismic wave shown in Figure 8, the frequency component of 0.305 Hz and 0.742 Hz is discovered to be the dominant frequencies of the response which is the first two natural frequencies of OWT, respectively. Meanwhile, the influence of the second order natural frequency is discovered to be more prominent.

So the reduction of the nacelle acceleration under such seismic case achieved by the TMD with the tuning frequency of the second natural frequency of the OWT is presented in order to investigate the effectiveness of TMD, such as the reducing of the structural responses achieved by TMD10–TMD12. On the other hand, in order to research the influence of the PGA of seismic waves on the control effect of TMD, El Centro seismic waves with different PGAs are also applied to the model. The reduction in the NAA time histories are illustrated in Figure 11, it can be seen that TMD10 reduces the NAA significantly under the El Centro seismic wave of PGA 0.1 g. With the increase in the PGA, the control effects of TMD decreases, as shown in Figure 11b; however, the reduction in NAA is still prominent.

Further details on the influence of the PGA on the control effect of TMD can be understood from Figure 12. Variations in the statistics of NAAs controlled by TMD01–03 and TMD10–12 under El Centro seismic waves with varying PGAs are shown in Figure 11, which tuning frequency is the fundamental and second order natural frequency of the OWT, respectively. The term reduction ratio in the Figures can be defined as,

$$\text{Reduction ratio} = \frac{(Statistics - Statistics_{\text{TMD}})}{Statistics} \times 100\% \tag{10}$$

where *Statistics* is the maximum or minimum value or the standard deviation of the structural responses of the OWT without TMD and $Statistics_{\text{TMD}}$ is the corresponding maximum or minimum value or standard deviation of the responses reduced by the TMD.

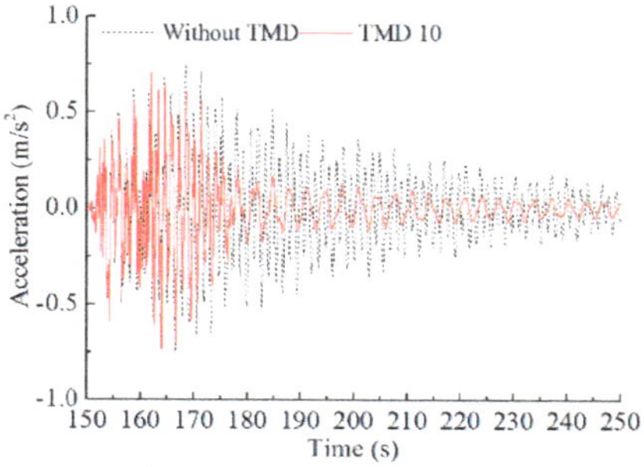

(**a**) El Centro (0.1 g) NAA TMD10

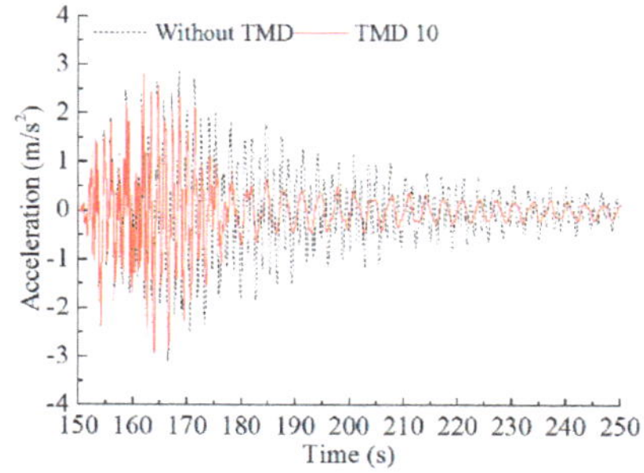

(**b**) El Centro (0.4 g) NAA TMD10

Figure 11. Reduction in nacelle acceleration (NAA) by TMDs under El Centro seismic wave.

Initially, the ineffectiveness of TMD01 can be observed from the slight reduction in the statistics, as shown in the Figure 12. TMD03 is found to be more effective in reducing the maximum values when the PGA of the El Centro seismic wave is 0.1 g, as shown in Figure 12a. However, the control effects of TMD03 decrease significantly with the increase in the PGA of the El Centro seismic wave; e.g., the reduction ratios of TMD03 for the maximum values of NAAs are 12.8% and 9% under the El Centro seismic waves of PGAs 0.1 and 0.4 g, indicating that the control effects of TMD03 weaken remarkably with the increasing of the PGA of El Centro seismic wave.

Further, Figure 12b demonstrates that TMD10–12 are more efficient in reducing the standard deviations of NAAs, and the maximum reduction ratio of the standard deviations of the responses exceeds 40%. Influence of the PGAs of seismic waves can also be discovered for such effective TMDs that the reducing ratio of TMD10 decreases to only 29% under the El Centro seismic wave with the maximum PGA.

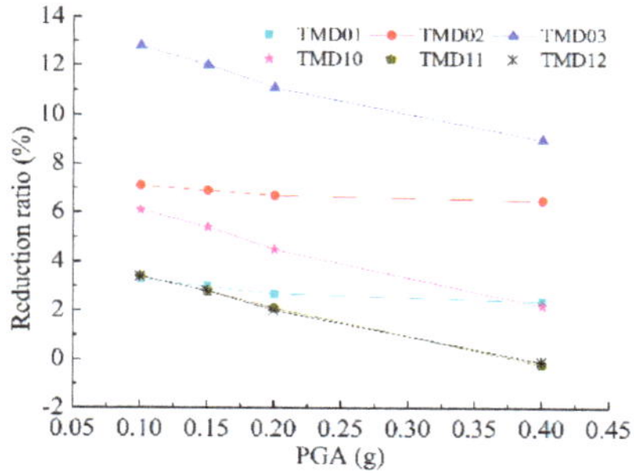

(**a**) Maximum values of NAAs

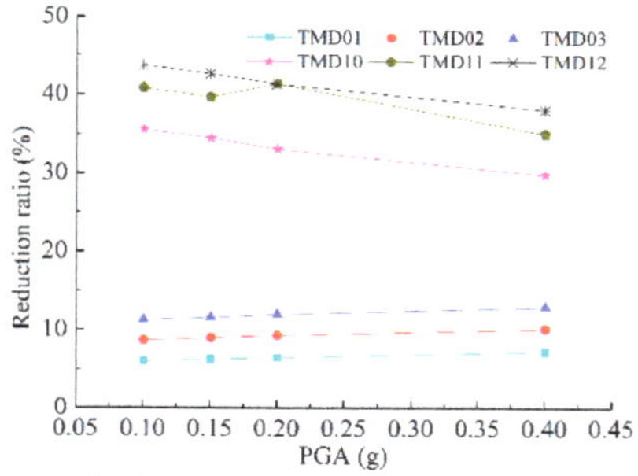

(**b**) Standard deviation of NAAs

Figure 12. Reduction ratios of NAAs by TMDs under El Centro seismic wave.

4.3.2. Control Effect of TMD on the Mudline Bending Moment

The influence of the PGAs on the control effect of bending moment is studied in this section. Figure 13 illustrates the reduction in the time histories of bending moments under El Centro seismic waves of PGAs 0.1 and 0.4 g.

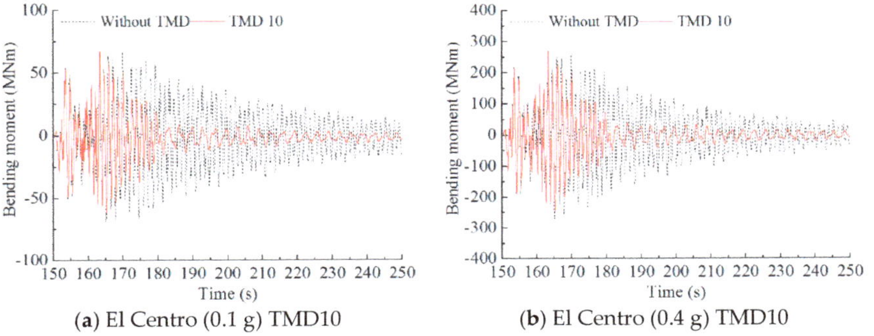

Figure 13. Reduction in bending moments by TMDs under El Centro seismic wave.

For TMD10, the bending moments under load case T10 (El Centro 0.1 g) and T56 (El Centro 0.4 g) is decreased significantly. Further detailed comparisons are depicted in Figure 14, based on the reduction in the statistics achieved. It can be observed that the control effects of TMD10 and 11 are sensitive to the variations in PGA. The reduction ratios of the maximum values is comparatively smaller than that of standard deviations, which can exceed 50%. Meanwhile, the reduction ratios of TMD01 decrease significantly under the El Centro seismic wave of the highest PGA, and the control effects cannot be improved even by increasing the mass of the TMDs under such seismic excitations.

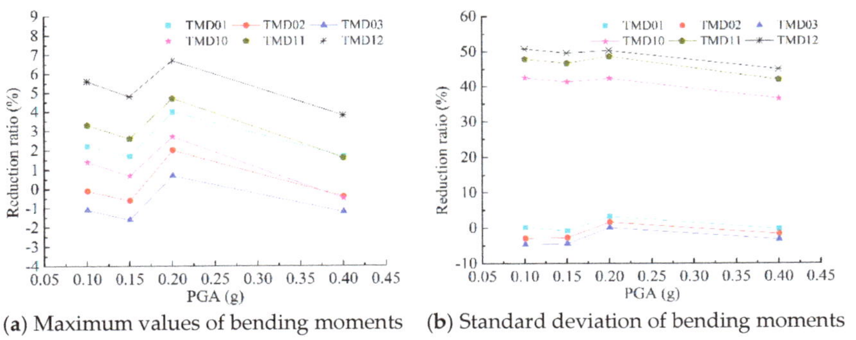

Figure 14. Reduction ratios of bending moments by TMD under El Centro seismic wave.

4.4. *Influence of the Seismic Excitations on Vibration Control*

4.4.1. Control Effect of TMD on NAA under Different Seismic Excitations

Figure 15 illustrates the reduction in NAAs by the TMD under different seismic excitations. It can be seen that the reduction in NAA by TMD01 is rather insignificant under the Northridge seismic waves, unlike that under the Chichi seismic waves, as shown in Figure 15a,c. Significant decrease in the NAA is achieved by TMD07 under the Northridge seismic wave, as shown in Figure 15b. The comparison of Figure 15c,d indicates that the control effects of TMD01 and 07 under the Chichi seismic wave are nearly identical.

From the above discussions, it can be seen that the type of the seismic wave influences the control effect of TMD significantly, which can be attributed to the characteristics in the frequency domains of the responses. For the NAA under the Northridge seismic wave, the dominant frequency of the response is the second natural frequency of the OWT. However, the nearly identical control effects of TMD01 and 07 under the Chichi seismic wave can be attributed to the comparable influence of the first two natural frequencies on the NAA under the Chichi seismic wave, as shown in Figure 8.

Figure 16 compares the effectiveness of TMD1–12 under the listed load cases of Table 5 based on the reductions in the maximum values and standard deviations of the bending moment time histories. As shown in Figure 16, using TMD, diverse control effects of the OWT under different types of seismic excitations are observed; however, the effectiveness of the TMD in reducing the standard deviations are reserved, especially for the Taft seismic wave. The reduction ratio of standard deviations of the NAAs under the Taft seismic wave nearly reaches to 55%. Although TMD02 and 03 marginally reduce the responses under the El Centro seismic wave, remarkable reduction ratios are obtained under the Chichi seismic wave. Furthermore, Figure 15 illustrates that TMD07–09 can reduce the statistics under the Taft and Northridge seismic waves significantly, while a more drastic reduction is found under the Chichi seismic wave.

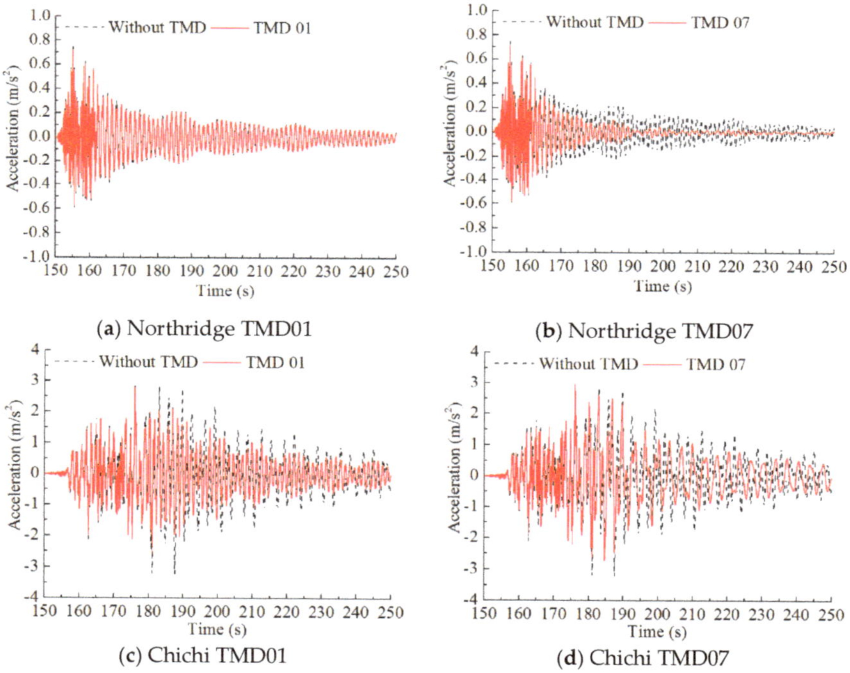

(a) Northridge TMD01 (b) Northridge TMD07 (c) Chichi TMD01 (d) Chichi TMD07

Figure 15. Reduction in NAAs by TMDs under different seismic excitations.

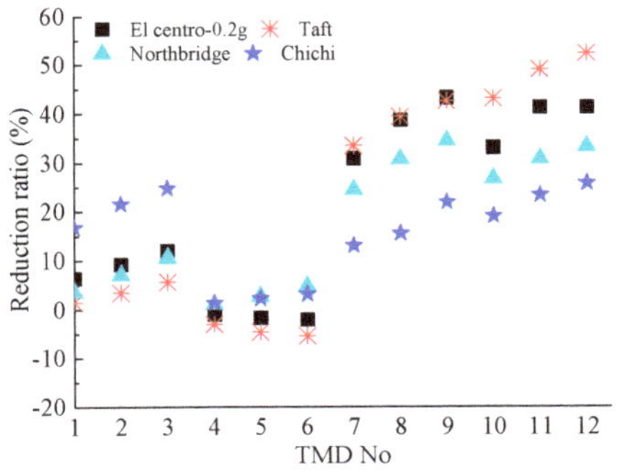

Figure 16. Reduction ratios of standard deviations of NAAs by TMDs under different earthquakes.

4.4.2. Variations in Bending Moments by TMD under Different Seismic Excitations

Reduction in mudline bending moments by TMD under different seismic excitations is shown in Figure 17. Trends similar to the control effects of TMD01 and 07 on the accelerations at the tower top can be found for the reduction in bending moments under the Chichi seismic wave. The obvious control effect of TMD10 on the bending moments under the Northridge seismic wave can be observed from Figure 17b, indicating that the tuning frequency of TMD07 is the dominant frequency of the response under that excitation. On the other hand, limited control effect of TMD01 is observed in Figure 17a due to the significant discrepancies between the tuning frequency of TMD and the dominant frequency of bending moments under the Northridge seismic wave. Furthermore, the reduction in the bending moments in Figure 17d is more distinct than the results in Figure 17c, especially in the periods of 190–250 s, indicating that under the Chichi seismic wave, more energy is captured by the second natural frequency of the OWT, as shown in Figure 9d, which is the tuning frequency of TMD07.

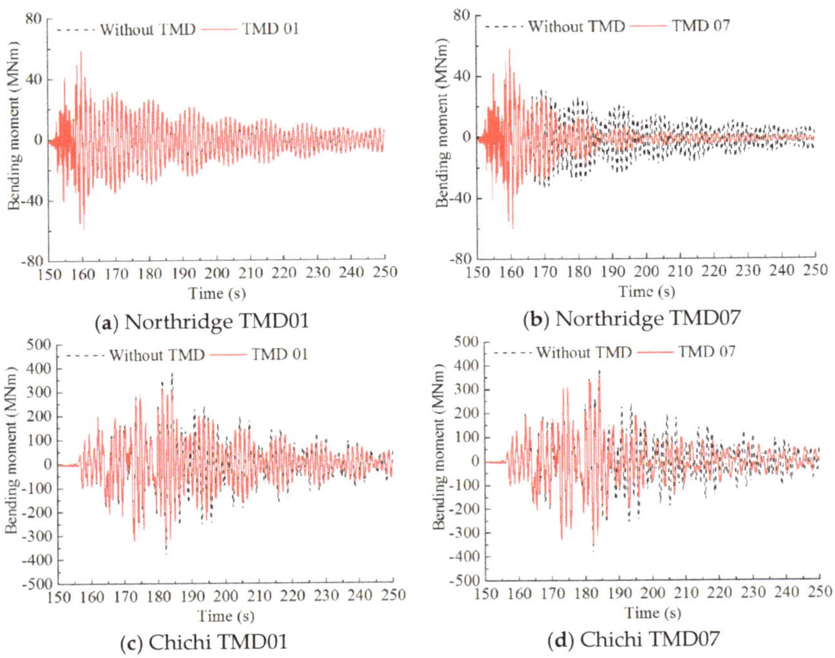

Figure 17. Reduction in bending moments by TMDs under different seismic excitations.

Variations in reduction ratios of bending moment statistics by TMD under different seismic excitations are displayed in Figure 18. From the figure, it can be seen that TMD10–12 are more effective in reducing the standard deviations of the bending moments under the Taft and Northridge seismic waves, and the maximum reduction ratio exceeds 60%. Reduction ratios decrease significantly under the Chichi seismic wave because the first two natural frequencies are both dominant frequencies of the bending moments. Only one TMD designed with one tuned frequency such as the first or the second order frequency of the OWT is installed in the OWT, which is not effective for controling the response with more than one dominant frequency.

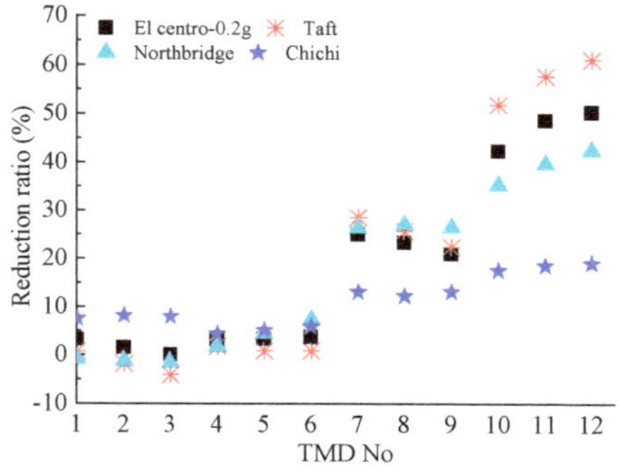

Figure 18. Reduction ratios of standard deviations of bending moments under earthquakes by TMD.

4.5. Influence of TMD's Mass Ratio on Vibration Control

4.5.1. Influence of TMD's Mass Ratio on the Reduction in NAAs

TMDs with different mass ratios such as 1% (TMD04, 07, 10), 2% (TMD05, 08, 11) and 3% (TMD06, 09, 12), are also considered in the study to analyze the effects of mass ratios on their control effects. Figure 19 illustrates the reduction in NAAs by TMD10 and 12 under the Taft and Chichi seismic waves. From the Figure, it can be seen that the improvements in the control effects obtained by increasing the mass of the TMD are trivial, though the mass ratio of TMD12 is increased threefold than that of TMD10, as listed in Table 4.

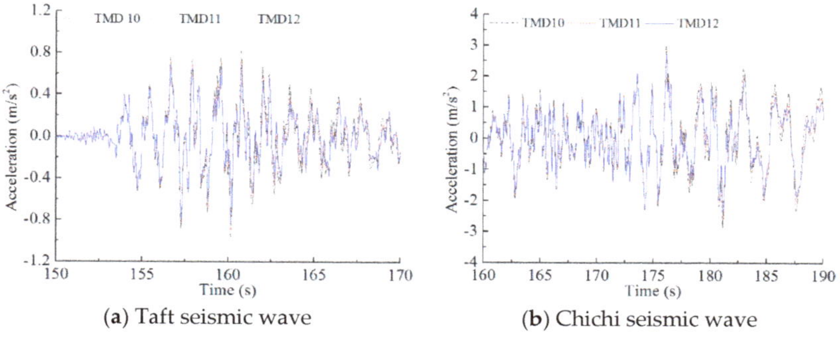

(**a**) Taft seismic wave (**b**) Chichi seismic wave

Figure 19. Reduction in NAAs by TMDs under Taft and Chichi seismic waves.

Figure 20 shows the variation of reduction ratio with respect to mass ratio of TMD. For the effective TMD, it is found that the achieved reduction ratios can be improved by increasing the mass ratios, however, an increase in the mass ratio of TMD cannot result in a proportional increase in the reduction ratio. From Figure 20b, the reduction ratio of the standard deviations of NAAs under the Taft seismic wave is 43.2% for TMD10 with 1% mass ratio and 52.4% for TMD12 with 3% mass ratio. Increment in the reduction ratio is only 9.2%, while the mass of the TMD is triple. For the ineffective TMD, such as for TMD04–06; it can be seen that improvements cannot be obtained in the reduction ratios by an increase in the mass ratios, such as for the reduction in the maximum values of NAAs shown in Figure 20a.

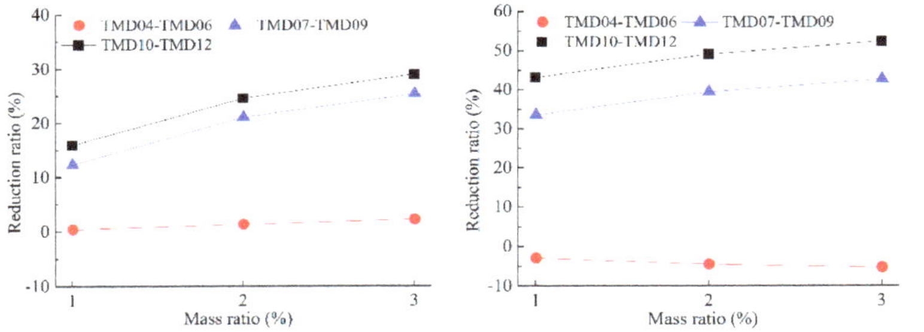

(**a**) Maximum values of NAAs (**b**) Standard deviations of NAAs

Figure 20. Reduction ratios of statistics of the NAAs under Taft seismic waves.

4.5.2. Influence of TMD's Mass Ratio on the Reduction in Bending Moments

Influence of TMD with increased mass on the reduction in bending moments under the Taft and Chichi seismic waves can be observed from Figure 21. The results similar to the reduction in NAAs under the same load cases can be obtained for the reduction in bending moments. The expected control effects cannot be achieved only by increasing the mass ratios of the TMD, especially under the Chichi seismic wave.

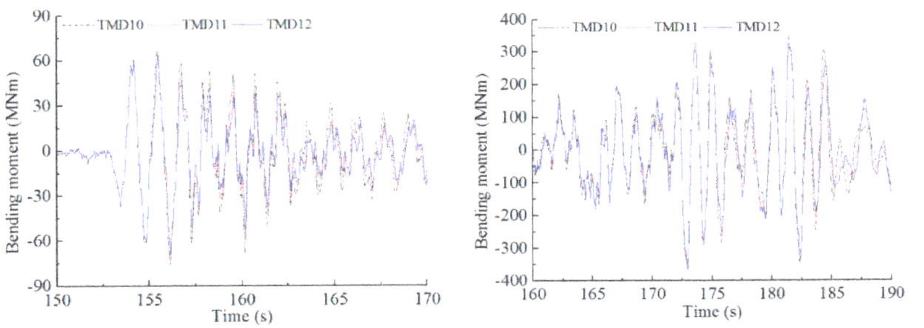

(**a**) Taft seismic wave (**b**) Chichi seismic wave

Figure 21. Reduction in bending moments by TMDs under Taft and Chichi seismic waves.

The influence of TMD's mass ratio on the reduction in bending moments can be observed from Figure 22. Remarkable control effects of TMD10–12 on the standard deviations of bending moments under the El Centro, Taft, and Northridge seismic waves can be found. The reduction ratio of the standard deviations by TMD10 and 12 under the Taft seismic wave were 51.8% and 61%, respectively. However, the achieved increment in the reduction ratio of bending moments was only 9.2% for a triple increase in the mass ratio of TMD.

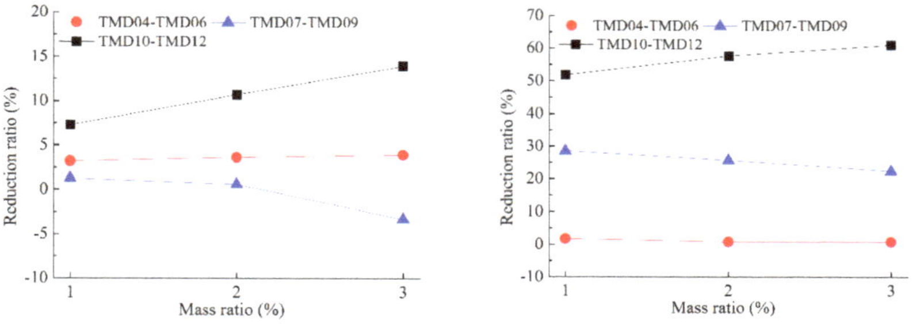

(a) Maximum values of bending moment (b) Standard deviations of bending moment

Figure 22. Reduction ratios of statistics of bending moments under Taft seismic waves.

4.6. Influence of TMD's Location on Vibration Control

4.6.1. Influence of TMD's Location on the Reduction of NAAs

As shown in Figure 23, TMD08 and 11 are selected to demonstrate the influence of TMD's location on its control effects. It can be observed that the TMD mounted at the nacelle (TMD08) or tower base (TMD11) decreases the NAAs under the Northridge seismic wave remarkably, as shown in Figure 23a. Meanwhile, according to Figure 22b, the control effects are limited by comparing with the reducing of the NAAs under the El Centro seismic wave, but the comparable control effects of the selected TMD08 and 11 under the Chichi seismic wave still can be discovered. So the nacelle and tower base would be the alternative optimal locations for the TMD to control the NAAs under such seismic waves.

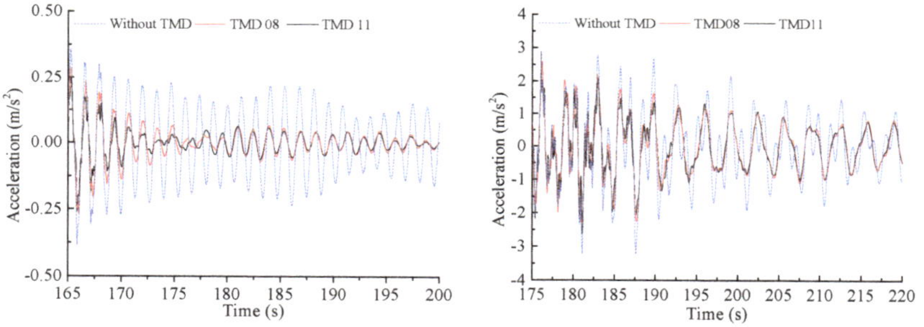

(a) Northridge seismic wave (b) Chichi seismic wave

Figure 23. Reduction in NAAs by TMDs under Northridge and Chichi seismic waves.

The reduction ratios of standard deviations of NAAs are illustrated in Figure 24. From Figure 24a, the reduction ratios of the standard deviations for the TMD installed at tower base or nacelle under the Northridge seismic wave are close for the corresponding cases such as TMD07 and 10, TMD08 and 11, TMD09 and 12. So the control effects are approximately same for the selected TMD's locations. According to Figure 24b, the reduction ratios of the standard deviations are different. For the corresponding cases, the reduction ratios achieved by TMD08 and 11 under such seismic case are about 15.6% and 23.4%, respectively. Hence, the tower base is a more appropriate location than the nacelle for mounting the TMD under the Chichi seismic wave.

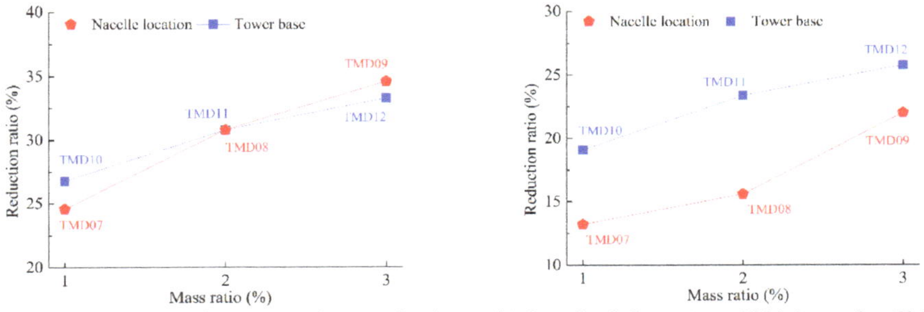

(**a**) Standard deviations of NAAs under Northridge (**b**) Standard deviation of NAAs under Chichi

Figure 24. Reduction ratios of the NAAs under Northridge and Chichi seismic waves.

4.6.2. Influence of TMD's Location on Reduction in Bending Moments

The influence of TMD's location on the reducing of the bending moment under the Northridge and Chichi seismic wave is shown in Figure 25. According to the comparisons shown in Figure 25a,b, it can be seen that approximately identical with the reducing of the nacelle accelerations under the selected seismic waves, the comparable control effects of TMD 08 and 11 on the bending moment under such seismic cases are also observed.

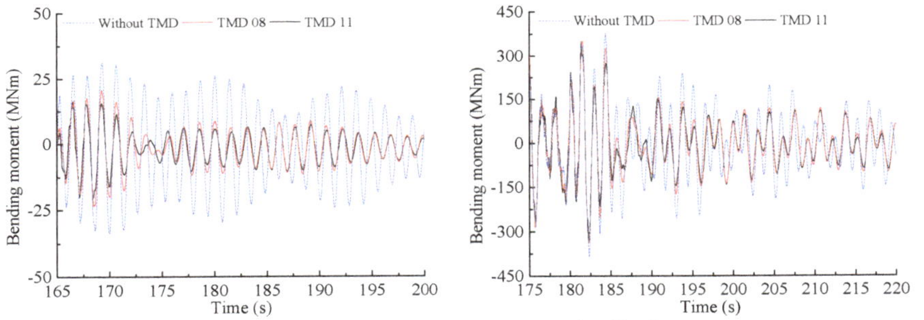

(**a**) Northridge seismic wave (**b**) Chichi seismic wave

Figure 25. Reduction in bending moments by TMDs under Northridge and Chichi seismic waves.

Meanwhile, the reduction ratios of the relevant statistics under such seismic cases by TMDs with varying locations are shown in Figure 26. From Figure 26a, the reduction ratios of the standard deviations of the response under the Northridge seismic wave achieved by TMD08 and 11 are about 27.2% and 39.7%, respectively. The reduction ratios of the standard deviations achieved by TMD09 and 12 are about 26.6% and 42.4%, respectively. Thus, the tower base is the more feasible location for the control of the mudline bending moment by TMD under such seismic wave.

Similarly, under the Chichi seismic wave, the location of TMD07–09 improved the reduction in the bending moment, as shown in Figure 26b. The reduction ratios of standard deviations achieved by TMD11 is about 18.6%, which increased approximately 6.2% by changing the location of TMD from nacelle for TMD08 to the tower base. So the locations of the TMD influence the control effects significantly.

Therefore, both the nacelle and tower base are the appropriate places to mount the TMD to control the motion of the OWT under earthquakes. Furthermore, the control effects of the OWT response by TMDs mounted at the tower base are better than the ones at the nacelle.

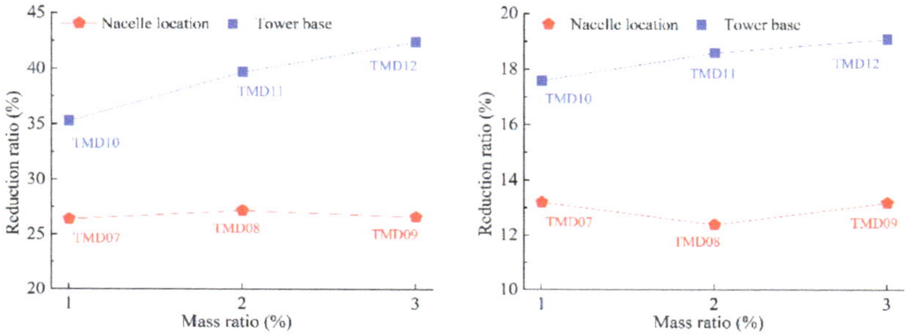

(**a**) Standard deviations under Northridge (**b**) Standard deviations under Chichi

Figure 26. Reduction in bending moments under Northridge and Chichi seismic waves.

4.7. Influence of TMD's Tuning Frequency on Vibration Control

4.7.1. Influence of TMD's Tuning Frequency on the Reduction in NAAs

The tuning frequency of TMD is another essential parameter that should be emphasised in the design of TMD. TMDs listed in Table 4 are selected to perform the investigation on the influence of TMD's tuning frequency on its control effects. From the table, the TMDs are divided into two categories by the tuning frequencies. One is that the tuning frequency is the fundamental frequency of the OWT such as TMD01–03 and TMD07–09; the other is that the TMDs designate the second order natural frequency of the OWT as the tuning frequency such as TMD04–06 and TMD10–12.

The influence of the tuning frequency on the control effects of OWT under earthquakes are shown in Figure 27. The significant discrepancies of the control effects between the TMDs with different tuning frequencies is observed. From Figure 27a, the control effects of TMD03 under El Centro seismic wave improved significantly by changing its tuning frequency from the fundamental frequency of the OWT to the second order frequency, and the reduction ratios of the standard deviations achieved by TMD03 was increased from 12% to about 41.3%. Approximate improvements can also be discovered from TMD04–06 and TMD10–12 under such seismic waves. So an effective method to improve the limited control effects TMD01–06 is to adjust the tuning frequencies of such TMDs from the fundamental frequency to the second order natural frequency of the OWT.

From Figure 27b,c, the reduction ratio of the standard deviations under the Northridge seismic wave achieved by TMD03 and 09 were about 10.7% and 33.3%, respectively; the reduction ratio of the standard deviations under the Taft seismic wave achieved by TMD03 and 09 were about 5.8% and 42.7%, respectively. On the other hand, different from the above introduced seismic cases, it can be discovered the comparable control effects of TMD01–03 and TMD07–09 on the NAAs under the Chichi seismic wave, as shown in Figure 27d. So both the first two natural frequencies of the OWT are the feasible tuning frequencies under such seismic case.

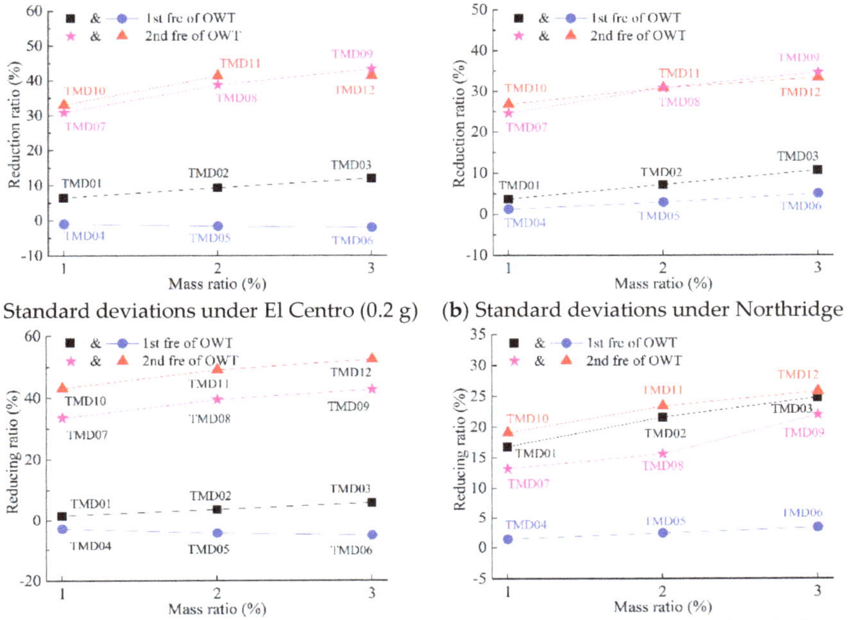

(**a**) Standard deviations under El Centro (0.2 g) (**b**) Standard deviations under Northridge

(**c**) Standard deviations under Taft (**d**) Standard deviations under Chichi

Figure 27. Reduction ratios of NAAs under El Centro and Northridge seismic waves.

4.7.2. Influence of TMD's Frequency on the Reduction in Bending Moment

Figure 28 shows the influence of the TMD's tuning frequency on the reduction of the bending moment under the different seismic cases. From the Figure, the prominent control effects of the TMDs are discovered when the second natural frequency of the OWT is designated as the tuning frequency. Reduction ratios of standard deviations achieved by TMD12 and 06 under the El Centro seismic wave were about 50.3 and 3.8%, respectively, as shown in Figure 28a. Although an optimum location and prominent mass ratio are selected in the design of TMD06, the control effects on the bending moments are trivial due to the discrepancy in the tuning frequency. If the tuning frequency of TMD06 is adjusted from the fundamental frequency of the OWT to the second natural frequency, the increments in the reduction ratio can reach 45%. Same tuning frequencies can also be applied to the TMD under the Northridge seismic wave, according to Figure 28b. It can be seen that the reduction ratios of bending moments achieved by TMD12 and 06 are 42.4% and 7.5%, respectively. Meanwhile, the same results can be obtained for TMD01–03 and TMD07–09 under the El Centro, Northridge and Taft seismic wave.

On the other hand, it should also be noted that limited control effects of TMD are observed under the Chichi seismic wave by comparing with reducing of the responses under the other seismic cases, as shown in Figure 28d. From Figure 9d, it can be seen that in addition to the second natural frequency, the influence of the fundamental frequency of the OWT is also non-negligible under such seismic load cases. Hence, a TMD with single tuning frequency cannot effectively control the vibration of the OWT under the Chichi seismic wave.

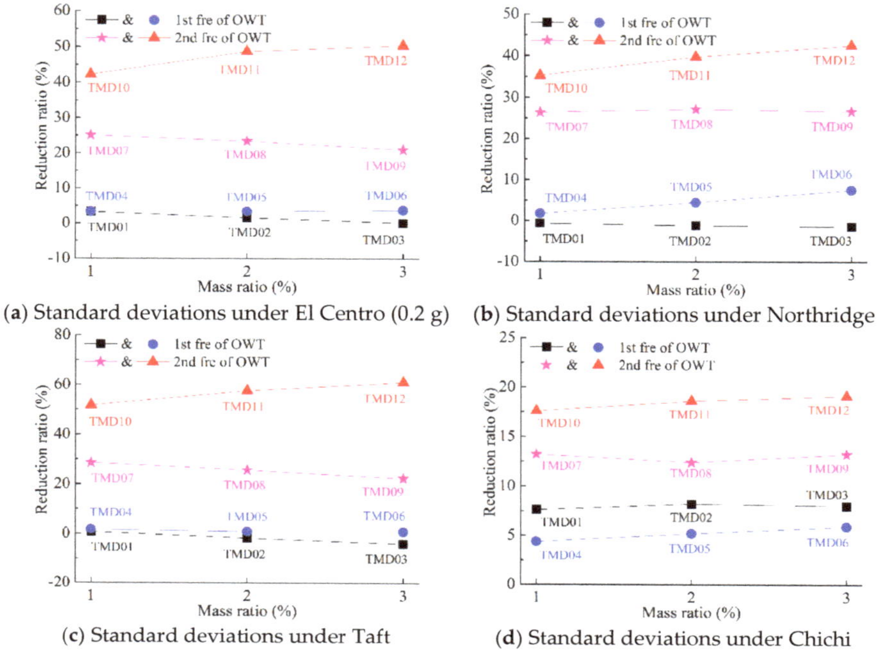

(**a**) Standard deviations under El Centro (0.2 g) (**b**) Standard deviations under Northridge

(**c**) Standard deviations under Taft (**d**) Standard deviations under Chichi

Figure 28. Reduction ratios of bending moments under seismic excitations.

5. Conclusions

Based on the integrated analysis model, the structural responses and dynamic characteristics of a Pentapod OWT under seismic excitations are studied. The influence of the rotor and the activation of higher-order modes of the OWT under seismic excitations can be observed. Further, a TMD is mounted on the OWT to control the structural responses under different seismic excitations. From the research, the following conclusions can be drawn:

(1) From the results that the mudline bending moments are influenced by the mode of the rotor blade, the influence of the rotor system on the motion of the support system cannot be neglected in the seismic analysis of OWTs. Thus, it is of necessity to establish the integrated model of the OWT to obtain reasonable structural responses under earthquakes.

(2) For the tower accelerations, the seismic excitations can stimulate higher-order frequency components and it can become the dominant frequency for the structural responses, especially for the seismic waves with abundant frequencies around the natural frequencies of OWTs, such as the Northridge seismic wave.

(3) Using TMD, the influences of the PGA and type of seismic excitation on the control effects of the OWT are investigated. It can be observed that these parameters of the input seismic wave can influence the control effects of TMD remarkably.

(4) The parameters of TMD, such as the tuning frequency and location, are more essentials than the mass ratio in the design of TMD. The higher-order dominant frequencies are proved as the more effective tuning frequencies for reducing the structural responses of the OWT under seismic excitations; e.g., TMD tuned with the second natural frequency of the OWT can achieve more effective control than the TMD tuned with the fundamental frequency.

(5) The locations of TMD installed in nacelle or tower base are validated as the appropriate locations to control the nacelle accelerations and bending moments of the OWT.

(6) The studies prove that the control effects of the TMD can be improved by increasing the mass of the TMD only when an effective tuning frequency and location are adopted.
(7) Limited control effects of TMDs are observed when the structural responses are composed of multi-dominant frequencies, such as the nacelle accelerations and bending moments under the Chichi seismic wave. Additional tuning frequencies or more TMDs may be necessary to achieve significant control effects under such load cases.

Author Contributions: W.W.: Writing and data collection; X.L.: Data analysis, interpretation and revising; Z.P.: Figures and tables; Z.Z.: Numerical analysis and revising.

Funding: This research was funded by National Natural Science Foundation of China [Grant No. 51879040].

Acknowledgments: The authors would like to acknowledge the support from the National Natural Science Foundation of China (Grant No. 51879040).

Conflicts of Interest: The authors declare no conflict of interest.

References

1. Bazeos, N.; Hatzigeorgiou, G.D.; Hondros, I.D.; Karamaneas, H.; Karabalis, D.L.; Beskos, D.E. Static, seismic and stability analyses of a prototype wind turbine steel tower. *Eng. Struct.* **2002**, *24*, 1015–1025. [CrossRef]
2. Witcher, D. Seismic analysis of wind turbines in the time domain. *Wind Energy* **2005**, *8*, 81–91. [CrossRef]
3. Häenler, M.; Ritschel, U.; Warnke, I. Systematic modelling of wind turbine dynamics and earthquake loads on wind turbines. In Proceedings of the European Wind Energy Conference and Exhibition 2006, EWEC 2006, Athens, Greece, 27 February–2 March 2006.
4. Zhao, X.; Maisser, P. Seismic response analysis of wind turbine towers including soil-structure interaction. *Proc. Inst. Mech. Eng. Part K J. Multi-Body Dyn.* **2006**, *220*, 53–61. [CrossRef]
5. Riziotis, V.A.; Madsen, H.A. *Wind Energy Systems: Optimising Design and Construction for Safe and Reliable Operation*; Woodhead Publishing Limited: Shaston, UK, 2010.
6. Prowell, I. An Experimental and Numerical Study of Wind Turbine Seismic Behavior. Ph.D. Thesis, University of California, San Diego, CA, USA, 2011.
7. Prowell, I.; Elgamal, A.; Uang, C.M.; Enrique Luco, J.; Romanowitz, H.; Duggan, E. Shake table testing and numerical simulation of a utility scale wind turbine including operational effects. *Wind Energy* **2014**, *17*, 997–1016. [CrossRef]
8. Zheng, X.Y.; Li, H.; Rong, W.; Li, W. Joint earthquake and wave action on the monopile wind turbine foundation: An experimental study. *Mar. Struct.* **2015**, *44*, 125–141. [CrossRef]
9. Wang, W.H.; Gao, Z.; Moan, T.; Li, X. Model Test of an Offshore Bottom Fixed Pentapod Wind Turbine under Seismic Loads. In Proceedings of the International Conference on Ocean, Offshore and Arctic Engineering, Busan, Korea, 19–24 June 2016.
10. Wang, W.H.; Gao, Z.; Moan, T.; Li, X. Model Test and Numerical Analysis of a Multi-pile Offshore Wind Turbine under Seismic, Wind, Wave and Current Loads. *J. Offshore Mech. Arct. Eng.* **2016**, *139*, 031901. [CrossRef]
11. Penzien, J.; Kaul, M.K.; Berge, B. Stochastic response of offshore towers to random sea waves and strong motion earthquakes. *Comput. Struct.* **1972**, *2*, 733–756. [CrossRef]
12. Kim, D.H.; Lee, S.G.; Lee, I.K. Seismic fragility analysis of 5 MW offshore wind turbine. *Renew. Energy* **2014**, *65*, 250–256. [CrossRef]
13. Nuta, E.; Christopoulos, C.; Packer, J.A. Methodology for seismic risk assessment for tubular steel wind turbine towers: application to Canadian seismic environment. *Can. J. Civ. Eng.* **2011**, *38*, 293–304. [CrossRef]
14. Sadowski, A.J.; Camara, A.; Málaga-Chuquitaype, C.; Dai, K. Seismic analysis of a tall metal wind turbine support tower with realistic geometric imperfections. *Earthq. Eng. Struct. Dyn.* **2017**, *46*, 201–219. [CrossRef]
15. Stewart, G.; Lackner, M.A. Offshore wind turbine load reduction employing optimal passive tuned mass damping systems. *IEEE Trans. Control Syst. Technol.* **2013**, *21*, 1090–1104. [CrossRef]
16. Dinh, V.N.; Basu, B. Passive control of floating offshore wind turbine nacelle and spar vibrations by multiple tuned mass dampers. *Struct. Control Health Monit.* **2015**, *22*, 152–176. [CrossRef]

17. Colwell, S.; Basu, B. Tuned liquid column dampers in offshore wind turbines for structural control. *Eng. Struct.* **2009**, *31*, 358–368. [CrossRef]
18. Jonkman, J.M.; Buhl, M.L., Jr. *FAST User's Guide*; Technical Report No. NREL/EL-500-38230; National Renewable Energy Laboratory: Golden, CO, USA, 2005.
19. Jonkman, J.; Butterfield, S.; Musial, W.; Scott, G. *Definition of A 5-MW Reference Wind Turbine for Offshore System Development*; Technical Report No. NREL/TP-500-38060; National Renewable Energy Laboratory: Golden, CO, USA, 2009.
20. DNV GL. *DNV GL-ST-0126: Support Structures for Wind Turbines*; DNV GL: Oslo, Norway, 2018.
21. DNV GL. *DNV GL-ST-0437: Loads and Site Conditions for Wind Turbines*; DNV GL: Oslo, Norway, 2016.
22. IEC. *IEC 61400-1: Wind Turbine-Part 1: Design Requirements*; IEC: Geneva, Switzerland, 2012.
23. Cava, L.W.; Lackner, M.A. *Theory Manual for the Tuned Mass Damper Module in FAST v8*; Technical Report; University of Massachusetts: Amherst, MA, USA, 2015.
24. Zhou, F.L. *Structural Motion Control in Civil Engineering*; Seismological Press: Beijing, China, 1997.
25. Connor, J.J. *Introduction to Structural Motion Control*; Prentice Hall: Upper Saddle River, NJ, USA, 2003.

Journal of
Marine Science
and Engineering

MDPI

Article

Experimental Investigations on Hydrodynamic Responses of a Semi-Submersible Offshore Fish Farm in Waves

Yunpeng Zhao [1], Changtao Guan [2], Chunwei Bi [1,*], Hangfei Liu [1] and Yong Cui [2]

1 State Key Laboratory of Coastal and Offshore Engineering, Dalian University of Technology, Dalian 116024, China
2 Qingdao Key Laboratory for Marine Fish Breeding and Biotechnology, Yellow Sea Fisheries Research Institute, Chinese Academy of Fishery Sciences, Qingdao 266071, China
* Correspondence: bicw@dlut.edu.cn

Received: 13 June 2019; Accepted: 20 July 2019; Published: 23 July 2019

Abstract: A series of physical model experiments was performed to investigate the hydrodynamic responses of a semi-submersible offshore fish farm in waves. The structural configuration of the fish farm primarily refers to that of the world's first offshore fish farm, Ocean Farm 1, developed by SalMar in Norway. The mooring line tension and motion response of the fish farm were measured at three draughts. The study indicated that the tension on the windward mooring line is greater than that on the leeward mooring line. As the wave height increases, the mooring line tension and motion responses including the heave, surge, and pitch exhibit an upward trend. The windward mooring line tension decreased slightly with increasing draught. The existence of net resulted in approximately 42% reduction in mooring line tension and approximately 51% reduction in surge motion. However, the heave and pitch of the fish farm increased slightly with the existence of net. It was found that the wave parameters, draught, and net have noticeable effect on the hydrodynamic response. Thus, these factors are suggested to be considered in structural designs and optimization to guarantee the ability of the fish farm to resist destruction and ensure safety of workers during intense waves.

Keywords: semi-submersible; offshore fish farm; hydrodynamic response; mooring line tension; motion response

---

1. Introduction

Owing to the limitations of ecological environments and space resources in near-shore areas; offshore aquaculture is becoming growing in the aquaculture industry worldwide. High-density polyethylene (HDPE) net cages are increasingly being used in the aquaculture industry. Generally, an HDPE net cage could be moored in a single or array configuration, which is flexible in industrial applications. However, the volume for aquaculture typically ranges from 10,000 m^3 to 30,000 m^3, not to mention the deformation induced by currents and waves. So far, more than 20 countries and regions around the world have been devoted to aquaculture field. For instance, France and Norway have cooperated to build an aquaculture vessel with 270 m length and the drainage can reach 10^5 t. Spanish companies have designed an aquaculture platform, which can resist to waves of 9 m. Norway, as one of the most developed fisheries countries in the world, has already been leading the world in the field of farming equipment design. Recently, a new type of offshore fish farm, Ocean Farm 1 [1] (see Figure 1), which is mainly made up of steel, with a volume of 250,000 m^3 was developed in Norway and built in China. This kind of fish farm may represent the first step toward a new era in offshore aquaculture and address key issues related to sustainable growth in the aquaculture industry. As the aquaculture is marching to the open sea, the power supply of the electrical equipment on the fish farm becomes a weakness of development. The floaters and columns for a fish farm could be ideally used as a wave energy converter platform. The aquaculture platform of sufficient dimensions can support several wind turbines and realize self-power supply of the offshore fish farm. Overall, both the converter platform of wave energy and offshore wind turbine can be combined with this kind of new type of offshore aquaculture facility. Due to its superior performance, the offshore fish farm considered is accepted and widely used all over the world. Hence, fundamental research on the offshore fish farm is very critical, especially in the initial stage.

Figure 1. Main structure and the location of connection of the ocean Farm 1.

In open seas, a fish farm is easily subjected to strong currents, severe waves, and storm surges, all of which can cause significant destruction. Therefore, studying the hydrodynamic responses of a net cage is important and urgent. In the past few decades, a considerable amount of research including experimental and numerical studies has been carried out to investigate the hydrodynamic responses of offshore net cages. Fredriksson et al. [2] analyzed both the motion response and mooring line tension of a net cage in waves and currents. Lader et al. [3] conducted a series of physical model experiments to study the forces and deformation of a circular net cage with different weights in a uniform flow. Li et al. [4] adopted a numerical model to analyze the influences of a sinker weight on the deformation and uniform load on the net cage; the results of the numerical simulation agreed well with the experimental data. Moe et al. [5] calculated the distribution of loads in the net cage owing to current, weights, and gravity based on commercial explicit finite element software. Tsukrov et al. [6] conducted experimental studies of drag forces on copper alloy net panels and empirical values for normal drag coefficients are proposed for various types of copper netting. Xu et al. [7] applied a statistical approach to determine the motion and tension transfer functions in irregular waves. Fu et al. [8] studied an extended three-dimensional hydro-elasticity theory to predict the dynamic response of 5×2 floating fish farm collars in waves. Zhao et al. [9] performed a number of physical model experiments to study the main mooring line tensions and flow velocity magnitudes when the current flowed through multiple net cages. Zhou et al. [10] presented hydrodynamic characteristics of knotless nylon netting normal to free stream and the effect of inclination angle on drag coefficient were discussed. Kristiansen et al. [11] investigated the mooring line loads on an aquaculture net cage that was analyzed by model tests and numerical simulations. Huang et al. [12] investigated the dynamic deformation of the floating collar of a net cage under a combination of waves and currents. Yao et al. [13] developed a novel hybrid volume approach to model the current loads on a net cage by considering the fluid–structure interaction. Bi et al. [14] investigated waves propagating through net cages with different levels of biofouling and studied them numerically using a three-dimensional computational fluid dynamics model. In addition, as for the submersible fish cage, Kim et al. [15] used the computational fluid dynamic software to analyze the flow field characteristics of a submersible abalone aquaculture cage and adopted a finite element model based on Morison equation to study the hydrodynamic responses of the moored containment structure. Shainee et al. [16,17] conducted numerical simulations and experimental model tests to examine the submergence characteristics of a self-submersible SPM cage system in both regular and random waves, respectively. Xu et al. [18] analyzed the hydrodynamic behavior of a self-submersible single-point mooring gravity cage in combined wave–current by numerical simulation.

Although the hydrodynamic responses of the net cage have been studied extensively, studies on new types of offshore fish farms are scarce. Li et al. [19] conducted a preliminary study on the global responses and mooring line loads of a vessel-shaped offshore fish farm in various waves. However, the effect of waves on the hydrodynamic responses of fish farms is an important research issue. Thus, it is essential to analyze the interaction mechanism between fish farms and waves. Unlike traditional net cages, the semi-submersible offshore fish farm is a new type of aquaculture facility that is unique in the sense that draught of the fish farm can be regulated by pontoons. When the waves and winds are severe, the draught of the fish farm can be increased appropriately to avoid destruction; when the fish farm need to be maintained, it can ascend to the sea surface by controlling the water of the pontoon. In addition, owing to the existence of net, the hydrodynamic response is different from the traditional offshore engineering structure. In this study, a series of laboratory experiments were performed in the wave–current flume to study the hydrodynamic characteristics of a semi-submersible offshore fish farm. The physical model of the fish farm, similar to Ocean Farm 1 [1], consists of a primary frame, four mooring lines, nets, pontoons, and sinkers. However, as we do not know the detailed information of Ocean Farm 1, the structural parameters of the fish farm in this study are designed according to our own understanding. The objective of this study is to analyze the hydrodynamic responses of this kind of semi-submersible offshore fish farm and provide positive suggestions for its structural optimization. The present research is based purely on scientific interest and has no commercial purpose.

This paper is organized as follows. The laboratory experiments, including the physical model, experimental setup, experimental conditions, and data analysis methods, are presented in Section 2. Section 3 presents the experimental results of the mooring line tension and motion response in different draughts. A detailed discussion including an analysis on the wave parameters, draught, and net damping is presented in Section 4. Finally, the conclusions are presented in Section 5.

2. Laboratory Experiment

In the open sea, a semi-submersible fish farm can be easily subjected to extreme waves. In order to investigate the hydrodynamic responses of a fish farm in regular waves, a series of physical model experiments were conducted. According to the actual sea conditions, prototype structure size, and experimental conditions, the model size, experimental layout, and wave conditions can be designed. All the laboratory experiments were performed in a wave–current flume at the State Key Laboratory of Coastal and Offshore Engineering, Dalian University of Technology, Dalian, China.

2.1. Physical Model

The physical model (see Figure 2) consists of the primary frame, net, weight, and mooring systems. The draught of the fish farm can be adjusted by changing the volume of ballast water in the pontoon to determine the working station. In the experiment, it is difficult to completely meet the same similarity law for each part of fish farm, so in order to reduce the influence of the scale effect on the structure, especially the simulation accuracy of the net, the study adopts different similar scales for the net and frame structure. The full scale can be designed by the gravity similarity. The frame and net system adopted two geometric scales, which were 1:120 and 3:40, respectively. More details about the similarity law and the calculation of full-scale values can be found in previous research [20,21].

Figure 2. Physical model of the semi-submersible offshore fish farm.

2.1.1. Frame System

The frame system is the primary structure that bears the environment load and constitutes the fish living space with the net. The frame system of the fish-farm model is made up of plexiglass and includes column, arc pipe, brace, and pontoon (see Figure 3). In accordance with the dimensions of the experimental facilities and the tested wave conditions, the diameter of the fish-farm model is set to 1 m. The detailed parameters of the fish farm in this study are presented in Table 1. The prototype values of the fish farm are also shown with a geometric similarity scale of 1:120, which is roughly similar to the dimensions of Ocean Farm 1 [1]. The deformation of the primary frame of fish farm is small and has a negligible effect on the hydrodynamic responses of the fish farm; therefore, the elastic similarities need not be considered.

Figure 3. Primary frame of the fish farm.

Table 1. Structure parameters of the fish farm in this study.

Component	Parameter	Prototype (m)	Model (m)
Upper arc pipe	Length	24.00	0.20
	Diameter	1.92	0.016
	Thickness	0.017	0.0023
Middle arc pipe	Length	24.00	0.20
	Diameter	0.96	0.008
	Thickness	0.015	0.0021
Lower arc pipe	Length	24.00	0.20
	Diameter	1.92	0.016
	Thickness	0.017	0.0023
Thick column	Length	36.00	0.30
	Diameter	3.60	0.03
	Thickness	0.025	0.0025
Thin column	Length	33.60	0.28
	Diameter	2.40	0.02
	Thickness	0.030	0.0031
Inclined column	Length	36.00	0.30
	Diameter	0.96	0.008
	Thickness	0.015	0.0021
Central column	Length	36.00	0.30
	Diameter	3.60	0.03
	Thickness	0.025	0.0025
Pontoon	Column Length	7.20	0.06
	Height	6.00	0.05
	Diameter	12.00	0.10
	Thickness	0.040	0.005
Brace	Length	60.00	0.50
	Diameter	1.90	0.016
	Thickness	0.017	0.0023
Upper Brace	Length	60.00	0.50
	Diameter	1.90	0.016
	Thickness	0.017	0.0023

2.1.2. Net System

The net system was designed with three parameters: The overall size, mesh size, and net diameter. To avoid an extremely small net diameter with the model scale of 1:120 such that the model net can hardly be attained, the experimental model primarily considered the drag force to guarantee that the force on the theoretical net is equal to that on the model net (see Figure 4). In the traditional net cage, the whole net system is flexible structure and exhibits a large deformation in regular waves. However, the effect on the deformation of net combined with the fish farm is slight. In the experiment, the net system is designed based on our previous research [20]. The geometric similarity of the net is as follows:

$$\frac{a_1}{d_1} = \frac{a_2}{d_2} \tag{1}$$

where a_1 and d_1 are prototype mesh size and prototype net diameter, respectively; a_2 and d_2 are the mesh size and net diameter of the model, respectively.

The prototype mesh size and net diameter are 5 cm and 3.75 mm, respectively. Therefore, it can be calculated that the ratio of the mesh size to the net diameter should be 40/3, such that the drag force on the theoretical net is the same as that on the equivalent experimental net. Through preliminary calculations, the mesh size is 8 mm and the net diameter is 0.6 mm in the experiment.

To guarantee the gravity similarity, the net system is designed based on a previous research study [20]. The gravity similarity of the net is as follows:

$$\Delta W = (\frac{1}{\lambda'} - \frac{1}{\lambda}) \times (\frac{\pi d_p^2}{4a_p\mu_1\mu_2} \times 10^4) \times (\rho_n - \rho) \times q \times S \tag{2}$$

$$\lambda' = d_p/d_m \tag{3}$$

where Δw is the corrected weight of the net; λ' is the small scale of the net that can be calculated by d_p/d_m; λ is the geometric scale; d_p is the prototype net diameter; d_m is the model net diameter; ρ_n is the net material density; ρ is the water density; a_p is the prototype mesh size; q is the solidity ratio of the net; S is the hanging ratio area of the model net; μ_1 and μ_2 are 0.707 and 0.707, respectively.

Some differences in weight exist between the theoretical net and the equivalent net. To guarantee the weight similarity, it is calculated that the weight of the equivalent net should be increased by 6.7 g, which can be adjusted by the weight system. Owing to the weak elasticity and stiffness of the polyethylene net, the elasticity similarity of the net will not be considered.

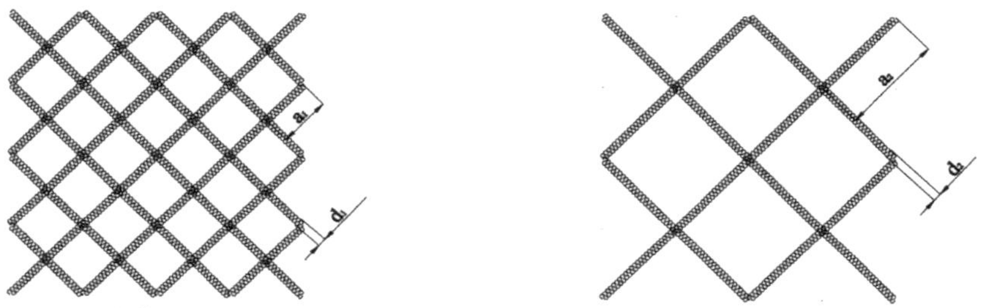

(**a**) Theoretical net (**b**) Equivalent net

Figure 4. Net system similarity.

2.1.3. Weight System

To overhaul and operate the fish farm, the draught of the fish farm can be controlled by a weight system (see Figure 5a) to attain a specified state in practical applications. In the experiment, combining the actual state with experimental conditions, the draught can be adjusted by injecting the water into the pontoon (see Figure 5b) and by a sinker (see Figure 5c). In addition, three draughts exist, including those of 7 cm, 28 cm, and 36 cm, representing the maintenance condition and two working conditions of the fish farm, respectively. Meanwhile, the maintenance condition can be reached depending on its own weight; no additional weight is required. However, the working state 1 can be obtained by hanging weights at the bottom of column, injecting water into the pontoon and adding sand to the top box, and the working state 2 needs to continue to add sand into the box to achieving the draught of 36 cm. The detailed conditions can be presented in Table 2.

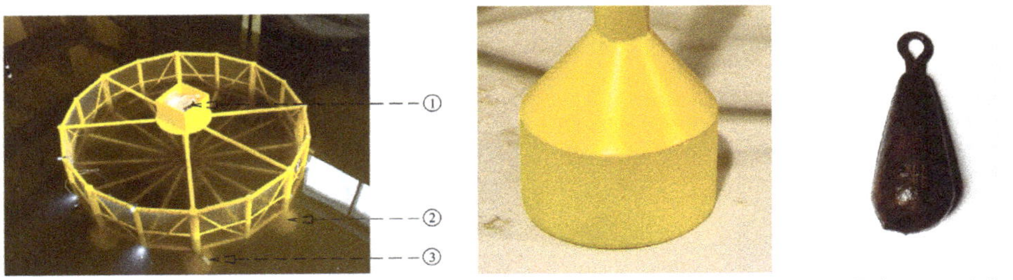

a) Integral layout (b) Pontoon (c) Balance weight

Figure 5. Weight system. ① Adding sand to the top box; ② injecting water into the pontoon; ③ hanging weights at the bottom of column.

Table 2. Weight system of the fish farm.

Model Conditions	Prototype Draught (m)	Model Draught (cm)	Weights (kg)
Maintenance state	8.4	7.0	0
Working state 1	33.6	28	4.02
Working state 2	43.2	36	6.59

2.1.4. Mooring System

The mooring system is an important part of the fish farm and is crucial for the fish farm to operate steadily in the sea. In the experiment, a four-point anchor form was adopted in the fish farm and the initial tension could be loaded in the mooring line. In the practical application, the wave loads on the mooring line are smaller than those on the fish farm; therefore, the elastic similarity was considered primarily in the experiment such that the model mooring line is similar to that of the prototype and to reduce test errors. The elastic part of the anchor is modeled by the spring, and the anchor rope can be fixed by lead blocks at the bottom. The relationship between the force and the calculated value of elongation of the mooring line can be calculated by Formula (4). The results can be listed in the Table 3.

$$F_m = \frac{C_p d_p^2 (\Delta S/S)^n}{\lambda^3} \tag{4}$$

where F_m is the model cable tension; C_p is the elastic coefficient of prototype cable, which is $26.97 \times 10^4 MPa$; d_p is the diameter of prototype cable, which is 120 mm; $\Delta S/S$ is the elongation rate of prototype cable; n is the coefficient, which is 1.5; λ is the geometric scale.

Table 3. Force and calculated value of elongation.

Fm (N)	1	2	3	4	5	6	7	8
ΔS/S	0.34%	0.54%	0.71%	0.85%	0.99%	1.12%	1.24%	1.36%

As listed in Table 3, the relationship between the force and the calculated value of elongation of the mooring line is close to linear. Therefore, the elasticity of the mooring line can be simulated by the spring in the experiment. To obtain a suitable spring, a variety of springs were tested (see Table 4). It can be concluded that the mooring line of the model can be simulated by a polyethylene rope and spring. The mooring line was 3.0 m long in the physical model experiment.

Table 4. Elongation rate of spring under different loads.

Load (N)	Elongation (cm)	Length of Mooring Line (cm)	Elongation Percentage (%)
1	1.2	300	0.4
2	1.6	300	0.53
3	2.0	300	0.67
4	2.4	300	0.80
5	2.9	300	0.97
6	3.3	300	1.10
7	3.7	300	1.23
8	4.1	300	1.37

The comparison between the calculated elongation of the mooring line and the measured values is shown in Figure 6. It is indicated that the error is small; therefore, the method of simplifying the mooring line is feasible.

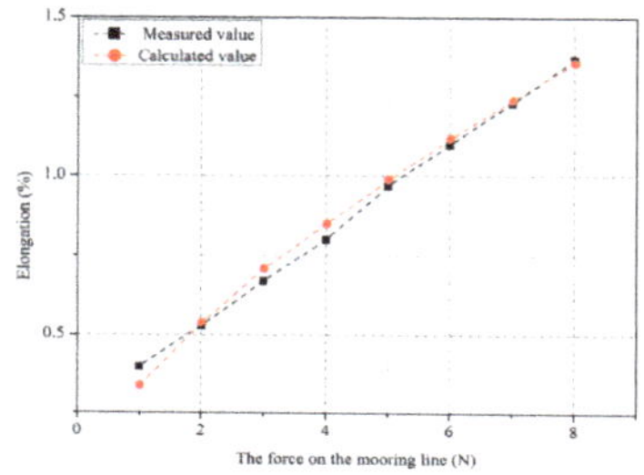

Figure 6. Comparison on mooring line elasticity of calculated values and measured value.

2.2. Experimental Setup

The wave–current flume is 69 m long, 2 m wide, and 1.8 m deep, and the water depth was 1 m during the experiments. In addition, the flume is equipped with a servo motor-driven, piston-type wave-maker capable of producing regular and irregular waves. At the end of the flume, wave absorbers are installed to mitigate the wave reflection. Both sides of the flume at the working section are smooth glass to reduce viscous dissipation owing to the boundaries. In this flume, both waves and currents can be generated. However, only regular waves were generated in this physical model experiment.

Figure 7 shows a sketch of the experimental setup. The fish farm model was moored by four mooring lines in its equilibrium position. Each mooring line primarily comprises one 2.9 m polyethylene rope whose diameter is 1.0 mm and one 0.1 m stainless spring. The linear density of the rope is 0.003 kg/m and the mass of the spring is 0.01 kg. To measure the forces acting on the mooring lines, four load cells were connected to the windward and leeward mooring lines.

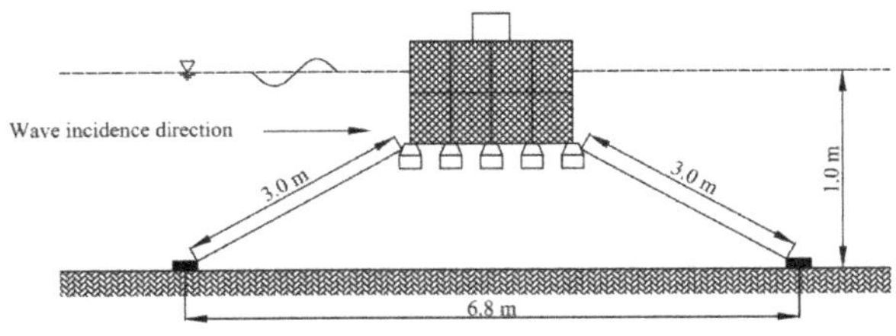

(**a**) Plan view

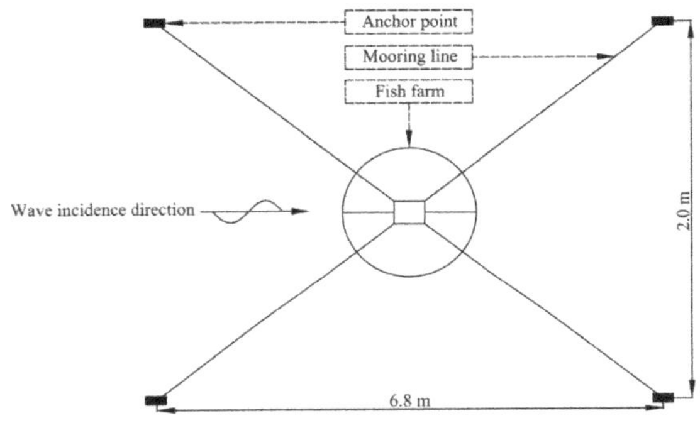

(**b**) Elevation view

Figure 7. Experimental layout.

To obtain the motion response, a charge-coupled device (CCD) camera was used to record the motion trajectory of the diodes that can be fixed on the fish farm, and the motion responses of the fish farm can be calculated by a self-developed software DUT-FlexSim. The motion response can be collected by the CCD high-speed acquisition camera that is arranged in the observation area of the flume (see Figure 8). To guarantee the accuracy of data acquisition of the motion response, the surrounding light should be maintained dark. Four light-emitting diode (LED) bulbs are arranged at the top and bottom of the fish farm as tracing points (see Figure 9). The tracing points along the x-axis direction are called the front point, and the tracing points on the back side are called the back point.

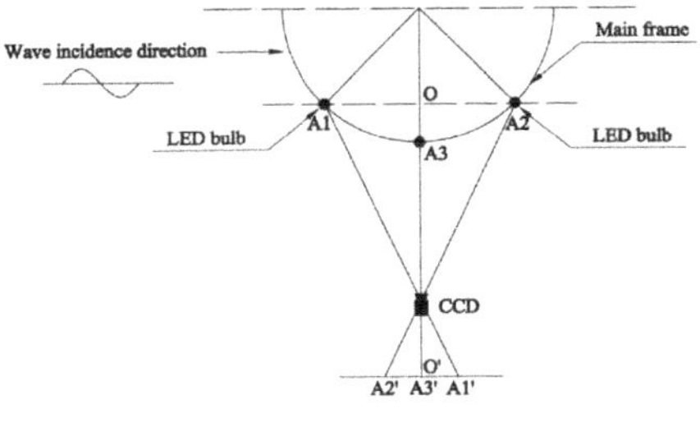

Figure 8. Charge-coupled device (CCD) high-speed camera setup.

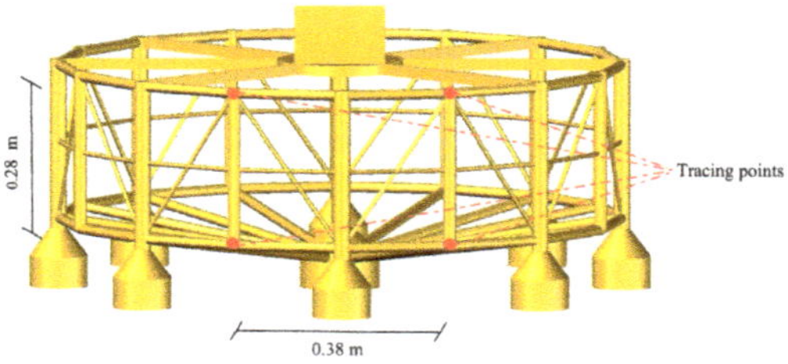

Figure 9. Location of tracing points.

2.3. Experimental Conditions

In the physical model experiment, the hydrodynamic response of fish farm can be measured in regular waves. According to the model scale, the experimental water depth can be designed to be 1.0 m and its corresponding prototype water depth is 120 m. The mooring line tension and motion response can be measured by the load cells and CCD, respectively. To guarantee the accuracy of data collection, the measurements were performed thrice, and the sampling rate was 0.02 s. Three draughts exist, including those of 7 cm, 28 cm, and 36 cm. In addition, the regular waves were designed as per the specifications in Table 5.

Table 5. Wave conditions.

Wave Case No.	Prototype Value		Model Value	
	Wave Height (m)	Wave Period (s)	Wave Height (m)	Wave Period (s)
1	7.2	10.95	0.06	1.0
2	7.2	13.15	0.06	1.2
3	7.2	15.34	0.06	1.4
4	9.6	10.95	0.08	1.0
5	12	10.95	0.10	1.0
6	12	13.15	0.10	1.2
7	12	15.34	0.10	1.4

2.4. Data Analysis Method

In the experiment, the water surface elevations were recorded by capacitance-type wave gauges arranged along the center line of the wave flume. The absolute accuracy of these wave gauges is approximately ±1 mm. Before initiating any measurements, the wave gauges were examined for soundness, cleaned if necessary, and subsequently calibrated. In addition, a computer control system developed by the Beijing Hydraulic Research Institute was used for collecting the data of free surface elevations with multiple channels. To collect accurate data, the time series of wave elevation at each measurement point was recorded with a sampling rate of 50 Hz and a stable data over a period of 10 s was chosen for data analysis; subsequently, the wave height at a measurement point was the average value of the corresponding time series.

Water-resistant load cells with a capacity of 10 N were used to measure the forces on the mooring lines, and the specified accuracy of the load cell was 0.1 N. Each measurement was performed thrice to diminish the impact of random and bias errors. Data sampling was conducted over a period of 20 s and the final experimental value was the average value of the three measurements. Figure 10 is the time series of mooring lines tensions when the wave height is 10 cm and the wave period is 1.4 s in a draught of 36 cm. The pretension of mooring line was measured before each group of tests began, and then the total force on the mooring line can be obtained by the load cells. Therefore, the mooring line tension in pure waves can be calculated as follows.

$$F_w = F_t - F_{pre} \tag{5}$$

where F_w is the average value of maximum mooring line tensions in pure waves; F_t is the average value of maximum mooring line tensions which can be measured by water-resistant load cells; F_{pre} is the pre-tension of mooring line. The windward and leeward pre-tension are 1.91 N and 1.87 N, respectively.

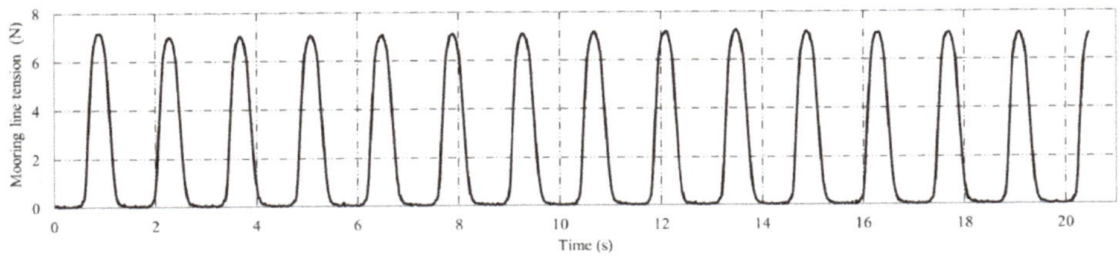

Figure 10. The time series of mooring lines tensions.

3. Results

In an actual operation of the fish farm, the mooring line tension is the key factor affecting its stability. In addition, the motion response contributes significantly to the safety of the workers. Therefore, the mooring line tension and the motion response can be measured and analyzed in different draughts. The maximum force on the mooring line and the maximum motion response were obtained from time intervals in duration of approximately 15 wave periods from the measured time series.

3.1. Mooring Line Tension

Wave period and wave height are two important factors for the mooring line tension. Figures 11 and 12 show the mooring line tension in different wave periods and wave heights, respectively. The primary part of the fish farm is above the sea level, and only nine pontoons of the fish farm are underwater with the draught of 7 cm for maintenance condition, which is easily subjected to the wave load. On the contrary, most of the fish farm is submerged beneath the free surface when the draughts are 28 cm and 36 cm.

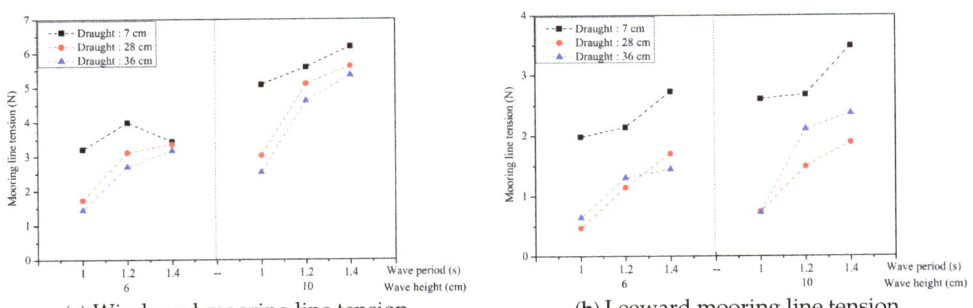

(**a**) Windward mooring line tension (**b**) Leeward mooring line tension

Figure 11. Mooring line tension in different wave periods.

Because of the symmetry of the fish farm model and the mooring lines, the average value of the tension in the mooring lines at symmetrical positions was used for force analysis. The maximum force on the mooring line can be measured by the average value of the two mooring lines in the symmetrical position, and the pretension can be loaded in the mooring line before the experiment begins. It can be observed from Figure 11 that the force on the mooring line is closely proportional to the wave period. Furthermore, as the draught increases, the mooring line tension exhibits a downward trend. It is apparent from Figure 12 that the windward and leeward mooring line tensions exhibit an upward trend with increasing wave height.

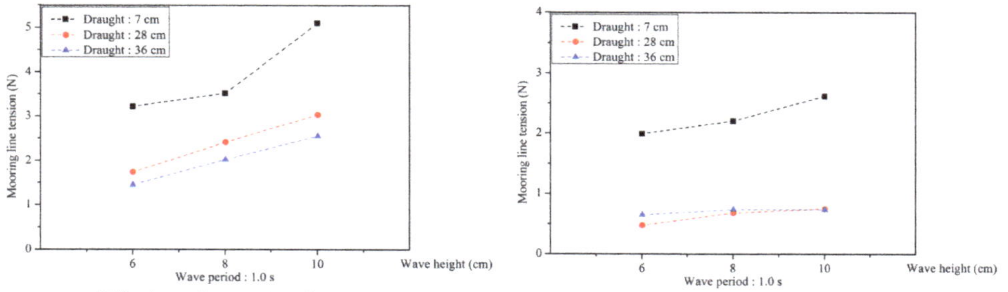

(**a**) Windward mooring line tension (**b**) Leeward mooring line tension

Figure 12. Mooring line tension in different wave heights.

3.2. Motion Responses

In the semi-submersible offshore fish farm, not only the mooring line tension, but also the motion responses are necessary to be analyzed for the fish farm. It is because the motion response is critical to the fish living in the fish farm and human activities on the fish farm. Smaller motion response can provide a stable living space for fish and avoid damage to fish in severe currents and waves, and further guarantee the quality of fish and the economic income of fishermen. The fish farm exhibits a small deformation under waves and is approximately a rigid structure. Therefore, only the motion responses of the fish farm require analysis, which include three aspects: The heave, surge, and pitch. In the present experiment, the fish farm model was fitted with a tracking point on the upstream side and the back side of the fish farm such that the motion trajectory can be analyzed by tracking the two points.

Figure 13 shows trajectory of the fish farm at H = 10 cm and T = 1.4 s in the draughts of 28 cm and 36 cm. As shown in the Figure 13, the trajectory of the windward side is oblique and elliptical, while the trajectory of the leeward side resembles an egg shape. Meanwhile, it is apparent from the motion track of the fish farm that the heave motion and surge motion exhibit some attenuations with increasing draught; this is consistent with the gradual attenuation of the mooring line tension on the upstream side.

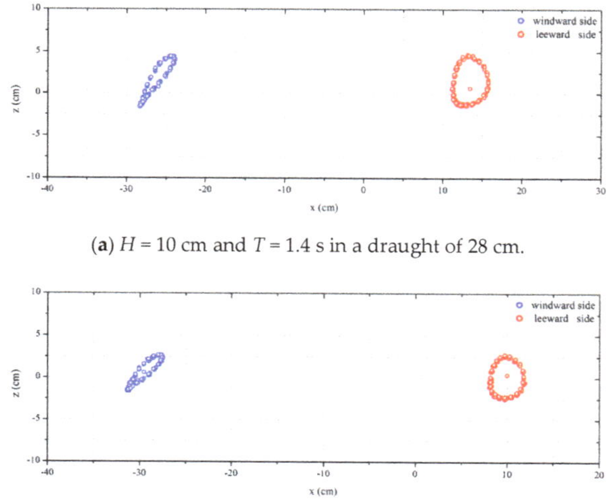

(**a**) H = 10 cm and T = 1.4 s in a draught of 28 cm.

(**b**) H = 10 cm and T = 1.4 s in a draught of 36 cm.

Figure 13. Motion trajectories of the fish farm.

Figure 14 shows the motion response of the fish farm in different wave periods. It can be observed that the motion response is positively correlated with the wave period in different draughts. However, when the draught is 7 cm, there is no obvious correlation between the surge or pitch and the wave period. In addition, it is apparent that as the draught increases, the motion response including the heave, surge, and pitch motion turn to decrease. It can be concluded that the maximum motion response can be attained in the draught of 7 cm, followed by those in the draughts of 28 cm and 36 cm. Meanwhile, it is obvious from Figure 15 that the heave, surge, and pitch motions are proportional to the wave height and reaches their peak values at the wave height of 10 cm when the wave period is 1 s.

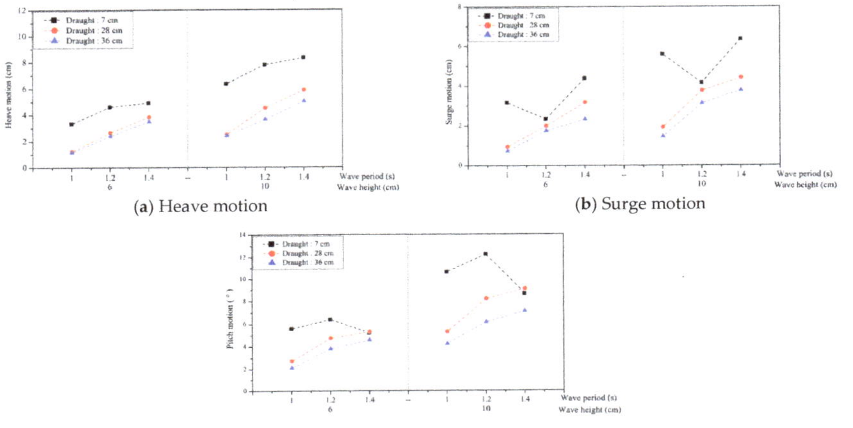

(**a**) Heave motion (**b**) Surge motion

(**c**) Pitch motion

Figure 14. Motion response in different wave periods.

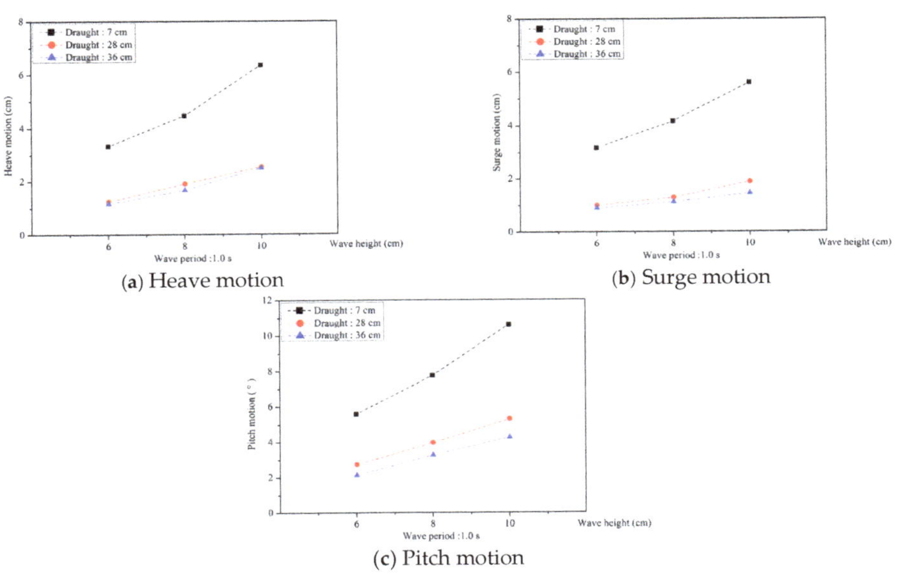

(**a**) Heave motion (**b**) Surge motion

(**c**) Pitch motion

Figure 15. Motion response in different wave heights.

4. Discussion

The semi-submersible offshore fish farm is a new type of aquaculture facilities unlike the traditional net cage. The new fish farm has the advantages including the resistance to severe waves, small deformation, and convenient towing. To analyze the hydrodynamic response of the fish farm in detail, the wave parameters, draughts, and net damping should be considered in the research.

4.1. Effects of Wave Parameters

In the open sea, the fish farm is easily subjected to the wave action; therefore, the wave heights and wave periods are two important factors affecting the hydrodynamic response. As the wave heights increase, the mooring line tension and motion response exhibit an upward trend. This is because the area that can be loaded by waves gradually turn to increase with the wave heights increasing so that the motion amplitude will become larger and the mooring line will be strained, which will produce a larger mooring line tension. In addition, the results above indicate that when the draught is 7 cm, no obvious correlation exists among the mooring line tension, motion response, and wave periods. This is due to the complex pontoon that can be submerged. The surface area of the pontoon is not linearly proportional to the height along the vertical of the water surface; further, a gradual change exists in the connection between the pontoon and the column.

Taking the draught of 28 cm as an example, when the period changes from 1.0 s to 1.2 s and 1.4 s, the increment of windward force on the mooring line is from 80% to 7.6% in the wave height of 6 cm and the increment of leeward force on the mooring line is from 141% to 49% in the wave height of 6 cm. It is apparent that the trend of growth is gradually becoming slower with the increasing of wave periods. Meanwhile, the windward and leeward peak values can reach 3.37 N and 1.71 N, respectively, whose prototype values are 5823.36 kN and 2021.76 kN, respectively. In addition, when the wave height is 10 cm, the fluctuation range of the heave value ranges from 191.13% to 30.17% at T = 1.0 s–1.4 s; the fluctuation range of the surge value ranges from 96.15% to 16.80% at T = 1.0 s–1.4 s; the fluctuation range of the pitch value ranges from 54.48% to 10.79% at T = 1.0 s–1.4 s. It can be concluded that the increase rate of the motion response decelerates with increasing wave period. However, the whole effect on the fluctuation value of the fish farm became more evident with wave period increasing; further, when the period increases to a certain extent the surge value of the fish farm tends to be stable.

The comparison of the windward and leeward forces of the mooring line tension can be represented in Figure 16, where the wave height and wave period are 6.0 cm and 1.4 s, respectively. It can be concluded that the mooring line tension in the line 2 is less than that in line 1. Using the draught of 28 cm as an example, it can be calculated that the maximum force is attenuated by 61.7% and 70.8% from the windward side to the leeward side at H = 6 cm and H = 10 cm, respectively. It can be explained that the wave-ward lines are more loaded due to the non-linearity and drift forces when the wave reaches the fish farm. On the contrary, less external force exists on the leeward mooring line than that in the windward. Therefore, the wave loads on the fish farm acts primarily on the windward mooring line and the leeward force is relatively smaller.

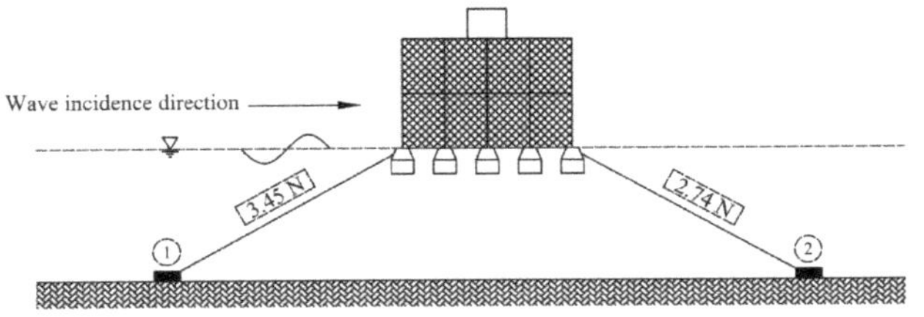

(**a**) Draught of 7 cm.

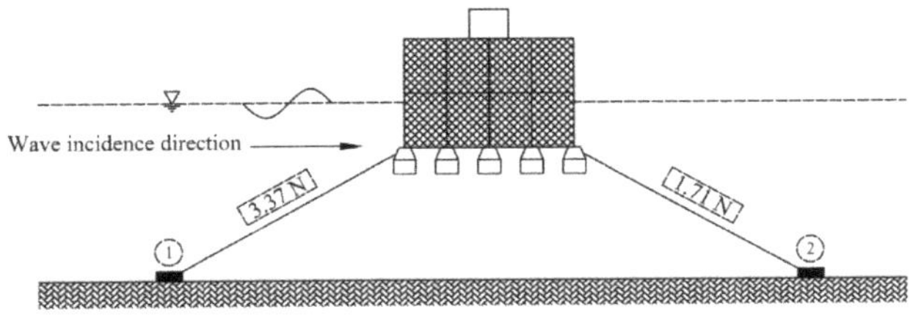

(**b**) Draught of 28 cm.

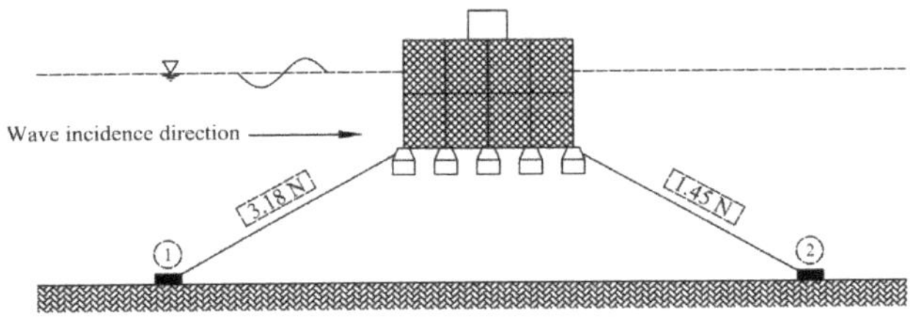

(**c**) Draught of 36 cm.

Figure 16. Mooring line tensions of the fish farm.

4.2. Effects of Draught

To overhaul and run the fish farm, different draughts can be adjusted by the weight system. With the change in draught, the force on the mooring line and the motion response will present different variation tendencies. It can be observed from Figure 17 that as the draught continues to increase, the windward force on the mooring line decreases. This is a desirable phenomenon that will provide a technical method for the security of the fish farm in severe waves. In addition, it can be calculated that the maximum reduction reaches 49.8% when the wave height is 10 cm. Meanwhile, it is apparent from Figure 18 that the motion amplitude exhibits a downward trend with increasing draught, and that the maximum attenuation including the heave, surge, and pitch motion can reach 75.6%, 73.9%, and 59.8%, respectively, when the wave height is 10 cm. This phenomenon can arise because the sloping bottom and the pontoons around the bottom contribute to the vast majority of volume of the fish farm. Therefore, a considerable wave force is produced by the interaction between the waves and structures. With the draught increases, the primary volume of the fish farm is submerged to a certain water depth and interacts with the lower velocity of the water particles. Meanwhile, the primary parts that interact with the waves are the columns in the circumference and the net chamber around the free surface that experience significantly weaker external forces. Thus, the wave force acting on the fish farm decreases gradually. In practical implication, the draught of the fish farm can be increased appropriately to maintain the stability of the fish farm and to avoid severe wave loads from acting on the fish farm.

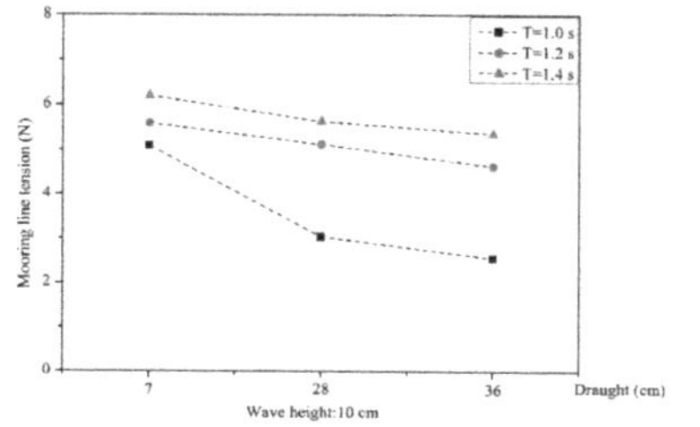

Figure 17. Windward mooring line tension in different draught.

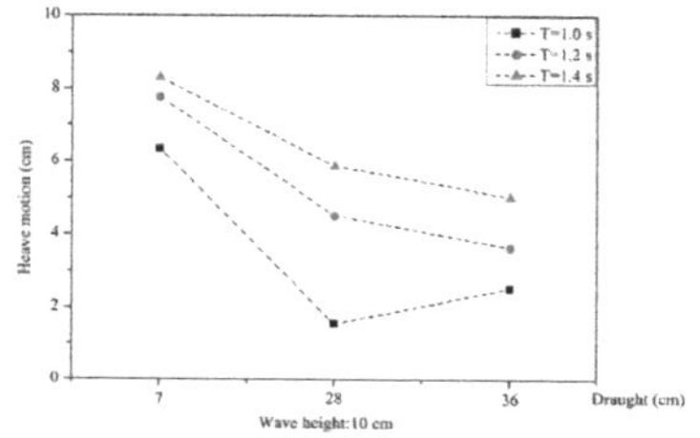

(**a**) Heave motion

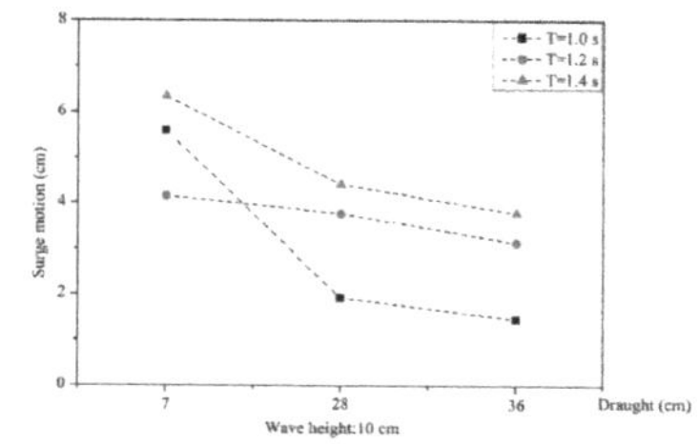

(**b**) Surge motion

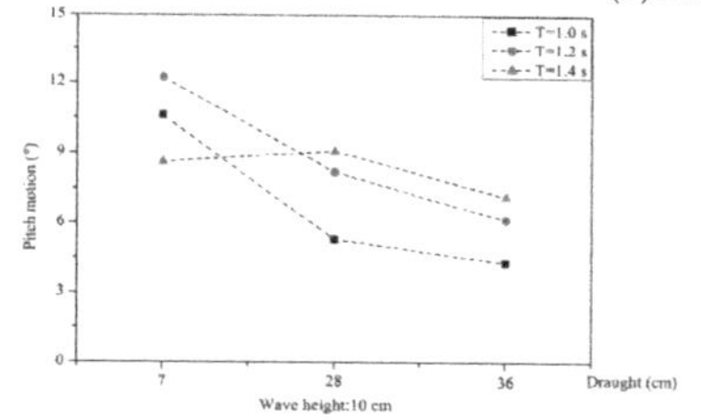

(**c**) Pitch motion

Figure 18. Motion response in different draughts.

4.3. Effects of Net Damping

From comparing the new semi-submersible offshore fish farm and a traditional semi-submersible offshore platform, it is apparent that the primary difference is whether a net exists. The existence of net will pose certain effects on the mooring line tension and motion response of the fish farm. Figure 19 shows the comparison of the windward mooring line tension with or without net in the draught of 36 cm. It can be observed that the mooring line tension in different regular wave decreases owing to the existence of nets, and the maximum reduction can reach 42%. The phenomenon can be explained by the shielding effect and damping effect of the net [22,23]. When the wave passes the net, the wave energy will be attenuated, and the wave height reduces significantly, such that the wave loads on the leeward fish farm exhibit a downward trend. Meanwhile, the fish farm will produce a relative motion along the direction of the wave propagation so that the fluid will exhibit a greater damp effect on the net and produce a reverse force. Compared with the no-net state, the horizontal motion amplitude will be restrained to a certain extent (see Figure 20) and the mooring line tension decreases. Overall, it can be concluded that the mooring line tension will exhibit a downward trend for the fish farm with a net compared with no net.

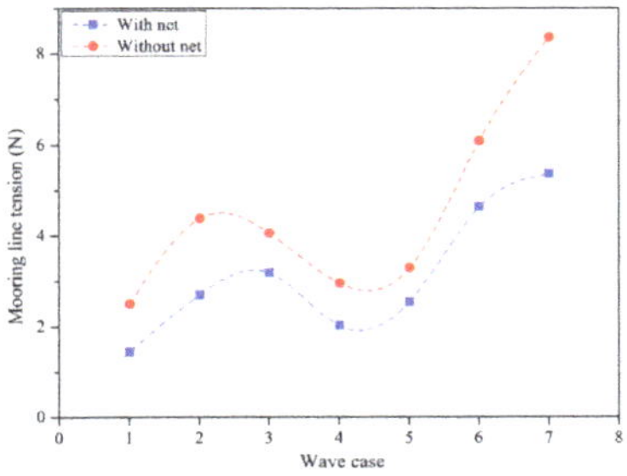

Figure 19. Comparison of mooring line tension.

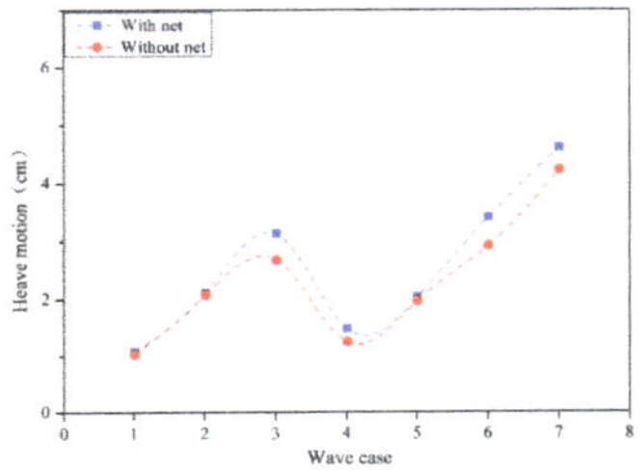

(**a**) Heave motion

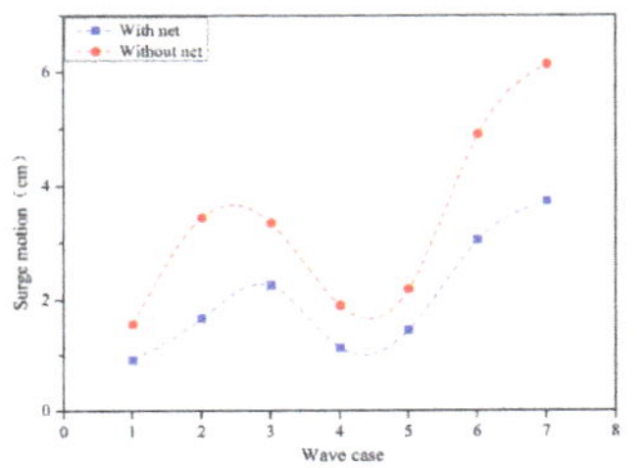

(**b**) Surge motion

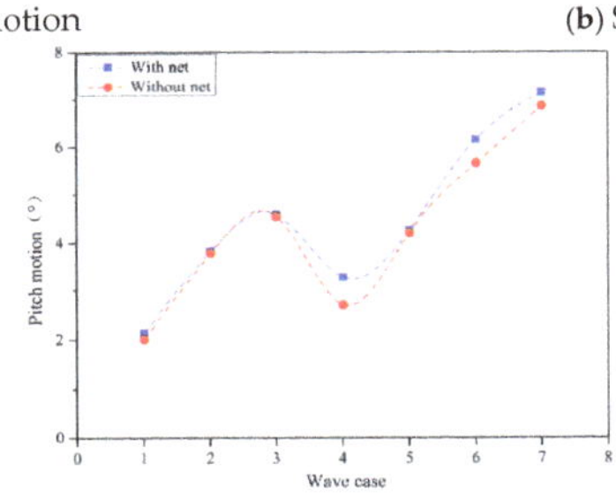

(**c**) Pitch motion

Figure 20. Comparison of motion response.

As for different wave conditions, the change trend of surge motion of the fish farm is the same as that of the mooring line tension. The surge motion will reduce for the fish farm without net. This is because the fluid damping will produce an important impact on the net. Overall, the surge motion of the fish farm with net is less than that without net, and the maximum difference can reach 51%. In addition, the heave motion and pitch motion of the fish farm with net are larger than that without net, and the maximum difference can reach 19% and 22%, respectively. It can be concluded from Figure 20 that the existence of net affects the pitch motion less but affects the surge motion more.

In the paper, the hydrodynamic response was studied and analyzed in the pure waves. Due to the existence of waves, the motion response of the fish farm will be significantly affected. However, in the actual sea conditions, waves and currents always exist at the same time and have an interaction with each other. The existence of water flow is equivalent to adding an external load on the fish farm, which will have a greater impact on the anchor rope force. Therefore, in the next work, the hydrodynamic characteristics of the fish farm under the combined action of waves and currents will be considered, which will provide suitable suggestions for the application of practical projects.

5. Conclusions

In the present study, the hydrodynamic responses of a semi-submersible offshore fish farm in regular waves were investigated at various draughts using physical model experiments. The mooring line tension and the motion response of the fish farm were analyzed. The conclusions are summarized as follows:

(1) The mooring line tension and motion response were closely proportional to the wave period and wave height. Nevertheless, when the draught was 7 cm, no obvious correlation existed between the hydrodynamic response and wave periods. In addition, the force on the windward mooring line was higher than that on the leeward; further, the upstream anchor lines endured most of the external loads on the fish farm. Therefore, the upstream anchor lines should be optimized and materials with better mechanical properties should be chosen in the actual operation.

(2) As the draught increased, the mooring line tension and motion response exhibited a slight downward trend. Although the area of interaction between the fluid and structure increased, the range of wave action was near the sea surface; subsequently, the effect became weaker gradually with increased draught. In addition, owing to the damping effect of fluid, the motion of the fish farm was limited and the force on the mooring line exhibited a slight downward trend. The experimental results provided a good method to avoid severe waves.

(3) The existence of net affected the hydrodynamic characteristics of the fish farm significantly. Compared with the fish farm without net, the mooring line tension exhibited an obvious increase and the reduction in the peak value could reach 42%. Because of the shielding effect and damping effect of the net, the surge motion of the fish farm exhibited a maximum attenuation of approximately 51%. However, the heave motion and pitch motion increased to some extent. It can be concluded that the net affected the surge degree the most but affected the heave and pitch degrees the least.

(4) Pure waves have a great influence on the motion response of fish farm and is a critical factor affecting the stability of fish farm, so it is important to design a suitable parameter to decrease or avoid the large motion of the fish farm. However, in actual sea conditions, the presence of water flow will increase the anchor rope force and affect the waves. Therefore, in the following work, the hydrodynamic response of the fish farm under the waves and currents will be studied to provide some experimental guidance for the practical engineering.

Author Contributions: Data curation, C.B., H.L. and Y.C.; formal analysis, C.G., H.L. and Y.C.; funding acquisition, Y.Z. and C.G.; investigation, Y.C.; methodology, C.B.; resources, C.G.; software, Y.C.; supervision, Y.Z.; writing—original draft, Y.Z., C.B., and H.L.; writing—review and editing, Y.Z., C.G., C.B., and H.L.

Funding: This research was funded by the National Natural Science Foundation of China (NSFC), grant number 31872610, 51822901, 51609035, 51579037 and 31772898; and Fundamental Research Funds for the Central Universities, project nos. DUT19ZD206, and DUT18RC (3) 076.

Conflicts of Interest: The authors declare no conflict of interest.

References

1. SalMar. Offshore Fish Farming: A New Era in Fish Farming is on its Way. 2018. Available online: https://www.salmar.no/en/offshore-fish-farming-a-new-era/?tdsourcetag=s_pcqq_aiomsg (accessed on 4 September 2018).
2. Fredriksson, D.; Swift, M.; Eroshkin, O.; Tsukrov, I.; Irish, J.; Celikkol, B. Moored Fish Cage Dynamics in Waves and Currents. *IEEE J. Ocean. Eng.* **2005**, *30*, 28–36. [CrossRef]
3. Lader, P.F.; Enerhaug, B. Experimental Investigation of Forces and Geometry of a Net Cage in Uniform Flow. *IEEE J. Ocean. Eng.* **2005**, *30*, 79–84. [CrossRef]
4. Li, Y.C.; Zhao, Y.P.; Gui, F.K.; Teng, B.; Dong, G. Numerical simulation of the influences of sinker weight on the deformation and load of net of gravity sea cage in uniform flow. *Acta Oceanol. Sin.* **2006**, *25*, 129–141.
5. Moe, H.; Fredheim, A.; Hopperstad, O. Structural analysis of aquaculture net cages in current. *J. Fluids Struct.* **2010**, *26*, 503–516. [CrossRef]

6. Tsukrov, I.; Drach, A.; Decew, J.; Swift, M.R.; Celikkol, B. Characterization of geometry and normal drag coefficients of copper nets. *Ocean Eng.* **2011**, *38*, 1979–1988. [CrossRef]
7. Xu, T.J.; Dong, G.H.; Zhao, Y.P.; Li, Y.C.; Gui, F.K. Analysis of hydrodynamic behaviors of gravity net cage in irregular waves. *Ocean Eng.* **2011**, *38*, 1545–1554. [CrossRef]
8. Fu, S.; Moan, T. Dynamic analyses of floating fish cage collars in waves. *Aquac. Eng.* **2012**, *47*, 7–15. [CrossRef]
9. Zhao, Y.P.; Bi, C.W.; Chen, C.P.; Li, Y.C.; Dong, G.H. Experimental study on flow velocity and mooring loads for multiple net cages in steady current. *Aquac. Eng.* **2015**, *67*, 24–31. [CrossRef]
10. Zhou, C.; Xu, L.; Hu, F.; Qu, X. Hydrodynamic characteristics of knotless nylon netting normal to free stream and effect of inclination. *Ocean Eng.* **2015**, *110*, 89–97. [CrossRef]
11. Kristiansen, T.; Faltinsen, O.M. Experimental and numerical study of an aquaculture net cage with floater in waves and current. *J. Fluids Struct.* **2015**, *54*, 1–26. [CrossRef]
12. Huang, X.H.; Guo, G.X.; Tao, Q.Y.; Hu, Y.; Liu, H.Y.; Wang, S.M.; Hao, S.H. Numerical simulation of deformations and forces of a floating fish cage collar in waves. *Aquac. Eng.* **2016**, *74*, 111–119. [CrossRef]
13. Yao, Y.; Chen, Y.; Zhou, H.; Yang, H. Numerical modeling of current loads on a net cage considering fluid–structure interaction. *J. Fluids Struct.* **2016**, *62*, 350–366. [CrossRef]
14. Bi, C.W.; Zhao, Y.P.; Dong, G.H.; Xu, T.J.; Gui, F.K. Numerical study on wave attenuation inside and around a square array of biofouled net cages. *Aquac. Eng.* **2017**, *78*, 180–189. [CrossRef]
15. Kim, T.; Lee, J.; Fredriksson, D.W.; Decew, J.; Drach, A.; Moon, K. Engineering analysis of a submersible abalone aquaculture cage system for deployment in exposed marine environments. *Aquac. Eng.* **2014**, *63*, 72–88. [CrossRef]
16. Shainee, M.; Decew, J.; Leira, B.; Ellingsen, H.; Fredheim, A. Numerical simulation of a self-submersible SPM cage system in regular waves with following currents. *Aquac. Eng.* **2013**, *54*, 29–37. [CrossRef]
17. Shainee, M.; Leira, B.; Ellingsen, H.; Fredheim, A. Investigation of a self-submersible SPM cage system in random waves. *Aquac. Eng.* **2014**, *58*, 35–44. [CrossRef]
18. Xu, T.J.; Dong, G.H.; Li, Y.C.; Guo, W.J. Numerical study of a self-submersible single-point mooring gravity cage in combined wave-current flow. *Appl. Ocean Res.* **2014**, *48*, 66–79. [CrossRef]
19. Li, L.; Jiang, Z.; Ong, M.C. A preliminary study of a vessel-shaped offshore fish farm concept. In Proceedings of the ASME 2017 36th International Conference on Ocean, Offshore and Arctic Engineering, Trondheim, Norway, 25–30 June 2017.
20. Li, Y.C.; Gui, F.K.; Zhang, H.H.; Guan, C.T. Simulation criteria of fishing nets in aquaculture sea cage experiments. *J. Fish. Sci. China* **2005**, *12*, 179–187. (In Chinese)
21. Li, Y.C. 2010. Hydrodynamic behavior of net cages in the open sea. In *Handbook of Coastal and Ocean Engineering*; World Scientific: Singapore, 2010.
22. Bi, C.W.; Zhao, Y.P.; Dong, G.H.; Xu, T.J.; Gui, F.K. Experimental investigation of the reduction in flow velocity downstream from a fishing net. *Aquac. Eng.* **2013**, *57*, 71–81. [CrossRef]
23. Bi, C.W.; Zhao, Y.P.; Dong, G.H.; Cui, Y.; Gui, F.K. Experimental and numerical investigation on the damping effect of net cages in waves. *J. Fluids Struct.* **2015**, *55*, 122–138. [CrossRef]

Journal of
Marine Science and Engineering

MDPI

Article

Study on the Optimal Wave Energy Absorption Power of a Float in Waves

Zhengxiao Luan [1], Guanghua He [1,2,*], Zhigang Zhang [1], Penglin Jing [1], Ruijia Jin [3,4], Baolei Geng [3] and Chaogang Liu [1]

1 School of Naval Architecture and Ocean Engineering, Harbin Institute of Technology, Weihai 264209, China
2 Institute of Ship and Ocean Engineering Hydrodynamics, Shandong Institute of Shipbuilding Technology, Weihai 264209, China
3 Tianjin Research Institute for Water Transport Engineering, M.O.T., Tianjin 300456, China
4 College of Engineering, Ocean University of China, Qingdao 266100, China
* Correspondence: ghhe@hitwh.edu.cn

Received: 6 July 2019; Accepted: 10 August 2019; Published: 13 August 2019

Abstract: The utilization of ocean renewable energy, especially wave energy, is of great significance in ocean engineering. In this study, a three-dimensional numerical wave tank was established to simulate the wave-float interaction based on the Reynolds-averaged Navier–Stokes equations and the Realizable K-Epsilon Two-Layer turbulence model was applied. Firstly, convergence studies with respect to the mesh and time step were carried out and confirmed by the published analytical and numerical data. Then, the resonance condition of a particular float was solved by both numerical and analytical methods. The numerical and the analytical results are mutually verified in good agreements, which verify the reliability of the analytical process. Furthermore, a wave energy converter (WEC) consisting of a single float without damping constant was adopted, and its hydrodynamic performance in different wave conditions was investigated. It was found that the damping factor can affect the motion response of the float and the wave force it receives. Under a certain wavelength condition, the WEC resonates with the wave, at which the wave force on the float, displacement of the float and other parameters reach a maximum value. Finally, the influence of linear damping constant on the power take-off (PTO) was studied. The results show that the damping factor does not affect the wave number turning point of the optimal damping constant.

Keywords: damping constant; resonance; power generation; wave energy; hydrodynamic analysis; numerical wave tank

---

1. Introduction

Energy and environment are key issues for humans and their development. Due to the problems of environmental pollution and energy deficit, the development and utilization of renewable energy have received strong attention by more and more countries [1,2]. Benefiting from the advantages of wide distribution, large reserves, renewable and no pollution, wave energy has received extensive attention. Barstow et al. [3] reported that wave energy can provide more than 2 TW, which is the same order of the world's electricity consumption.

Wave energy is usually captured by motions of a wave energy converter (WEC) [4]. While development of modern wave energy converter dates back to 1799 [5], the technology did not receive worldwide attention until the 1970s, when an oil crisis occurred and Stephen Salter published a notable paper about the technology in Nature in 1974 [6]. Early research mainly focused on the oscillating water column WECs, but the device is (1) expensive to build, (2) has low conversion efficiency (10–30%) and (3) high cost of power generation [7–9]. At present, the main focus is on the oscillating float-type (point absorption) WECs [10–12]. The float-absorber-type WEC reflects a simple but robust technology, which consists of buoys or floating bodies to capture the wave's motion. A point absorber with a direct drive power take-off (PTO) system is the most efficient and beneficial in converting the low-speed oscillating motion of ocean waves [13]. Bhatta et al. [14] carried out numerical study on the cylindrical shape point absorption WEC, and analyzed the hydrodynamic performance under different degrees of freedom such as surge, heave and pitch by using the mathematical technique method. Zheng et al. [15,16] studied the diffraction and radiation of a square floating structure under the condition of three-dimensional wave incidence, and analyzed the influence of different incident wave directions on the force and hydrodynamic performance of the WEC by using the eigenfunction expansion matching method. Shen et al. [17] used a semi-analytical method to solve the parameters of the wave excitation force and additional mass of the cylindrical WEC. Vantorre et al. [18] employed a software named AQUA+ to study floating structures with cylindrical, conical and hemispherical shapes at the bottom. The numerical results show that the WEC energy capture efficiency of the tapered bottom structure is higher. However, Backer et al. [19] carried out their research by boundary element method and shows that the difference in WEC energy capture efficiency between different shapes of float structures is small.

Consequently, the analysis of motion response of WEC in waves and its hydrodynamic performance is especially important. For hydrodynamic studies, analytical and semi-analytical methods have been used for wave-structure interaction, especially for simple geometries [20,21]. Potential flow theory is another widely used numerical approach. Since potential flow solvers are not easy to take the viscous effect into account, which may affect the accuracy of numerical results [22–24]. Recently, CFD (Computational Fluid Dynamics) method has been widely used to simulate the complicated interaction between strong nonlinear waves and floating bodies. Yu et al. [25] studied a two-body floating-point absorber (FPA) WEC system, and analyzed the hydrodynamic response and the power absorption performance of the system in regular waves. The study showed that the nonlinear effects could significantly decrease the power output and the motion of the FPA system, particularly in larger waves. However, this study does not show the relationship between WEC power absorption and resonance condition. Gallizio et al. [26] studied the dynamics of an inertial wave energy converter by coupled a CFD method with a dynamic model of the power generation system. Coe et al. [27] predicted extreme loading in a two-body WEC using a combination of a time-domain model based on linearized potential theory and CFD method based on unsteady Reynolds-averaged Navier–Stokes equation. CFD is considered to be a good choice to analyze the hydrodynamic performance of WEC especially for real sea situation [28,29]. Recently, a commercial CFD solver, STAR-CCM+ becomes popular and powerful in ocean engineering [30–32]. In the present paper, the STAR-CCM+ was employed to analyze a WEC's motion in waves.

Firstly, a three-dimensional numerical wave tank was established by STAR-CCM+, then convergence studies with respect to mesh and time step were verified. The problem of a truncated column in regular waves was simulated and compared with a potential flow solver DIFFRACT and a CFD tool OpenFOAM [33] and available experimental data from MOERI [34]. Secondly, the resonance condition of a particular float was solved not only by numerical method but also by analytical method. The numerical results and the analytical results were mutually verified. The velocity and displacement of a float without damping constant were investigated against various wave numbers. Finally, the relationship between the optimal damping constants and wave number was studied. Some conclusions and future research directions were then considered.

2. Numerical Schemes and Validation

2.1. Free-Surface Capturing

The STAR-CCM+ can solve the Reynolds-averaged Navier–Stokes equations with/without fluid compressibility based on finite volume mesh method. The prism layer mesher, surface remesher and trimmed model are applied in the mesh continuum model. An implicit unsteady time-marching model is used in the physics continuum model, where the governing equations are discretized over a computational domain. The Realizable K-Epsilon Two-Layer turbulence model is applied with a two-layer all-y+ wall treatment model.

The volume of fluid method is applied to capture the water free surface. This modeling approach assumes that the mesh resolution is sufficient to resolve the position and the shape of the interface between multiple phases [35]. The distribution of phases and the position of the interface are described by the fields of phase volume fraction α_i. The volume fraction of phase i is defined as:

$$\alpha_i = \frac{V_i}{V} \tag{1}$$

where V is the volume of a cell, and V_i is the volume of phase i in the cell. The volume factions of all phases in this cell are satisfied the following relationship:

$$\sum_{i=1}^{N} \alpha_i = 1 \tag{2}$$

where N is the total number of phases, $N = 2$ in the present study.

2.2. Numerical Wave Tank

Convergence studies of meshes and time steps for the multi-phase-flow numerical wave tank were first carried out to confirm the accuracy and efficiency of the present numerical model. The wave forcing method was implemented in the numerical wave tank, as shown in Figure 1. The 3D Navier–Stokes equations were solved in one domain, within the blue zones. No forcing was applied within the inner zone (3D Navier–Stokes), but within the outer zone (Forcing zone) the forcing source term was activated along boundaries of the computational domain. This method forces the solution of the discretized Navier–Stokes equations towards the theoretical solution over a specified distance, and it reduces the computational load by using a reduced-size solution domain. This forcing also eliminates problems that are associated with reflections of surface waves at boundaries, owing to the damping feature of the gradual forcing [35].

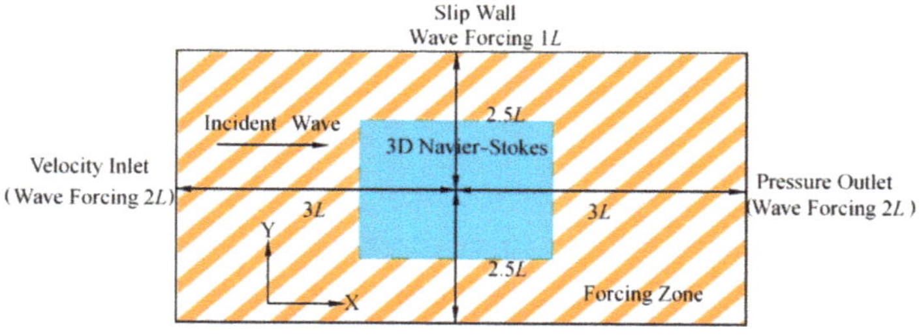

Figure 1. Numerical wave tank (top view).

The lengths of upstream, downstream and forcing zone in the numerical wave tank were set as 3 L, 3 L, 2 L (L is the incident wavelength), respectively, while the width of the tank is 2.5 L, and forcing zone was set as 1 L. The water depth is 1.5 L. An extruder mesh was used to reduce the computational time and to provide additional numerical dissipation for energy consumption. The conditions of upstream and top boundaries are the velocity inlet, while the downstream is the pressure outlet, and the side and bottom is the slip wall.

2.2.1. Mesh Convergence

The systematic verification of the grid spacing in x-direction, Δx, and in z-direction, Δz, at the water free surface was carried out. The parameters of incident wave were set the same as that in Sun et al. [33], which are shown in Table 1. In Table 1, H is wave height, L is wave length, k is wave number, A is wave amplitude.

Table 1. Wave parameters.

H/L	k	L (m)	H (m)	A (m)
1/10	0.082	76.44	7.644	3.822

Three types of mesh, named Mesh A, Mesh B, and Mesh C, were studied. Mesh size of the free surface was 25% of the base size in x-direction (wavelength direction), while it was 12.5% of the base size in z-direction (wave height direction). The specific size and quantity of the meshes are shown in Table 2. The wave probes were located at distances of 1 L, 3 L from the entrance of the wave. The origin of the coordinate was at the center of mass of the truncated cylinder.

Table 2. Meshes size and quantity.

Mesh	Base Size (m)	Number of Cells per Wave Length	Number of Cells for Wave Height H	Total Number
A	4.7	65	13	768,385
B	3.6	85	17	1,258,783
C	3.1	105	20	1,569,229

Figures 2 and 3 show the wave elevations at different locations with three type of meshes. It can be seen from Figure 2b that when the waves reach the location of 3 L from the left entrance boundary, the amplitude of the waves is reduced by about 5% compared to the incident one, which is caused by physical and numerical dissipations. The difference of the wave elevation between the incident wave height and the measured wave height at 3 L location calculated by these three types of meshes is 2.96%, 2.44% and 1.85%, respectively. For balancing the numerical accuracy and computational cost, Mesh B is mainly used in the following studies otherwise specified.

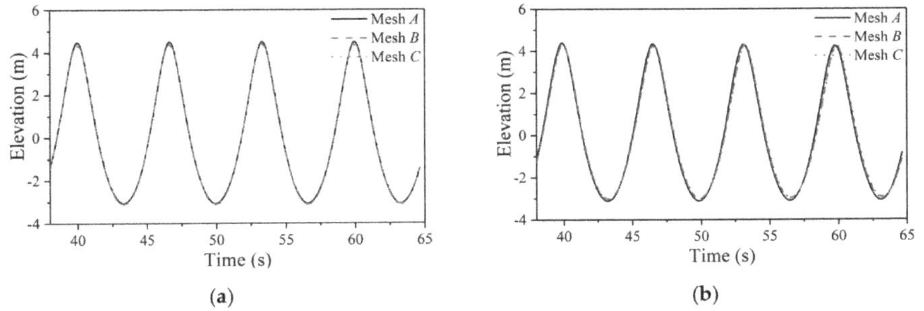

Figure 2. Wave elevation at different positions from the left entrance boundary: (**a**) 1 L from the left entrance boundary; (**b**) 3 L from the left entrance boundary.

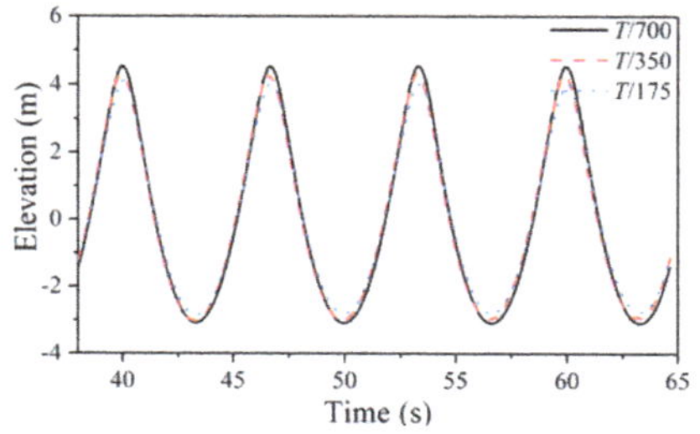

Figure 3. Wave elevations at a distance of 3 L from the left entrance boundary by using different time steps.

2.2.2. Time-Step Convergence

The time-step convergence is verified in this section. The same wave conditions as Sun et al. [33] were studied. Mesh B with L (85) and H (17) was used, and then the time steps were set to $\Delta t = T/175$, $T/350$ and $T/700$. The free-surface elevation at 3 L is shown in Figure 3. From Figure 3, it can be seen that the wave curves simulated by the time steps of $T/350$ and $T/700$ are basically the same, considering that the time required by the case of $T/700$ is significantly higher than that of the $T/350$. The time step of $T/350$ is employed in the subsequent calculations. Figure 4 shows the wave elevation in the calculation domain at $t = 10\ T$ (T is wave period). It can be seen that the wave propagates stably from the upstream to the downstream.

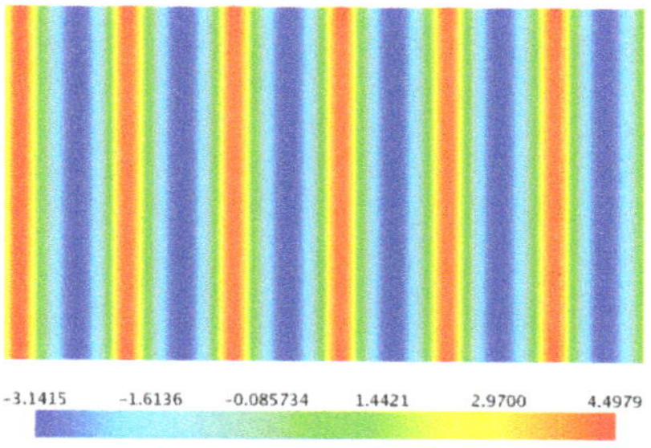

Figure 4. Top view of wave-surface elevation in the computational domain at the time instant of 10 T.

3. Free-Surface Elevation around a Truncated Column

3.1. Setting of Numerical Model

Based on the above validated numerical wave tank, the free-surface elevation around a column in strongly-nonlinear waves were studied and compared with the corresponding experimental data from MOERI [34]. The settings of numerical model are shown in Figure 5a. We used half of the computational domain in the following simulations to reduce the computational load benefit from the symmetry of body on the x-z plane. The radius of the column is $r = 8$ m, while the draft is 24 m. The positions of wave probes are shown in Table 3. The layout of the wave probes is shown in Figure 5b. The conditions of incident waves are shown in Table 1.

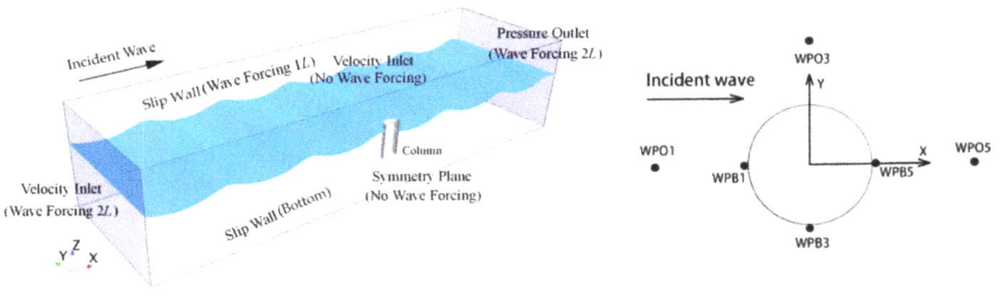

(a) (b)

Figure 5. Numerical model: (**a**) numerical wave tank; (**b**) layout of wave probes.

Table 3. Locations of wave probes.

Wave Probes	x (m)	y (m)
Wpb1	−8.2	0.0
Wpb3	0.0	−8.20
Wpb5	8.2	0.0
Wpo1	−16.0	0.0
Wpo3	0.0	16.0
Wpo5	16.0	0.0

The meshes of the numerical wave tank and column are shown in Figure 6. The STAR-CCM+ extruder meshing model [35] was used in the width direction and the downstream, which not only increases the numerical dissipation of waves, but also saves the calculation time.

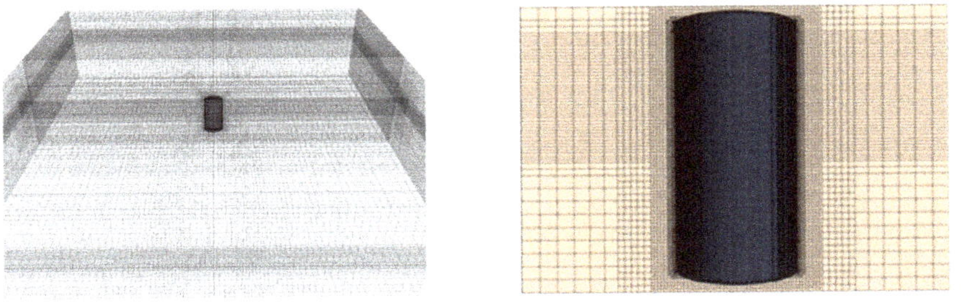

(a) (b)

Figure 6. Meshes for the numerical wave tank and column: (**a**) numerical wave tank; (**b**) column.

3.2. Comparison of Free-Surface Elevations

The RAOs (Response Amplitude Operator) of the wave-surface elevation around the truncated column under an incident wave of $k_0r = 0.657$ (k_0r is the dimensionless wave number, where $k_0 = 2\pi/L$ and L is wave length, r is the radius of column) was studied and compared with Sun et al. [33]. An average value of free-surface elevations at wave probes in 10 stable wave periods were obtained and plotted in Figure 7. The RAOs (1st harmonics) of the free-surface elevations were calculated by Fast Fourier Transform (FFT) and compared with the experimental data from MOERI [34]. The comparison results are shown in Tables 4 and 5, in which the data of the potential flow solver DIFFRACT, and CFD tool OpenFOAM are from Sun et al. [33].

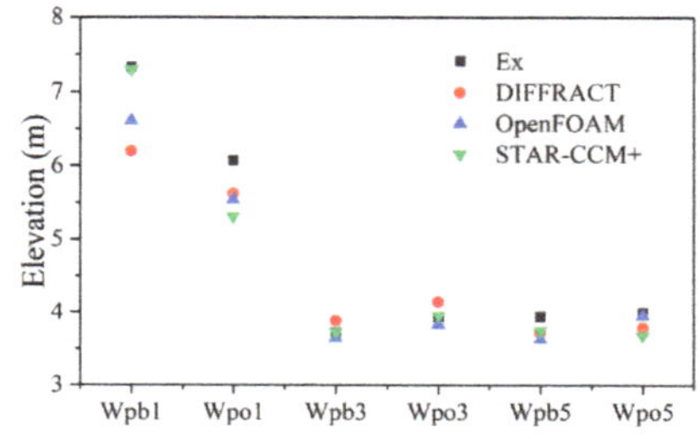

Figure 7. Free-surface elevations at different wave probes.

Table 4. RAOs of free-surface elevations at wpb1, 3 and 5.

	Wpb1	Difference	Wpb3	Difference	Wpb5	Difference
Experiment	1.918	-	0.975	-	1.031	-
DIFFRACT	1.622	−15.4%	1.017	4.3%	0.972	−5.7%
OpenFOAM	1.731	−9.7%	0.955	−2.1%	0.951	−7.8%
STAR-CCM+	1.902	−0.834%	0.976	0.103%	0.9814	−4.7%

Table 5. RAOs of surface elevations at wpo1, 3 and 5.

	Wpo1	Difference	Wpo3	Difference	Wpo5	Difference
Experiment	1.588	-	1.030	-	1.046	-
DIFFRACT	1.471	−7.4%	1.084	5.2%	0.993	−5.1%
OpenFOAM	1.450	−8.7%	1.005	−2.4%	1.034	−1.1%
STAR-CCM+	1.389	−12.5%	1.035	0.485%	0.9633	−7.9%

The RAOs (1*st* harmonics) was calculated as follows:

$$RAO^{1st} = \frac{\zeta^{(1)}}{A} \tag{3}$$

where, $\zeta^{(1)}$ is the 1*st* harmonics of the free-surface elevations, A is the amplitude of the incident waves.

From Tables 4 and 5, it can be seen that the wave elevations at the probes are in good agreement with the experimental data—excepting wpo1 probe, which has a slight discrepancy—which verifies the accuracy of numerical model.

Figure 8 shows snapshots of wave elevations around the column in one wave period. From Figure 8, two different types of scattered waves are obviously seen. One is a concentric circular wave field, called Type-1 scattered-wave field, and the other is related to the motion of water surrounding the column surface, called Type-2 scattered-wave field. Type-1 scattered-wave field is primarily in the upstream direction, while Type-2 wave by passing the surface of the column and propagating to downstream and merging to form a water mound behind the column. These two types of scattered-wave fields were also reported by Swan and Sheikh [36] in their experimental measurements.

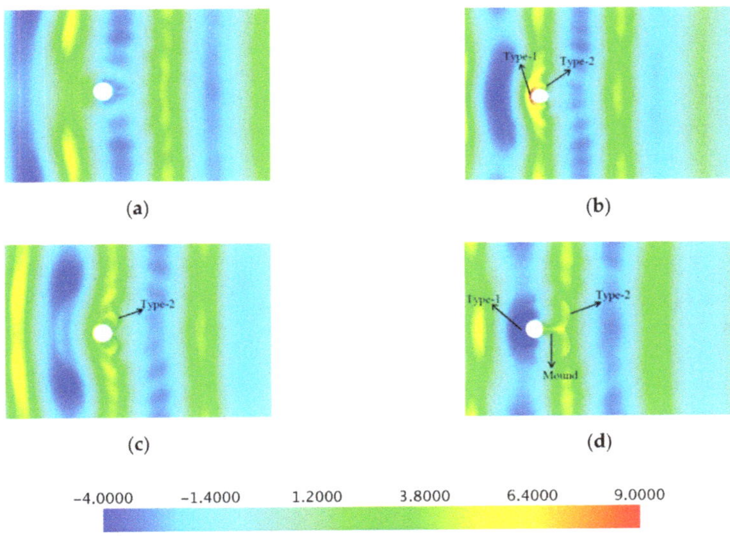

Figure 8. Contours of the free-surface elevations (m): (**a**) $t = 1/4\ T$; (**b**) $t = 2/4\ T$; (**c**) $t = 3/4\ T$; (**d**) $t = T$.

4. A Single Float in Different Waves

4.1. Resonance Frequency of a Single-Float WEC

Wave conditions for the following simulations were selected based on actual wave conditions in the South China Sea [37]. The selected wave parameters are shown in Table 6, where r and D are the radius and diameter of a float, respectively. The dimension parameters of the float are shown in Table 7.

Table 6. Wave parameters (H = 1.2 m).

k_0r	L (m)	T (s)	H/L	L/D
0.21	76.5	7	0.016	15.3
0.33	47.5	5.5	0.025	9.5
0.42	37.5	4.88	0.032	7.5
0.52	30.21	4.36	0.04	6.04
0.62	25.5	3.99	0.047	5.1
0.85	18.48	3.37	0.065	3.7
1.02	15.4	3.05	0.078	3.1
1.2	13.09	2.78	0.092	2.6
1.4	11.22	2.53	0.107	2.2
1.6	9.82	2.33	0.122	1.964

Table 7. Basic parameters of the float.

Draft (m)	Radius (m)	Density (kg/m^3)	Mass (kg)	Added Mass (kg)
1.217	2.5	512.5	25,157.28	21,354.17

According to Michael and Cormick [38], the heaving displacement of a single-float WEC system can be represented by:

$$z(t) = \frac{(F_0/ = \rho g A_{wp})\cos(\omega t + \gamma - \sigma_z)}{\sqrt{\left(1-\omega^2/\omega_z^2\right)^2 + \left(2\Delta_z\omega/\omega_z\right)^2}} = Z_0\cos(\omega t + \gamma - \sigma_z) \quad (4)$$

where F_0 is the amplitude of wave force, ρ is the density of fluid, A_{wp} is the water plane area of the float, Z_0 is the motion amplitude, ω is the circular wave frequency, t is time in seconds, γ is a phase angle that depends on the wave force components, and σ_z is a phase angle that depends primarily on the dimensionless system damping factor Δ_z, ω_z is the natural heaving frequency. In this study, the float is symmetry about the x-z and y-z planes, so the phase angle γ is always equal to zero.

The natural heaving frequency can be represented by:

$$\omega_z = \frac{1}{\sqrt{\frac{m+m_w}{\rho g A_{wp}}}} \tag{5}$$

where m is mass of the float, m_w is added mass of the float. According to the data in Table 7, ω_z is equal to 2.059, therefore, the dimensionless wave number ($k_0 r$) corresponding to natural heaving frequency is equal to $k_0 r = 1.02$.

The partial derivative of Z_0 to ω equal to zero in Equation (4) can be obtained:

$$\frac{\partial Z_0}{\partial \omega} = 0 \Rightarrow \omega' = \omega_0 \sqrt{1 - 2\Delta_z^2} \tag{6}$$

where ω' is the circular frequency of resonance state considering the damping factor.

The damping factor Δ_z can be represented by:

$$\Delta_z = \sqrt{1 - \frac{q^2}{\omega_z^2}} \tag{7}$$

where q is the natural heaving frequency of the float under radiation damping.

The q can be represented by:

$$q = 2\pi / T_0 \tag{8}$$

where T_0 is the free vibration period of the float under radiation damping.

Figure 9 plots the free vibration curve of the WEC under radiation damping. The T_0 is equal to 3.32 s in this research, therefore the q is equal to 1.945 rad/s, the Δ_z is equal to 0.329, the ω' is equal to 1.823 rad/s. Therefore, the dimensionless wave number ($k_0 r$) corresponding to resonance condition considering damping factor (Δ_z) is equal to 0.85. In the following section, we utilize the numerical method to analyze the resonance state of float and to verify the reliability of the above process.

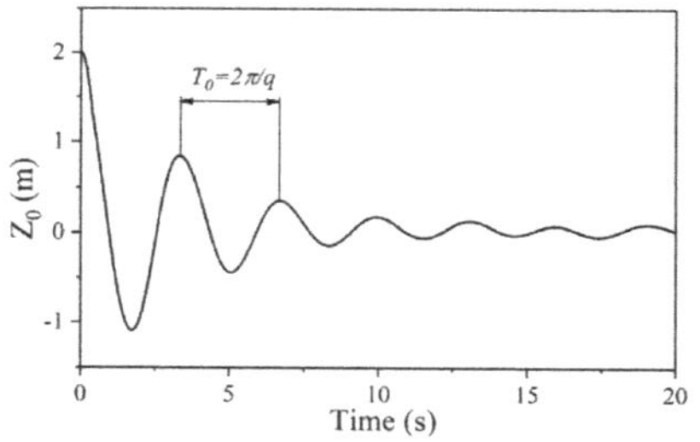

Figure 9. The free vibration curve of the wave energy converter under radiation damping.

4.2. Hydrodynamic Performance and Motion Response of the Float

Figure 10a shows the motion response of the float. From Figure 10a, it can be seen that when the wavelength is greater than the resonance one, the magnitude of heave motion of the float is equal to the amplitude of the incident wave. (A is constant, A = 0.6 m) The velocity of the float increases with the decrease of the incident wavelength. The maximum velocity of the float reaches 1.48 m/s when the resonance occurs. However, when the wavelength is smaller than the one at resonance condition, the magnitudes of displacement and velocity of the float begin to decrease, e.g., the maximum displacement of the float is only one third of the amplitude of the incident wave at k_0r = 1.4–1.6.

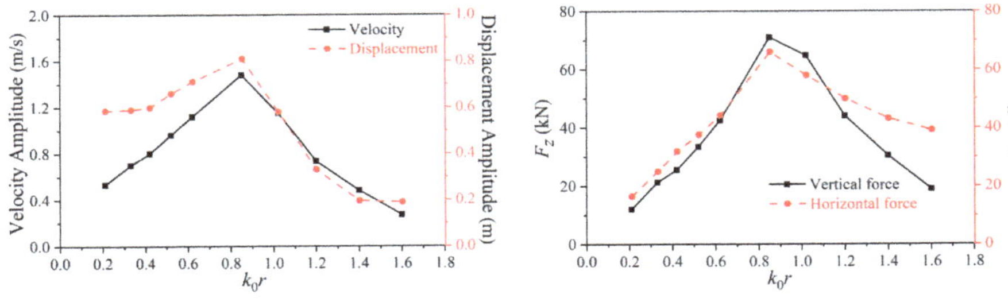

(**a**) (**b**)

Figure 10. Wave forces and motion responses of the float: (**a**) motion response; (**b**) wave force.

Figure 10b plots wave forces acting on the float in different sea conditions; where F_x and F_z are the horizontal and vertical wave forces, respectively. From Figure 10b, it can be seen that when the wavelength is greater than the resonance one, the horizontal and vertical wave forces increase as the wavelength decreases. The resonance (k_0r = 0.85) is turning point, when the incident wavelength is smaller than the resonance wavelength, the horizontal and vertical wave forces decrease as the wavelength decreases.

From the above analysis, we can see that the turning point from the numerical results is near k_0r = 0.85, while the corresponding analytical results at the resonance condition considering damping factor is also at k_0r = 0.85, which has already mentioned in section of '*Resonance Frequency of a Single-Float WEC*'. A good agreement between the numerical and analytical solutions is obtained, which verifies the reliability of the analytical process. Furthermore, we can find that the turning point of motion response and wave force is near k_0r = 0.85 rather than 1.02, which means that the damping factor can affect the motion response of the float and the wave force it receives.

4.3. *Influence of Linear Damping Constant on the PTO*

In this section the power generation of the float-type WEC, which considers the equivalent damping, is calculated. In the WEC system, the equivalent damping of the generator is the resistance generated by the electrical generator. The work done by this resistance is the electric energy generated by the generator. We simplify the equivalent damping of the generator as a damping force f_d acting on the float. The damping force, f_d, is defined as:

$$f_d = -C_{d,f}v \tag{9}$$

where $C_{d,f}$ is the linear damping constant, v is the vertical velocity of the float.

The average power of a purely heaving symmetric float in waves is represented by:

$$P_z = \frac{1}{T}\int_0^T f_d v dt \tag{10}$$

where f_d is the damping force, T is the wave period.

Generally, under certain sea conditions, the WEC system has an optimal linear damping constant, which maximizes the overall power generation of the system. Figure 11 plots the variation of power generated by the float via the linear damping constant for different wavenumbers.

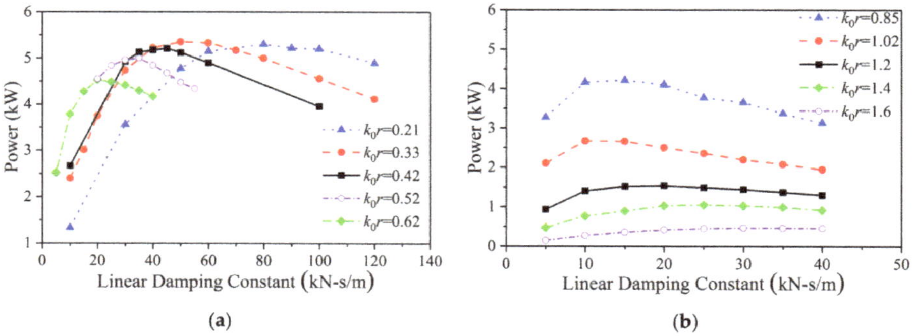

Figure 11. Power absorption performance of the WEC system: (**a**) k_0r = 0.21–0.62; (**b**) k_0r = 0.85–1.6.

From Figure 11, it can be seen that (1) the power generated by the float increases first and then decreases with the increase of the linear damping constant. (2) when the wavelength is larger than the resonance condition (k_0r = 0.85) as shown in Figure 11a, the value of the optimal linear damping constant increases with the increase of the wavelength; and the maximum of the power generated by the float slightly varies in long waves, which is basically concentrated at 5.0–5.5 kw; and the corresponding range of the optimal linear damping constants is from 35 kN-s/m to 80 kN-s/m. (3) However, when the wavelength is smaller than resonance condition, as shown in Figure 11b, the value of the optimal damping constant decreases as the wavelength increases; and the generation power is lower than the long wavelength, the maximum power is about 0.47–2.6 kw, and the corresponding range of the optimal linear damping constants is from 10 kN-s/m to 35 kN-s/m.

Figure 12 shows the optimal linear damping constant of WEC and the maximum power generated by the float with consideration of equivalent damping. The power generated by the float under the optimal linear damping constant decreases as the wavelength decreases. The optimal linear damping constants decrease first and then increase with the increases of wave number; the turning point is near k_0r = 1.02, which is coincident with the analytical natural heaving frequency without consideration of damping. This means that the damping factor does not affect the wave number turning point of the optimal damping constant.

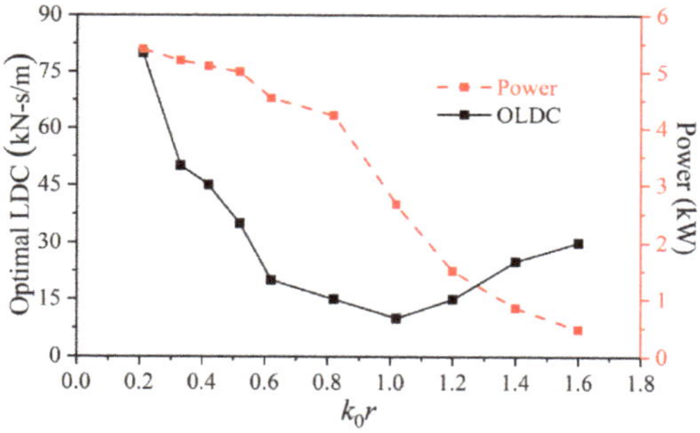

Figure 12. Optimal linear damping constants and maximum power.

5. Discussion

In this paper, a numerical model for calculating the hydrodynamic performance of a float was established. First, the numerical accuracy and reliability of the established multiphase flow numerical wave tank were verified by the convergence studies. Then the wave-surface elevations surrounding a truncated column in waves were simulated and compared with the corresponding experimental data from MOERI [34]. After, the resonance condition of a particular float was solved not only by numerical method but also by analytical method. The results are in good agreement, which verify the reliability of the analytical process. Then, the hydrodynamic forces and motion response of the float without the damping constants in different waves were evaluated, and the power absorption performance of the float considering the damping constants in different waves were studied.

The velocity amplitude and the displacement amplitude of a float without damping constant were investigated against various wave numbers. It was found that velocity amplitude and the displacement amplitude increase first and then decrease with the increases of wave number, with the turning point being near $k_0r = 0.85$ rather than 1.02. This means that the damping factor can affect the motion response of the float and the wave force it receives.

The relationship between the optimal damping constant and the wave number was obtained. When $k_0r > 1.02$, the optimal damping constant increases as the wave number increases. When $k_0r < 1.02$, the optimal damping constant decreases as the wave number increases. In other words, the turning point of optimal linear damping constants is near $k_0r = 1.02$; this value is the analytical natural heaving frequency without consideration of damping. This means that the damping factor does not affect the wave number turning point of the optimal damping constant.

The power of the float generator under the optimal damping constant decreases as the wave number increases. When $k_0r < 0.42$, the absorption power of the float at the optimal damping constant is basically the same (5.5 kW).

The study provided an analytical approach to solving the resonance condition of a specific float, and provided an insight into the hydrodynamic performance of a WEC system and its power generation with optimal linear damping constant in different waves.

Author Contributions: Conceptualization, G.H.; Data curation, C.L.; Formal analysis, Z.L.; Funding acquisition, G.H.; Investigation, R.J. and B.G.; Methodology, G.H.; Project administration, G.H.; Resources, G.H.; Software, Z.L.; Supervision, G.H.; Validation, Z.L., Z.Z. and P.J.; Writing—original draft, Z.L.; Writing—review & editing, G.H. and Z.Z.

Funding: This research was funded by National Natural Science Foundation of China (51579058), Marine Renewable Energy Fund Project in 2018, Application Demonstration of High Reliability Marine Energy Supply Equipment (GHME2018SF02).

Conflicts of Interest: The authors declare no conflicts of interest.

References

1. Najam, A.; Cleveland, C.J. Energy and Sustainable Development at Global Environmental Summits: An Evolving Agenda. *Environ. Dev. Sustain.* **2003**, *5*, 117–138. [CrossRef]
2. Dincer, I.; Rosen, M.A. A worldwide perspective on energy, environment and sustainable development. *Int. J. Energy Res.* **1998**, *22*, 1305–1321. [CrossRef]
3. Barstow, S.; Mørk, G.; Mollison, D.; Cruz, J. The Wave Energy Resource. In *Ocean Wave Energy*, 2nd ed.; Springer: Berlin/Heidelberg, Germany, 2008; Volume 4, pp. 93–130. Available online: https://link.springer.com/chapter/10.1007%2F978-3-540-74895-3_4 (accessed on 17 March 2019).
4. Muetze, A.; Vining, J.G. Ocean Wave Energy Conversion-A Survey. In Proceedings of the Conference Record of the 2006 IEEE Industry Applications Conference Forty-First IAS Annual Meeting, Tampa, FL, USA, 8–12 October 2006; Volume 3, pp. 1410–1417. [CrossRef]
5. Ross, D. *Power from Sea Waves*, 1st ed.; Oxford University Press: Oxford, UK, 1995; Available online: https://www.mendeley.com/catalogue/power-waves-2/ (accessed on 24 July 2019).
6. Salter, S.H. Wave power. *Nature* **1974**, *249*, 720–724. [CrossRef]
7. Oreun, A. Hydrodynamic characteristics of an oscillating water column device. *Appl. Ocean Res.* **1996**, *17*, 155–164. [CrossRef]
8. Curran, R.; Gato, L.M.C. The energy conversion performance of several types of Wells turbine designs. *Proc. Inst. Mech. Eng. Part A J. Power Energy* **1997**, *211*, 133–145. [CrossRef]
9. Toyota, K.; Nagata, S.; Imai, Y.; Setoguchi, T. Effects of Hull Shape on Primary Conversion Characteristics of a Floating OWC "Backward Bent Duct Buoy". *J. Fluid Sci. Technol.* **2008**, *3*, 458–465. [CrossRef]

10. Alves, M.; Traylor, H.; Sarmento, A. Hydrodynamic optimization of a wave energy converter using a heave motion buoy. In Proceedings of the 7th European wave and Tidal Energy Conference, Porto, Portugal, 11–13 September 2007; pp. 11–14. Available online: https://www.mendeley.com/catalogue/hydrodynamic-optimization-wave-energy-converter-using-heave-motion-buoy/ (accessed on 24 July 2019).
11. Eriksson, M.; Waters, R.; Svensson, O.; Leijon, M. Wave power absorption: Experiments in open sea and simulation. *J. Appl. Phys.* **2007**, *102*, 034105. [CrossRef]
12. Falcão, A.F. Modelling and control of oscillating-body wave energy converters with hydraulic power take-off and gas accumulator. *Ocean Eng.* **2007**, *34*, 2021–2032. [CrossRef]
13. Pastor, J.; Liu, Y. Hydrokinetic Energy Overview and it's Renewable Energy Potential for the Gulf of Mexico. In Proceedings of the 2012 IEEE Green Technologies Conference, Tulsa, OK, USA, 19–20 April 2012. [CrossRef]
14. Bhatta, D.D.; Rahman, M. On scattering and radiation problem for a cylinder in water of finite depth. *Int. J. Eng. Sci.* **2003**, *41*, 931–967. [CrossRef]
15. Zheng, Y.H.; Shen, Y.M.; You, Y.G.; Wu, B.J.; Jie, D.S. Wave radiation by a floating rectangular structure in oblique seas. *Ocean Eng.* **2006**, *33*, 59–81. [CrossRef]
16. Zheng, Y.H.; You, Y.G.; Shen, Y.M. On the radiation and diffraction of water waves by a rectangular buoy. *Ocean Eng.* **2004**, *31*, 1063–1082. [CrossRef]
17. Shen, Y.M.; Zheng, Y.H.; You, Y.G. On the radiation and diffraction of linear water waves by a rectangular structure over a sill. Part I. Infinite domain of finite water depth. *Ocean Eng.* **2005**, *32*, 1073–1097. [CrossRef]
18. Vantorre, M.; Banasiak, R.; Verhoeven, R. Modelling of hydraulic performance and wave energy extraction by a point absorber in heave. *Appl. Ocean Res.* **2004**, *26*, 61–72. [CrossRef]
19. Backer, G.D.; Vantorre, M.; Banasiak, R.; Beels, C.; Rouck, J.D. Numerical modelling of wave energy absorption by a floating point absorber system. In Proceedings of the 17th International Offshore (Ocean) and Polar Engineering Conference, ISOPE, Lisbon, Portugal, 1–6 July 2007; Available online: https://www.onepetro.org/conference-paper/ISOPE-I-07-186 (accessed on 24 July 2019).
20. MacCamy, R.C.; Fuchs, R.A. Wave forces on piles: A diffraction theory. In *Beach Erosion Board Office of the Chief Engineers*; Technical Memorandum; Department of the Army: Arlington County, VA, USA, 1954; Volume 69, pp. 1–17. Available online: https://www.researchgate.net/publication/235139845_Wave_forces_on_piles_A_diffraction_theory (accessed on 17 March 2019).
21. Taylor, E.R.; Hung, S.M. Second order diffraction forces on a vertical cylinder in regular waves. *Appl. Ocean Res.* **1987**, *9*, 19–30. [CrossRef]
22. Lu, L.; Chen, X.B. Dissipation in the gap resonance between two bodies. In Proceedings of the 27th International Workshop on Water Waves and Floating Bodies, Copenhagen, Denmark, 22–25 April 2012; Available online: https://www.mendeley.com/catalogue/dissipation-gap-resonance-between-two-bodies (accessed on 17 March 2019).
23. Song, R.; Zhang, M.; Qian, X.; Wang, X.; Dai, M.; Chen, J. A Floating Ocean Energy Conversion Device and Numerical Study on Buoy Shape and Performance. *J. Mar. Sci. Eng.* **2016**, *4*, 35. [CrossRef]
24. Hardisty, J. Experiments with point absorbers for wave energy conversion. *J. Mar. Eng. Techno.* **2012**, *11*, 51–62. [CrossRef]
25. Yu, Y.H.; Li, Y. Reynolds-Averaged Navier–Stokes simulation of the heave performance of a two-body floating-point absorber wave energy system. *Comput. Fluids* **2013**, *73*, 104–114. [CrossRef]
26. Gallizio, F.; Bergmann, M.; Bracco, G.; Giorcelli, E.; Iollo, A.; Mattiazzo, G.; Ponzetta, M. A two-way coupling CFD method to simulate the dynamics of a wave energy converter. In Proceedings of the OCEANS 2015-Genova, Genoa, Italy, 18–21 May 2015. [CrossRef]
27. Coe, R.G.; Rosenberg, B.J.; Quon, E.W.; Chartrand, C.C.; Yu, Y.; van Rij, J.; Mundon, T.R. CFD design-load analysis of a two-body wave energy converter. *J. Ocean Eng. Mar. Energy* **2019**, *5*, 99–117. [CrossRef]
28. Agamloh, E.B.; Wallace, A.K.; Jouanne, A.V. Application of fluid-structure interaction simulation of an ocean wave energy extraction device. *Renew. Energy* **2008**, *33*, 748–757. [CrossRef]
29. Westphalen, J.; Greaves, D.M.; Hunt-Raby, A.; Williams, C.J.K.; Taylor, P.H.; Hu, Z.; Omidvar, P.; Causon, D.M.; Mingham, C.G.; Stansby, P.K.; et al. Numerical simulation of wave energy converters using Eulerian and Lagrangian CFD methods. In Proceedings of the 20th International Offshore (Ocean) and Polar Engineering Conference, ISOPE, Beijing, China, 20–25 June 2010; Available online: https://www.researchgate.net/publication/286723909_Numerical_simulation_of_wave_energy_converters_using_Eulerian_and_Lagrangian_CFD_methods?ev=prf_high (accessed on 17 March 2019).

30. Maggi, A.; Moreno, E.C.; Stansby, P.; Taylor, R.E.; Maggi, A. Wave energy conversion with high capture width by the three-float line absorber M4. In Proceedings of the 1st International Conference on Renewable Energies Offshore, RENEW 2014, Lisbon, Portugal, 24 November 2014; Available online: https://www.mysciencework.com/publication/show/wave-energy-conversion-high-capture-width-three-float-line-absorber-m4-c1d4a01f (accessed on 17 March 2019).
31. Hann, M.R.; Greaves, D.M.; Raby, A. EGU General Assembly Conference Abstract, Physical measurements of breaking wave impact on a floating wave energy converter. In Proceedings of the EGU General Assembly Conference, Vienna, Austria, 7–12 April 2013; Available online: https://www.researchgate.net/publication/258784114_Physical_measurements_of_breaking_wave_impact_on_a_floating_wave_energy_converter (accessed on 17 March 2019).
32. Yan, S.; Qiu, H.; Wang, G.Y.; Li, C.; Lv, N.; Wang, D.H. Research and design of the scale parameter of a wave energy gathering device. *J. Dalian Marit. Univ.* **2017**, *43*, 38–42. [CrossRef]
33. Sun, L.; Zang, J.; Chen, L.; Chen, L.; Taylor, R.E.; Taylor, P.H. Regular waves onto a truncated circular column: A comparison of experiments and simulations. *Appl. Ocean Res.* **2016**, *59*, 650–662. [CrossRef]
34. Ocean Engineering Committee. *Final Report and Recommendations to the 27th ITTC*; Ocean Engineering Committee: Copenhagen, Denmark, 2014.
35. CD-adapco. *User Guide STAR-CCM+*; Version 14.02.010-R8; CD-adapco: Melville, NY, USA, 2019.
36. Swan, C.; Sheikh, R. The interaction between steep waves and a surface-piercing column. *Philos. Trans A Math. Phys. Eng. Sci.* **2015**, *373*, 41251–41257. [CrossRef] [PubMed]
37. Yi, F.; Feng, W.B.; Cao, H.J. Wave analysis based on era-interim reanalysis data in the south china sea. *Mar. Forecast.* **2018**, *35*, 44–51. [CrossRef]
38. McCormick, M.E. Wave energy conversion. In *Ocean Wave Energy Conversion*, 2nd ed.; Courier Corporation: New York, NY, USA, 2007; Volume 4, pp. 45–59.

Journal of
Marine Science and Engineering

MDPI

Article

Experimental Analysis on Dynamic Responses of an Electrical Platform for an Offshore Wind Farm under Earthquake Load

Zhen-Zhou Sun [1], Chun-Wei Bi [2,*], Sheng-Xiao Zhao [1], Guo-Hai Dong [2] and Hua-Feng Yu [1]

[1] Power China Huadong Engineering Corporation Limited, Hangzhou 310014, China
[2] State Key Laboratory of Coastal and Offshore Engineering, Dalian University of Technology, Dalian 116024, China
* Correspondence: bicw@dlut.edu.cn

Received: 29 June 2019; Accepted: 15 August 2019; Published: 18 August 2019

Abstract: Offshore wind power is gradually developing to more open sea. Considering the economy of power transmission, it will be an inevitable choice to adopt the extra-large electrical platform. The offshore electrical platform is easily affected by sudden extreme loads such as earthquake and high current loads. With a large volume of electrical equipment arranged on the deck, the offshore electrical platform is characterized as a top-heavy structure in the offshore wind farm. The dynamic effect of the structure will aggravate the vibration problem of the structure. In this paper, a physical model test was carried out to study the dynamic characteristics of the electrical platform of a 10,000-ton offshore converter station under seismic load. The acceleration response, displacement response and stress response of the offshore electrical platform under the typical direction of seismic action were obtained. The effect of the dry–wet environment, mode of seismic excitation, whipping effect and weak positions of electrical platform structure were analyzed. It was determined that the average damping ratio of the first-order mode of the electrical platform was 5.73% and 8.68% with and without water, respectively. The bidirectional seismic excitation was more dangerous to the structure than unidirectional excitation. The peak acceleration along the height of the platform showed a typical whipping effect.

Keywords: offshore electrical platform; dynamic response; earthquake load; laboratory experiment

1. Introduction

Wind power is the main form of wind energy utilization. Due to low cost, eco-friendly and significant benefits, wind power has drawn the world's attention as the fastest developing new energy with great potential. In recent years, the attention of the world wind power industry gradually turns to offshore wind power. Offshore wind farms, which initially invest several times more than onshore wind farms, have better wind resources because they are not affected by site and power limitations. Therefore, an offshore wind farm is far more profitable than an onshore one during the operating period of the project. At present, the offshore wind power is gradually developing to more open sea [1,2]. Considering the economy of power transmission, it will be an inevitable choice to adopt the extra-large electrical platform, such as offshore converter station, for the large-scale offshore wind power development in the future. The structure of offshore electrical platform is complex, especially under the combined action of loads caused by wind, wave, current and other environmental factors, and threatened by sudden extreme loads such as typhoon and earthquake. In addition, with a large volume of electrical equipment arranged on the deck, the offshore electrical platform is characterized as a top-heavy structure in the offshore wind farm. The dynamic effect of the structure will aggravate the vibration problem of the structure. The choice of the platform type is related to the structure weight, the geological condition and the marine hydrological parameters. When the total weight of the superstructure is less than 1000 t, single-pile foundation is usually adopted, while gravity foundation can be used when the geological conditions are good and the water depth is relatively shallow. However, when the water depth is large and the total weight of the superstructure exceeds 1000 t, jacket foundation should be considered. Most offshore wind farm and oil platform are located in the edge of the continental plates. Thus, a strong earthquake is one of the main fracture loads of marine engineering structures. A strong earthquake could lead to very serious damage, studies on seismic responses and collapse analysis of the offshore electrical platform is very necessary [3–5].

In recent years, scholars have carried out a lot of theoretical and experimental research on the structural response of fixed offshore platforms under seismic load. For design specifications for offshore platforms under seismic conditions, Bea et al. [6] analyzed the seismic response of the traditional steel jacket platform under strong earthquakes, and verified that American Petroleum Institute (API) related specifications for the design of an offshore platform structure could guarantee the offshore platform structure to have good anti-seismic performance. Chandrasekaran and Gaurav [7] analyzed the dynamic behavior of a triangular tension leg platform under distinctly high sea waves in the presence of both horizontal and vertical seismic excitations. Huang and Foutch [8] studied the effect of hysteresis models on the global collapse drift limit, seismic demand and capacity/demand ratio for moment resisting frame (MRF) structures under seismic loads. Yu et al. [9] developed a transient dynamic analysis method to study the dynamic characteristics of a jack-up platform in regular and random waves. Araki and Hjelmstad [10] proposed criteria for assessing the propensity for dynamic collapse of an elastoplastic structural system. Salem et al. [11] assessed the response modification factor of open steel platform subjected to seismic loads using both nonlinear static pushover analysis and nonlinear dynamic time history analysis. Besides, there are a number of studies on the dynamic characteristics of wind turbine under seismic load. Failla et al. [12] studied the effects of response-spectrum uncoupled analyses on seismic assessment of offshore wind turbine. Zhao et al. [13] conducted a fully coupled time domain simulation to investigate the dynamic characteristics of an ultra-large semi-submersible offshore wind turbine in typical sea states. Prowell et al. [14] carried out shake table tests on an actual wind turbine in both parked and operational conditions with parallel and perpendicular orientations of shaking direction. Zheng et al. [15] performed a series of shake table model tests of a monopile wind turbine foundation in dry flume, low and high calm water levels, with and without regular or random waves. Wang et al. [16] conducted both a physical model test and numerical analysis of a multi-pile offshore wind turbine under seismic, wind, wave and current loads. Sadowski et al. [17] presented an analysis of the seismic response of a steel support tower of the wind turbine modeled as a near-cylindrical shell structure with realistic axisymmetric weld depression imperfection. Wang et al. [18] developed an integrated analysis model to analyze the dynamic characteristics of a bottom-fixed offshore wind turbine under earthquakes. Although there are many researches on the dynamic response of the jacket platform and wind turbine, the research on the dynamic characteristics of this kind of top-heavy electrical platform under the earthquake load has not been reported.

The electrical platform in the offshore wind farm bears a volume of electrical equipment arranged on each deck. For the offshore converter station platform in open sea, the environmental loads such as wind, wave, current and earthquake acting on the overall structure are complex alternating and fluctuating loads, which are likely to gradually stimulate and form excessive vibration response. Coupled with the large size and weight of the upper block of the converter station, the platform is a typical top-heavy structure. The dynamic effect of the upper structure may aggravate the vibration problem of the structure. In this paper, a physical model test is carried out to study the dynamic characteristics of the electrical platform of a 10,000-ton offshore converter station under seismic load.

2. Materials and Methods

2.1. Experimental Model Design

Based on the hydroelastic similarity and the previous experiences on the physical model test, plexiglass was selected as the material for the experimental model [16]. In order to obtain the basic material parameters of plexiglass, the mechanical test was carried out for the selected plexiglass, which was used to make the experimental model. According to the results of the mechanical test, the density of plexiglass material used in the test model was 1201 kg/m^3, the static elastic modulus was 2.62 GPa, the Poisson ratio was 0.42 and the dynamic modulus was 3.91 GPa.

2.1.1. Experimental Model Similarity

Considering the dimensions of converter station, marine hydrological parameters and laboratory conditions, the geometric scale of this test model was 1:60. The model similarity criterion is the key of this research, and the structural vibration should satisfy the elastic similarity as follows [19]:

$$\lambda_\rho \cdot \lambda_A \cdot \lambda^3 \cdot \lambda_u \cdot \lambda_t{}^{-2} = \lambda_I \cdot \lambda_E \cdot \lambda_u \cdot \lambda^{-3}, \quad (1)$$

where λ_ρ is the density scale, λ is the geometric scale, λ_u is the deformation scale, λ_E is the elastic modulus, λ_t is the time scale, λ_A is the sectional area scale and λ_I is the sectional moment of the inertia scale.

For a structure dominated by bending vibration, the elastic similarity law can be rewritten as follows:

$$\lambda_t{}^2 = \lambda^4 \cdot \lambda_\rho \cdot \lambda_E{}^{-1} \cdot \lambda_r{}^{-2}, \quad (2)$$

where λ_r is the radius of the inertia scale.

Froude number is the ratio of inertial force to gravity. Thus, Froude number similarity represents the similarity relationship between the inertial force and gravity, which is expressed as follows [20]:

$$\lambda_v = \lambda^{0.5}, \quad (3)$$

where λ_v is the speed scale.

Considering acceleration scale is $\lambda_g = 1$, thus,

$$\lambda_t = \lambda^{0.5}. \quad (4)$$

Combining the Equations (2) and (4), and simultaneously satisfy elastic similarity and Froude number similarity, the hydroelastic similarity can be given as follow:

$$\lambda^3 \cdot \lambda_r{}^{-2} = \lambda_E. \quad (5)$$

The prototype structural material was steel, and its elastic modulus was 206 GPa. The model material was plexiglass, and its dynamic elastic modulus was 3.91 GPa. Therefore, the elastic modulus scale was 52.69. The ratio of the section radius of inertia was 64.027. The similarity relations of other parameters could be calculated according to the geometric scale and the sectional radius of the inertia scale. The similarity relations of basic parameters are summarized in Table 1. Some of the similarities relations related to the section took the platform leg as an example.

Table 1. The similarity relation of basic parameters for the model design. Taking the platform leg as an example.

Parameter	Similarity	Similar scale
Length	λ	60
Area	λ_A	1120.7
Volume	$\lambda \cdot \lambda_A$	67241
Density	$\lambda_\rho = 1$	1
Mass	$\lambda_\rho \cdot \lambda \cdot \lambda_A$	67241
Speed	$\lambda^{0.5}$	7.746
Acceleration	$\lambda_g = 1$	1
Time	$\lambda^{0.5}$	7.746
Frequency	$\lambda^{-0.5}$	0.129
Force	λ^3	60^3
Moment	λ^4	60^4
Moment of area	$\lambda_A \cdot {\lambda_r}^2$	4.59×10^6
Moment of mass	$\lambda \cdot \lambda_A \cdot {\lambda_r}^2$	2.76×10^8
Stress	$\lambda^4 \cdot \lambda_D \cdot {\lambda_A}^{-1} \cdot {\lambda_r}^{-2}$	176.3

2.1.2. Geometric Dimension Design of the Experimental Model

With reference to the prototype structure of the electrical platform, the geometric dimensions of each component of the test model were determined according to the geometric scale and the inertia radius scale. Schematic diagram of the electrical platform is shown in Figure 1. The parameters of the superstructure model are shown in Table 2. The lower jacket structure was mainly the cylindrical steel pipe. The length of the members and the outer diameter of the pile legs strictly followed the geometric scale; therefore, the hydrodynamic characteristics of the platform could be simulated accurately. The pipe section of the pile leg was calculated according to the area scale to achieve the accurate simulation of the elastic response of the structure (see Table 3).

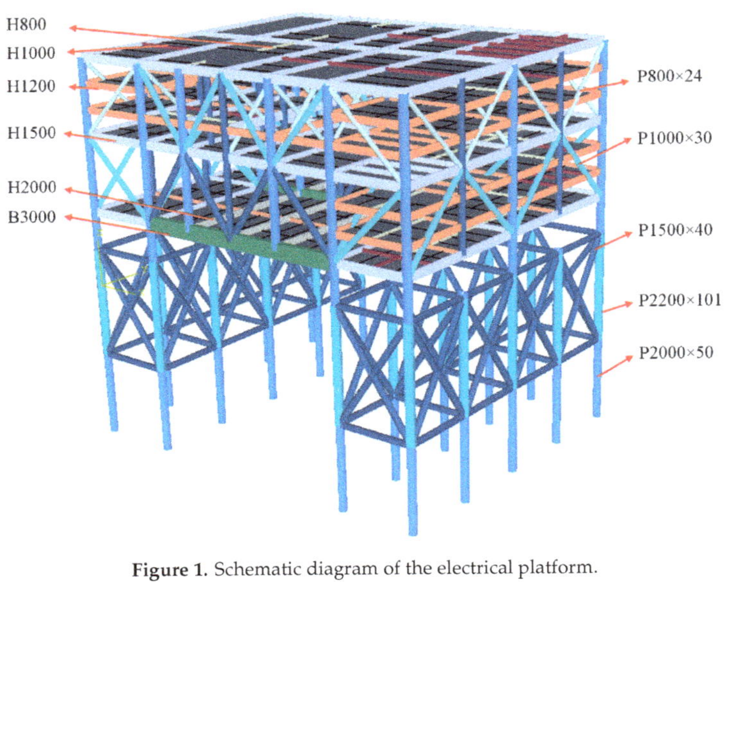

Figure 1. Schematic diagram of the electrical platform.

Table 2. Geometric dimension parameters of the superstructure model.

Member Type	Model Number	Sectional Dimension (mm)	Section for Model
Box beam	B3000	59.1 × 23	Rectangular section (Height × Width)
I-beam	H2000	45.5 × 8.6	
	H1500	34.2 × 6.8	
	H1200	27.6 × 6.4	
	H1000	22.8 × 5.1	
	H800	18.1 × 3.8	
Pipe	P2000 × 50	32 × 3	Pipe (Outside diameter × Thickness)
	P1500 × 40	25 × 2	
	P1000 × 30	22	Solid bar (Diameter)
	P800 × 24	18	

Table 3. Geometric dimension parameters of the jacket structure model.

Member Type	Model Number	Sectional Dimension (mm)	Section for Model
Pipe	P2200 × 101	35 × 5	Pipe (Outside diameter × Thickness)
	P2000 × 50	32 × 3	
	P1500 × 40	25 × 2	

2.1.3. Counterweight Design of the Experimental Model

Due to the difference of the prototype and model material, the weight distribution on the platform was different. Therefore, the sheet lead was used to balance each component of the model structure in this study. For the structural mass of the model component, the weight was scaled according to the mass scale. For the non-structural mass, such as the weight of the equipment, it was scaled as force. Thus, both gravity similarity and hydroelastic similarity were satisfied. Finally, the balance weight was 274.4 kg for the superstructure and 76.9 kg for the jacket model. The total weight of the platform model was 427.8 kg.

2.1.4. Equivalent Pile Design of the Experimental Model

The foundation of the electrical platform was the pile foundation. However, in the laboratory test, the soil data was difficult to determine. Coupled with the complexity of the seismic load, it was difficult to directly simulate the interaction between the pile and soil. Therefore, it was necessary to adopt a simplified approach instead. In this study, the equivalent pile method was used to simulate the pile–soil interaction of the converter station [21]. Firstly, the finite element model was established to simulate the nonlinear characteristics between pile and soil based on the p–y method. Then, taking the structural basic frequency as the constraint condition, the relationship between the equivalent pile length and pile diameter was determined by adjusting the equivalent pile length. In this study, the equivalent pile length of the electrical platform model was determined to be seven times the pile diameter. In the experiment, the 20 piles of the platform were fixed at the vibration table through two plexiglass plates.

2.2. Laboratory Setup

The laboratory experiments were conducted in a wave and current flume at the State Key Laboratory of Coastal and Offshore Engineering (SLCOE), Dalian University of Technology, Dalian, China. The flume was 21.6 m long, 5.0 m wide and 1.0 m deep. The flume was equipped with an elliptic underwater vibration table (see Figure 2) and earthquake simulating system. Basic parameters of the earthquake simulating system are shown in Table 4.

Based on the consideration of the structural symmetry, the laboratory experiment of the electrical platform using an underwater earthquake simulating system was carried out under the typical seismic propagation direction, where the normal direction was determined to be the most dangerous direction. The experimental arrangement is shown in Figure 3.

Figure 2. The elliptic underwater vibration table in the flume.

Table 4. Parameters of the earthquake simulating system.

Parameter	Value
Sizes of Shaking Table	4 m × 3 m
Max. Loading Mass	10,000 kg
Max. Horizontal Displacement	±75 mm
Max. Horizontal Velocity	50 cm/s
Max. Horizontal Acceleration	1.0 g
Max. Vertical Displacement	±50 mm
Max. Vertical Velocity	35 cm/s
Max. Vertical Acceleration	0.7 g
Operation Frequency Range	0–50 Hz

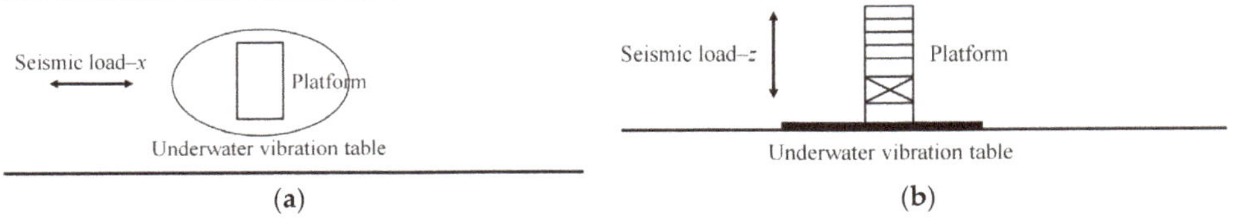

Figure 3. The experimental layout of the electrical platform model. (**a**) Top view and (**b**) side view.

In this experiment, strain data at some key positions of the platform structure was collected using some strain gauges (Figure 4) and a data acquisition instrument from the National Instrument (Figure 5). Waterproof and non-waterproof acceleration sensors (Figure 6) were adopted to measure the acceleration response at key positions on the platform structure and equipment. The measuring range of the acceleration sensor was 1.0 g, the accuracy was up to 0.3% and the measuring frequency was up to 1000 Hz. The displacement response of the structure was measured by the laser displacement meter, as shown in Figure 7.

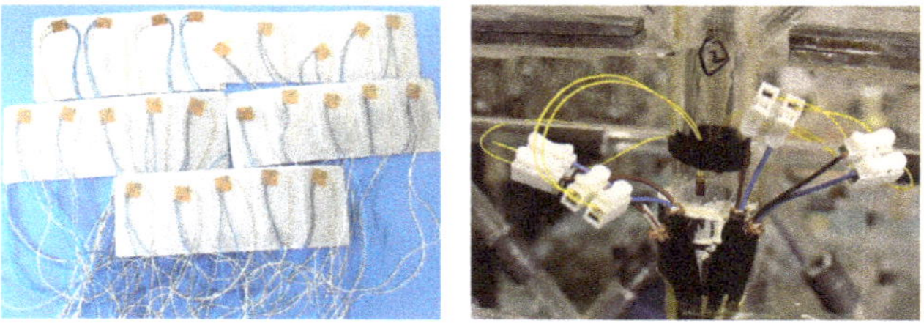

Figure 4. The experimental layout of the strain gauge.

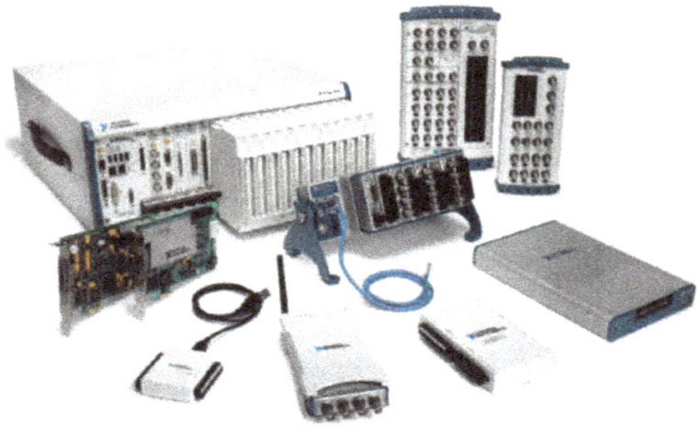

Figure 5. The National Instruments (NI) data acquisition instrument used in this experiment.

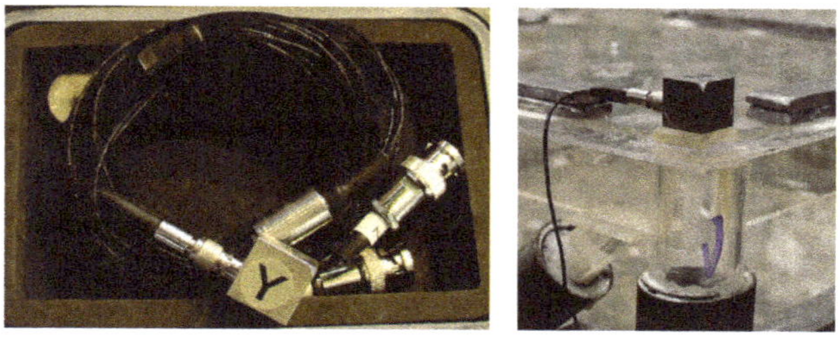

Figure 6. The experimental layout of the acceleration sensor.

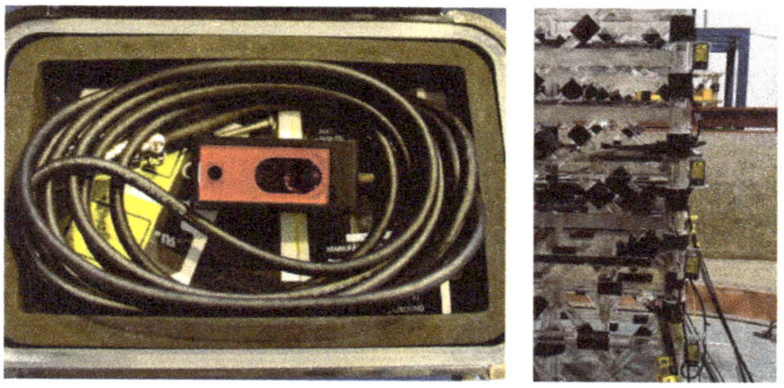

Figure 7. The experimental layout of the laser displacement meter.

2.3. Experimental Conditions

The water depth was 0.32 m during this experiment, and the corresponding prototype water depth was 19.4 m with the model scale of 1:60. There were no water waves and only the seismic load was applied during experiments. The dynamic response of the electrical platform structure excited by different seismic waves would be tested to obtain the time history curve of the dynamic response of the jacket electrical platform structure. The seismic loads in this experiment, presented in Table 5, included four recorded earthquake waves from the ground motion database of the Pacific Earthquake Engineering Research Center (PEER) and one artificial earthquake wave based on the spectrum of the API standard [21]. The four selected seismic spectrums were compared with the standard spectrum, and the overall error was within 20%.

Table 5. The seismic loads in the experiment.

Seismic Wave	Excitation Orientation	Water Level
API RSN169_IMVPALL RSN800_LOMAP RSN1116_KOBE RSN6988_DARFIELD	x z	Without level Mean water level

In order to eliminate the residual strain in the model, a sufficient time interval was set between two adjacent tests. To study the influence of water damping effect on the dynamic response of the platform structure, the dynamic response characteristics of the electrical platform under seismic load were tested with and without water respectively. The excitation direction of seismic wave could be divided into two working conditions: Unidirectional (x direction) and bidirectional (x and z direction). The ratio of the peak acceleration was x:z = 1:0.5 for the bidirectional seismic excitation.

2.4. Measuring-Point Arrangement

To measure the structural strain, seven key nodes were selected on the electrical platform and four strain gauges were arranged in an orthogonal form for each node. Eleven acceleration sensors were used, including nine on the electrical platform and two on the ground for reference. Along a pile leg, five displacement sensors were arranged from top to bottom to measure the displacement response of the structure under the earthquake load (Figure 8).

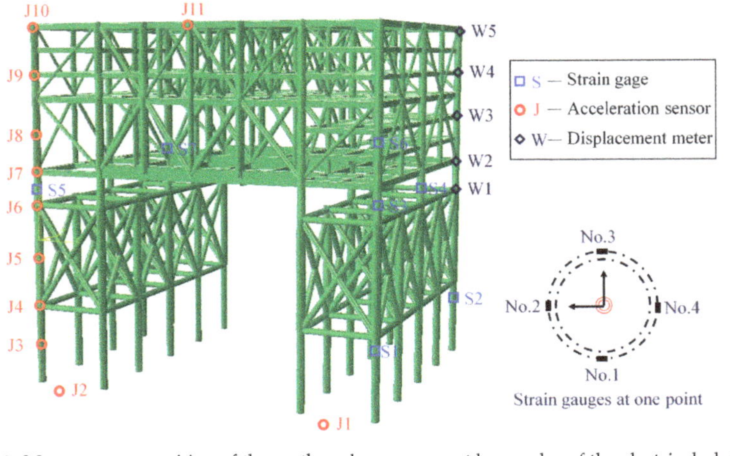

Figure 8. Measurement position of the earthquake response at key nodes of the electrical platform.

2.5. Experimental Uncertainties

The physical model of the electrical platform was made of plexiglass pipes and plates. Although the model size was carefully designed, there would inevitably be global and local dimension errors when manufacturing the experimental model. An uncertainty of ±1% in the dimensions of model height, width and pipe diameter might seem realistic.

The tested seismic wave was converted from the time series of the measured seismic wave. However, the time series of the output seismic wave acceleration still had some biases. Therefore, dimensionless results were analyzed and discussed in this study.

There will be some random error in the experiment. Due to various accidental factors, such as the measurement accuracy of the instrument, the measurement value might deviate from the true value. In this experiment, the same physical quantity was measured for several times to get the average value, so as to reduce accidental errors.

3. Results

3.1. Basic Dynamic Parameters of the Electrical Platform

In this section, an experimental study on the dynamic characteristics of the offshore electrical platform was conducted. By white noise excitation, the structural dynamic parameters such as structural frequency, mode of vibration and damping ratio were obtained.

3.1.1. Structural Frequency

According to the time-series data of the structural acceleration response collected by the acceleration sensor, the corresponding power spectral density of each point was calculated. Taking measurement point J4 for acceleration as an example, the time-series of acceleration response and power spectral density function at the pile leg with and without water are shown in Figure 9. The frequency corresponding to the first peak of the power spectral density function was the fundamental frequency in the normal (x) direction of the model structure. The frequency corresponding to the first peak of the power spectral density function was the fundamental frequency in the x-direction of the model structure.

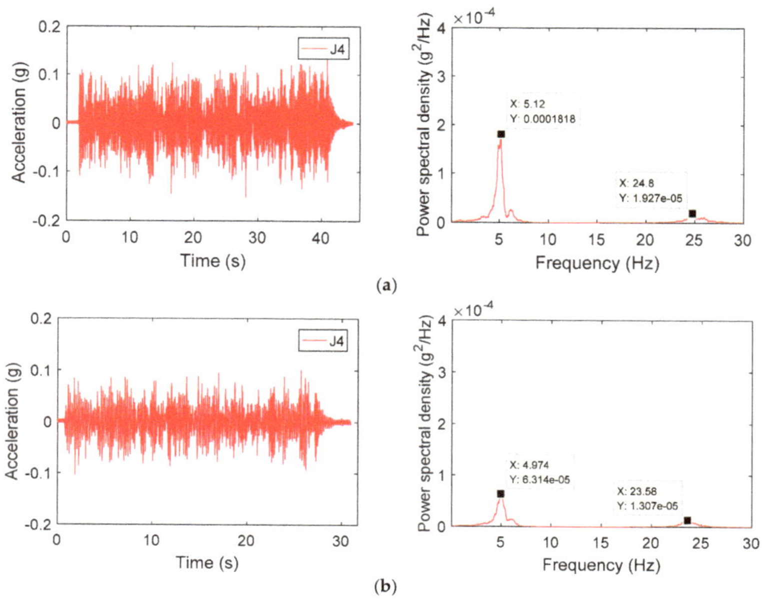

Figure 9. Time-series of the acceleration response and power spectral density function. (**a**) Without water and (**b**) with water.

The first-order mode frequencies of the electrical platform along the normal direction with and without water conditions were 5.12 Hz and 4.974 Hz, respectively. The results showed that the fundamental frequency of the structure with water was lower than that without water due to the water induced additional mass of the structure.

3.1.2. Mode of Vibration

The vibration modes of the electrical platform in with/without water states were analyzed under the excitation of white noise. The power spectral density function of acceleration response data of each measurement point was calculated. The structural mode of vibration is shown in Tables 6 and 7 for the electrical platform in with/without water states, respectively. The mode shape of the structure is shown in Figure 10. The mode showed a relatively high consistency between the presence and absence of water.

Table 6. Power spectrum peaks of the first and second mode in the without water condition.

Measuring Point	Height (m)	Amplitude of First Mode		Amplitude of Second Mode	
		Original Value (10^{-4})	Normalized Value	Original Value (10^{-7})	Normalized Value
J2	0	0.000	0.000	0.000	0.000
J3	0.22	0.882	0.286	−76.950	−0.339
J4	0.415	1.828	0.593	−191.400	−0.844
J5	0.56	1.800	0.583	−226.900	−1.000
J6	0.76	2.054	0.666	−87.250	−0.385
J7	0.92	2.523	0.818	0.259	0.001
J8	1.12	3.064	0.993	8.173	0.036
J9	1.306	3.079	0.998	13.470	0.059
J10	1.53	3.085	1.000	24.640	0.109

Table 7. Power spectrum peaks of the first and second mode in water condition.

Measuring Point	Height (m)	Amplitude of First Mode		Amplitude of Second Mode	
		Original Value (10^{-4})	Normalized Value	Original Value (10^{-7})	Normalized Value
J2	0	0.000	0.000	0.000	0.000
J3	0.22	0.307	0.295	−55.990	−0.364
J4	0.415	0.631	0.608	−131.600	−0.856
J5	0.56	0.619	0.596	−153.800	−1.000
J6	0.76	0.686	0.660	−57.870	−0.376
J7	0.92	0.846	0.814	0.630	0.004
J8	1.12	1.000	0.962	7.823	0.051
J9	1.306	1.038	0.999	11.470	0.075
J10	1.53	1.039	1.000	17.250	0.112

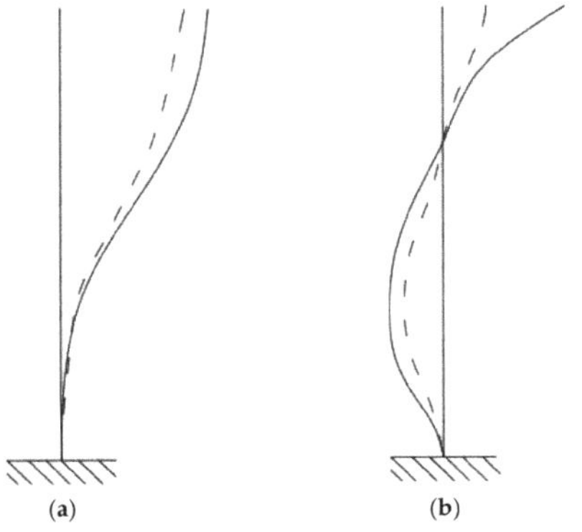

Figure 10. Structural mode of vibration. The solid line presents the results without water and the dotted line presents the results with water. (**a**) First-order mode of vibration and (**b**) second-order mode of vibration.

3.1.3. Damping Ratio

Taking acceleration data at J10 on the top of the platform as an example, the power spectral density functions are shown in Figure 11. For without water state, frequency peak of the first-order vibration mode of the power spectral density function was 3.058×10^{-4} g^2/Hz. The half peak was 1.529×10^{-4} g^2/Hz. The corresponding frequencies could be obtained from Figure 11a as $f_1 = 4.767$ and $f_2 = 5.347$, respectively. Thus, the damping ratio could be defined as follows:

$$\zeta = \frac{f_2 - f_1}{f_2 + f_1} = \frac{5.347 - 4.767}{5.347 + 4.767} = 5.73\%. \tag{6}$$

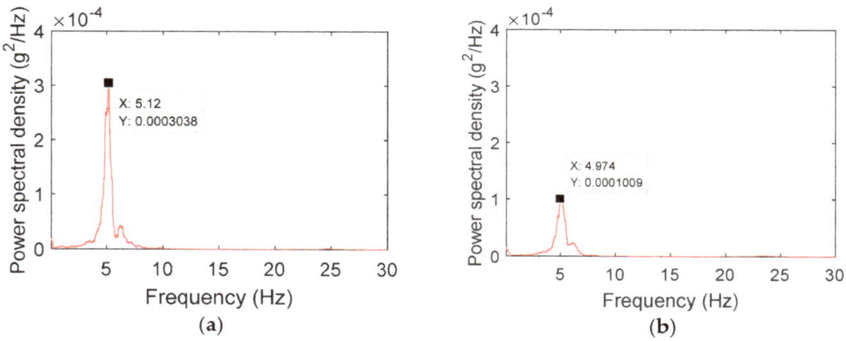

Figure 11. Power spectral density function of acceleration sensor J10. (**a**) Without water and (**b**) with water.

While, the damping ratio of the vibration mode was 8.68% for the with water state. The results for different measurement positions are present in Table 8. According to the experimental results, the average damping ratio of the first-order mode in the x-direction was 5.46% in the without water state, while it was 8.42% in the with water state. The damping ratio of the platform in water was larger than that without water. It was determined that the structural damping would increase due to the existence of water damping when the wind farm electrical platform was exposed in the open sea.

Table 8. Damping ratio of the first-order mode of the platform.

Measuring Point	J3	J4	J5	J6	J7	J8	J9	J10	J11	Mean Value
Without water	5.32%	5.35%	5.26%	5.65%	5.52%	5.36%	5.63%	5.73%	5.34%	5.46%
With water	7.83%	8.67%	8.25%	8.52%	8.66%	8.72%	8.66%	8.68%	7.82%	8.42%

3.2. Strain Response of the Electrical Platform

The time series of the strain response on the pile leg of the electrical platform showed a good following feature with the seismic excitation (see Figure 12). The strain at the normal positions of the pile leg was obviously larger than that at the tangential position on two sides. According to the strain analysis, the ratio of the four strain gauges at the same position in various test conditions remained basically unchanged. The strain ratios of the four strain gauges at measuring point S1 were averaged among different seismic excitations and the strain-peak ratios of strain gauges no. 1 to no. 4 were 0.745:0.177:1:0.215. The strain-peak ratios for measuring points S2, S3 and S4 were 1:0.428:0.747:0.153, 0.548:0.378:1:0.429 and 0.897:0.257:1:0.226, respectively.

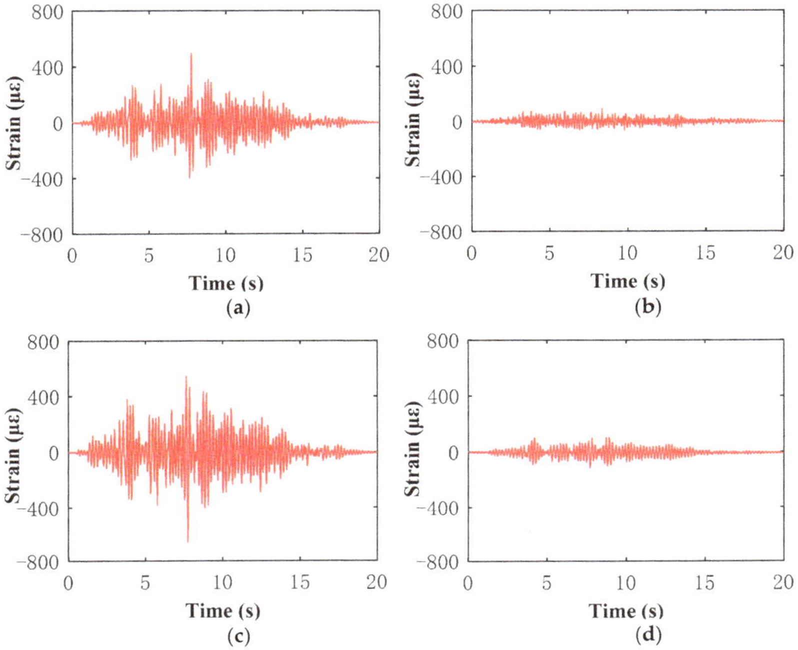

Figure 12. Time series of the strain response on measuring point S1. API (American Petroleum Institute) seismic wave with a peak ground acceleration of 0.25 g. (**a**) Strain at strain gauge no. 1, (**b**) strain at strain gauge no. 2, (**c**) strain at strain gauge no. 3 and (**d**) strain at strain gauge no. 4.

The structural strain increased with the increase of seismic excitation intensity (see Figure 13). When the peak acceleration of the input seismic wave was within the range of 0.2 g, the relationship between the structural strain response and the seismic acceleration was approximately linear. As the input acceleration of the foundation increased, the peak strain response of the structure changed nonlinearly due to the nonlinear interaction between water and structure.

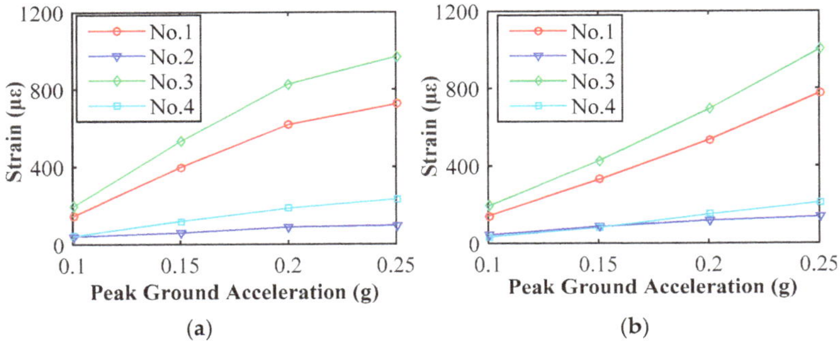

Figure 13. The peak value of the strain response on measuring point S1 for various seismic accelerations. (**a**) RSN169_IMVPALL and (**b**) RSN1116_KOBE.

According to the relationship between the stress and strain of the plexiglass material, the stress of the structure can be calculated according to the results of the strain. Combined with the stress similarity relation of model structure and prototype structure, the stress similarity scale of this model test was determined to be 176.31. Thus the stress of the structural model could be converted to that of the prototype structure. For the electrical platform, the structural strength of the upper structure and the upper and lower joint parts needed more attention. According to the test results, the maximum stress value at measuring points S3, S4, S5 and S6 reaches 115.8 MPa, 223.1 MPa, 137.4 MPa and 22.1 MPa, respectively, when the peak acceleration was 0.25 g. The results show that the maximum stress value on the prototype electrical platform structure was 223.1 MPa.

3.3. Motion Response of the Electrical Platform

The peak acceleration response of the structure was the maximum acceleration response of the structure under the action of corresponding seismic wave, which is an important index to reflect the seismic response of the structure. Overall, the acceleration response peak at different positions of the structure increased with increasing input seismic excitation. Due to the damping effect of water on the structure, the peak acceleration response of the structure varied nonlinearly with the input acceleration. The peak value of the acceleration responses for various seismic accelerations of the artificial and observed seismic wave is presented in Figure 14.

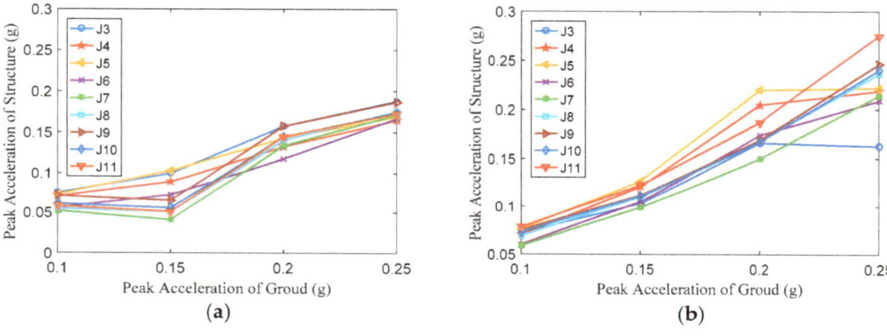

Figure 14. The peak value of the acceleration response for various seismic accelerations. (**a**) API and (**b**) RSN169_IMVPALL.

Under the action of different seismic waves, the peak accelerations at different positions on the structure were mainly concentrated around 5 Hz and 25 Hz, which was the 1st- and 2nd-order mode frequencies in the x-direction of the structure (see Figure 15). By comparing the acceleration response results in the frequency domain of the acceleration sensors arranged along the pile leg, it could be seen that the peak value of the acceleration frequency domain results of the jacket structure at the bottom of the platform not only appeared near the frequency of the first vibration mode of the structure, but also produced obvious high-frequency components. For the acceleration sensors arranged on the superstructure of the platform, the peak value of acceleration in the frequency domain was mainly concentrated near the 1st-order mode frequency of the structure. The low-frequency component excited was relatively obvious. The frequency of acceleration transited from high frequency to low frequency along the height direction.

The displacement response of the structure in the horizontal direction was measured by the laser displacement sensor. The time-history curve of displacement response arranged at the five measuring points of the jacket and the superstructure of the platform is shown in Figure 16. The results show that the electrical platform would produce a significant displacement response under the excitation of seismic load. For different positions of the structure, the response trends were consistent. Overall, the displacement response amplitude of the platform structure was larger than the input value of the vibration table. It was indicated that the displacement response amplitude of the electrical platform had a certain dynamic amplification effect under the excitation of a seismic load. In this paper, the peak displacement response of the platform structure under five different seismic waves was analyzed. The results show that the peak displacement response of the platform under the action of seismic wave RSN800_LOMAP was the largest. The maximum displacement response was found at the top of the platform with 5.17 mm, corresponding to the prototype value of 0.31 m.

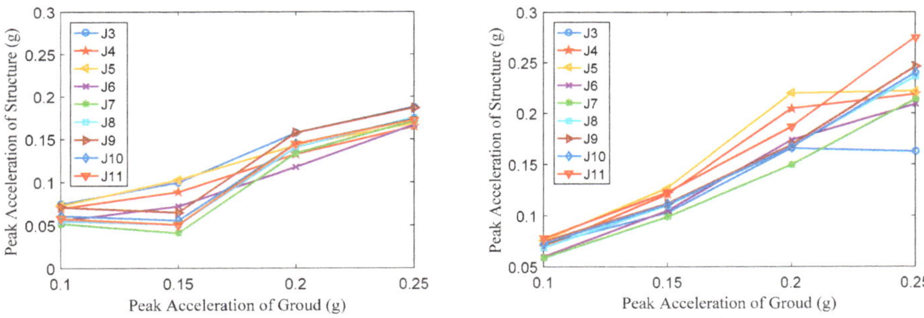

(a) (b)

Figure 15. The peak value of the acceleration response for various seismic accelerations. (**a**) API and (**b**) RSN169_IMVPALL.

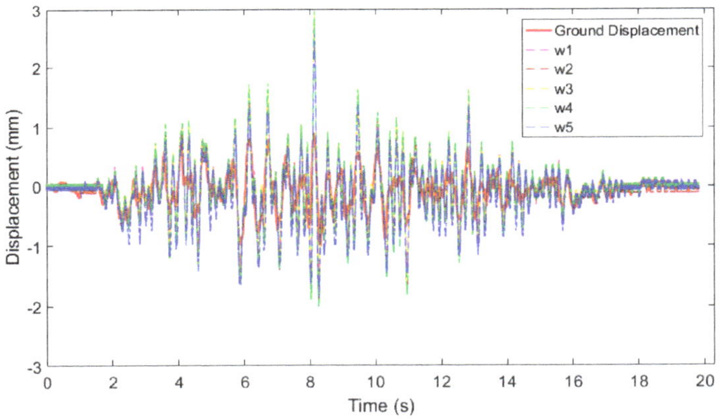

Figure 16. Time series of displacement response under the API seismic wave with a peak ground acceleration of 0.25 g.

4. Discussion

4.1. The Effect of Water Damping

The experimental results show that the damping of the structure under the normal water level condition was larger than that without water. In general, the peak value of strain of the structure in with water state was smaller than that without water. This was consistent with that of the platform structure that experienced larger damping ratio under the normal water level condition. It was indicated that the strain response of the structure under the action of earthquake was suppressed due to the influence of water damping. Under the action of different seismic waves, the structural strain response of the electrical platform showed the same variation of the strain peak. Therefore, only the results under the action of API seismic wave are present here (see Figure 17).

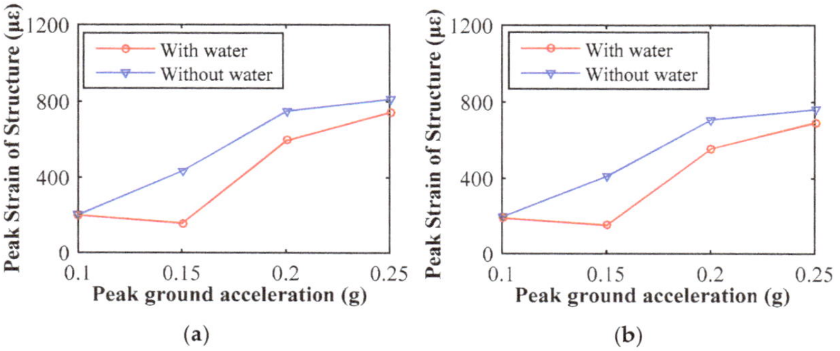

Figure 17. Structural strain response of the electrical platform in the with/without water condition under API seismic wave. (**a**) The strain response of S1 and (**b**) the strain response of S2.

Water damping had a significant influence on the acceleration response of the platform structure (see Figure 18). It could be seen from the distribution of structural peak acceleration at the representative positions that the peak acceleration of the structure under the condition of no water was greater than that with water. This was mainly due to the damping effect of water. It was indicated that the existence of water was beneficial to the stability of the structure under the action of an earthquake.

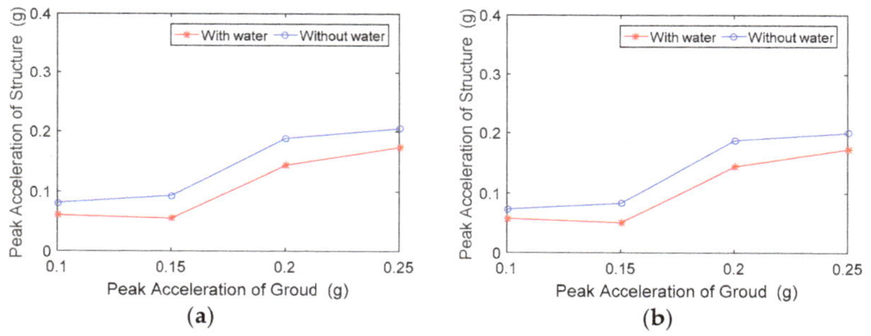

Figure 18. Structural peak acceleration of the electrical platform in the with/without water condition under the API seismic wave. (**a**) The acceleration response of J10 and (**b**) the acceleration response of J11.

4.2. The Effect of the Seismic Loading Mode

Taking the RSN1116_KOBE wave as an example, the dynamic response of the electrical platform under a unidirectional and bidirectional earthquake was compared (see Figure 19). With the increase of input peak acceleration, the strain response of the structure under unidirectional and bidirectional seismic action increased. In addition, the strain response peak value was approximately 35% larger on average in the case of bidirectional earthquake than that of a unidirectional earthquake. It can be considered that the bidirectional seismic load will cause stronger structural strain response, which is more harmful to the safety of the structure.

In the case of with and without water, the peak acceleration at different positions of the platform structure is shown in Figure 20. Under the action of the bidirectional seismic load, the acceleration response of the structure was greater than that under the unidirectional load. With the presence of water, the maximum increase of acceleration response of the structure, ~23.66%, occurred at measuring point J7. While in the dry mode, the maximum increase of the acceleration response of the structure, ~23.46%, occurred at measuring point J8. For both the with and without water conditions, the average increment of the acceleration response of the structure under a bidirectional seismic load was 18.86% and 15.87%, respectively, which showed a significant influence. For the same seismic wave, the variation trends of the acceleration response in the height direction were consistent under unidirectional and bidirectional loads for both the with and without water conditions. It was indicated that the presence of water would affect the acceleration amplitude, but not significantly affect the effect of the unidirectional and bidirectional seismic loads.

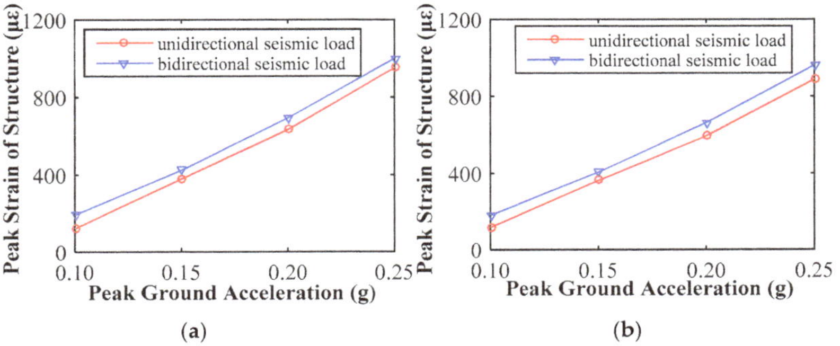

Figure 19. Structural strain response of the electrical platform under a unidirectional and bidirectional seismic wave. (**a**) The strain response of S1 and (**b**) the strain response of S2.

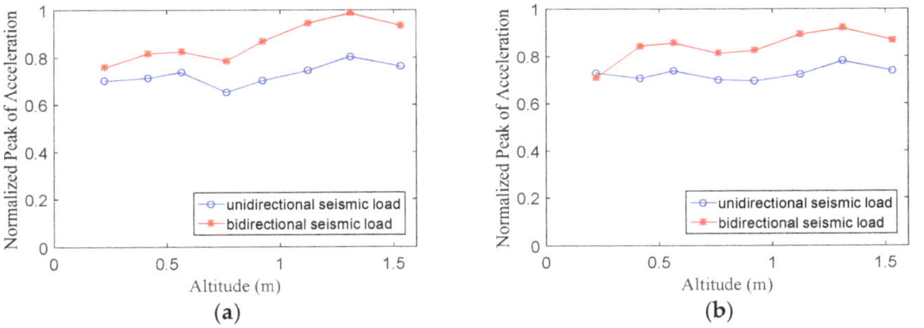

Figure 20. Structural peak acceleration of the electrical platform under a unidirectional and bidirectional seismic wave. (**a**) With water and (**b**) without water.

4.3. The Trend of Peak Acceleration in the Height Direction

According to the experiment results of acceleration at different locations, the peak acceleration in the height direction of the platform were obtained. A quadratic polynomial fitting using the least square method was conducted based on measured results at various heights along the pile leg for different seismic waves (see Figure 21). Under the excitation of different seismic waves, the distribution of the peak acceleration of the structure along the height of the platform decreased first and then increased.

Since the bottom of the structure was close to the seismic source, the structural response would follow the input of ground acceleration. However, the top of the platform would produce a certain whipping effect [22,23]. With the joint action of the above two characteristics, the peak acceleration of the structure decreased first and then increased along the height of the platform. Overall, this trend accords with the dynamic response characteristics of a structure under the earthquake load. For an offshore electrical platform of the wind farm, there is a number of important electrical equipment inside the superstructure. Therefore, reducing the whipping effect of the platform is extremely important for the security and stability of the system.

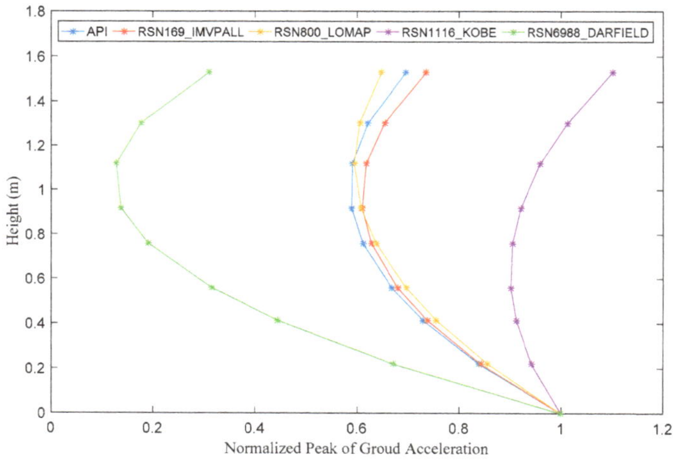

Figure 21. The variation trend of peak acceleration in the height direction of the platform by bidirectional seismic excitation in water.

5. Conclusions

In this study, a physical model test was carried out to investigate the dynamic characteristics of a 10,000-ton offshore converter station platform under seismic load. The acceleration response, displacement response and stress response of the offshore electrical platform under the typical direction of seismic action were obtained. The main conclusions were summarized as follows:

(1) The effect of the environment on the structural dynamic characteristics was analyzed. The average damping ratio of the first-order mode of the electrical platform was 5.73% and 8.68% in the conditions of with and without water, respectively. The damping ratio of the structure in water was greater than that in the dry mode. Due to the influence of water damping, the peak of the strain and acceleration of the platform structure in water were smaller than those in the dry mode.

(2) The unidirectional and bidirectional seismic excitations had a significant influence on the dynamic characteristics of the structure. At each location of the structure, the peak value of the strain response under bidirectional seismic excitation was generally larger than that under unidirectional seismic excitation. It could be determined that the bidirectional seismic excitation was more dangerous to the structure.

(3) The peak acceleration along the height of the platform structure decreased first and then increased, which showed a typical whipping effect. The dynamic response characteristics of the electrical platform under different seismic loads were generally consistent.

(4) The weak positions of electrical platform structure were found in the experimental study. According to the analyzed data, the strain on the pile leg at the mud surface was the largest. In addition, large strains and stresses were generated at the connections between the superstructure and the jacket. It was indicated that the connections were the weak positions of the electrical platform under a dynamic load.

Author Contributions: Conceptualization, Z.-Z.S. and S.-X.Z.; Data curation, C.-W.B.; Formal analysis, H.-F.Y.; Funding acquisition, S.-X.Z. and G.-H.D.; Investigation, Z.-Z.S. and C.-W.B.; Methodology, C.-W.B. and G.-H.D.; Project administration, S.-X.Z.; Resources, Z.-Z.S., S.-X.Z.; Supervision, S.-X.Z.; Writing—Original draft, C.-W.B.; Writing—Review & editing, Z.-Z.S., G.-H.D. and H.-F.Y.

Funding: This research was funded by National Natural Science Foundation of China (NSFC), grant nos. 51609035 and 31972843; Research Fund of Power China Huadong Engineering Corporation, project no. KY2018-ZD-03; and Fundamental Research Funds for the Central Universities, project no. DUT18RC(3)076.

Acknowledgments: The authors acknowledge the technical support from Power China Huadong Engineering Corporation Limited.

Conflicts of Interest: The authors declare no conflict of interest.

References

1. Sun, X.; Huang, D.; Wu, G. The current state of offshore wind energy technology development. *Energy* **2012**, *41*, 298–312. [CrossRef]
2. Harrison, R.; Hau, E.; Snel, H. *Large Wind Turbines: Design and Economic*; Wiley: Chichester, UK, 2000.
3. Butterfield, S.; Musial, W.; Jonkman, J.M.; Sclavounos, P. *Engineering Challenges for Floating Offshore Wind Turbines*; National Renewable Energy Laboratory (NREL): Golden, CO, USA, 2007.
4. Snyder, B.; Kaiser, M.J. Ecological and economic cost-benefit analysis of offshore wind energy. *Renew. Energy* **2009**, *34*, 1567–1578. [CrossRef]
5. Liu, J.; Thomas, E.; Manuel, L.; Griffith, D.; Ruehl, K.; Barone, M. Integrated system design for a large wind turbine supported on a moored semi-submersible platform. *J. Mar. Sci. Eng.* **2018**, *6*, 9. [CrossRef]
6. Bea, R.G.; Akky, M.R.; Audibert, J.M.E. Earthquake response of offshore platforms. *J. Struct. Div.* **1979**, *105*, 377–400.
7. Chandrasekaran, S.; Gaurav. Offshore triangular tension leg platform earthquake motion analysis under distinctly high sea waves. *Ships Offshore Struct.* **2008**, *3*, 173–184. [CrossRef]
8. Huang, Z.; Foutch, D.A. Effect of hysteresis type on drift limit for global collapse of moment frame structures under seismic loads. *J. Earthq. Eng.* **2009**, *13*, 939–964. [CrossRef]
9. Yu, H.; Li, X.; Yang, S. Dynamic analysis method of offshore jack-up platforms in regular and random waves. *J. Mar. Sci. Appl.* **2012**, *11*, 111–118. [CrossRef]
10. Araki, Y.; Hjelmstad, K.D. Criteria for assessing dynamic collapse of elastoplastic structural systems. *Earthq. Eng. Struct. Dyn.* **2015**, *29*, 1177–1198. [CrossRef]
11. Salem, Y.S.; Wang, L.; Aziz, G. Assessment of response modification factor of open steel platform structures subjected to seismic loads. In *Facing the Challenges in Structural Engineering*; Rodrigues, H., Elnashai, A., Calvi, G., Eds.; Springer: Cham, Switzerland, 2017; pp. 131–143.
12. Failla, G.; Santangelo, F.; Foti, G.; Scali, F.; Arena, F. Response-spectrum uncoupled analyses for seismic assessment of offshore wind turbines. *J. Mar. Sci. Eng.* **2018**, *6*, 85. [CrossRef]
13. Zhao, Z.X.; Li, X.; Wang, W.H.; Shi, W. Analysis of dynamic characteristics of an ultra-large semi-submersible floating wind turbine. *J. Mar. Sci. Eng.* **2019**, *7*, 169. [CrossRef]
14. Prowell, I.; Elgamal, A.; Uang, C.M.; Luco, E.J.; Romanowitz, H.; Duggan, E. Shake table testing and numerical simulation of a utility scale wind turbine including operational effects. *Wind Energy* **2014**, *17*, 997–1016. [CrossRef]
15. Zheng, X.Y.; Li, H.; Rong, W.; Li, W. Joint earthquake and wave action on the monopile wind turbine foundation: An experimental study. *Mar. Struct.* **2015**, *44*, 125–141. [CrossRef]
16. Wang, W.H.; Gao, Z.; Moan, T.; Li, X. Model Test and Numerical Analysis of a Multi-Pile Offshore Wind Turbine under Seismic, Wind, Wave and Current Loads. *J. Offshore Mech. Arct. Eng.* **2016**, *139*, 031901. [CrossRef]
17. Sadowski, A.J.; Camara, A.; Málaga-Chuquitaype, C.; Dai, K. Seismic analysis of a tall metal wind turbine support tower with realistic geometric imperfections. *Earthq. Eng. Struct. Dyn.* **2017**, *46*, 201–219. [CrossRef]
18. Wang, W.H.; Li, X.; Pan, Z.; Zhao, Z.X. Motion control of pentapod offshore wind turbines under earthquakes by tuned mass damper. *J. Mar. Sci. Eng.* **2019**, *7*, 224. [CrossRef]
19. Lin, G.; Zhu, T.; Lin, B. Dynamic model test similarity criterion. *J. Dalian Univ. Technol.* **2000**, *40*, 1–8. (In Chinese)
20. Chakrabarti, S. *Handbook of Offshore Engineering*; Elsevier Press: Amsterdam, The Netherlands, 2005.
21. API. *Recommended Practice for Planning, Designing, and Constructing Fixed Offshore Platforms-Working Stress Design*, 22nd ed.; American Petroleum Institute Recommended Practice 2A-WSD: Washington, DC, USA, 2007.

22. Yang, L.J.; Liu, D.W.; Guo, Z.L.; Li, J.; Dai, B.H. Engineering mechanics in whipping effect of high-rise building. *Appl. Mech. Mater.* **2014**, *540*, 173–176. [CrossRef]
23. Cao, Y.; He, E.; Furrer, M. Whipping effect of the seismic response of unbraced tied arch bridges. *IABSE Symp. Rep.* **2015**, *105*, 1–8. [CrossRef]

Journal of
Marine Science and Engineering

MDPI

Article

Numerical Study of the Interaction between Level Ice and Wind Turbine Tower for Estimation of Ice Crushing Loads on Structure

Ming Song [1], Wei Shi [2], Zhengru Ren [3] and Li Zhou [1,*]

1 School of Naval Architecture and Ocean Engineering, Jiangsu University of Science and Technology, Zhenjiang 212003, China
2 State Key Laboratory of Coast and Offshore Engineering, Deepwater Engineering Research Center, Dalian University of Technology, Dalian 116024, China
3 Department of Marine Technology, Norwegian University of Science and Technology, No-7451 Trondheim, Norway
* Correspondence: zhouli209@hotmail.com

Received: 8 October 2019; Accepted: 11 November 2019; Published: 1 December 2019

Abstract: In this paper, the interaction between level ice and wind turbine tower is simulated by the explicit nonlinear code LS-DYNA. The isotropic elasto-plastic material model is used for the level ice, in which ice crushing failure is considered. The effects of ice mesh size and ice failure strain on ice forces are investigated. The results indicate that these parameters have a significant effect on the ice crushing loads. To validate and benchmark the numerical simulations, experimental data on level ice-wind turbine tower interactions are used. First, the failure strains of the ice models with different mesh sizes are calibrated using the measured maximum ice force from one test. Next, the calibrated ice models with different mesh sizes are applied for other tests, and the simulated results are compared to corresponding model test data. The effects of the impact speed and the size of wind turbine tower on the comparison between the simulated and measured results are studied. The comparison results show that the numerical simulations can capture the trend of the ice loads with the impact speed and the size of wind turbine tower. When a mesh size of ice model is 1.5 times the ice thickness, the simulations can give more accurate estimations in terms of maximum ice loads for all tests, i.e., good agreement between the simulated and measured results is achieved.

Keywords: ice crushing; ice load; finite element method; wind turbine tower; numerical simulation

1. Introduction

With the growing renewable energy demands and the increasing concern about environmental pollution, the development of renewable energy harnessing has been paid more and more attention. As a sort of clean energy, wind energy has become the most promising renewable energy after decades of development. Compared with the land-based wind, offshore environment has more abundant wind energy resources with higher quality, and OWTs could avoid the problems of land acquisition and noise [1]. However, a key technology challenge for OWTs is operation in cold climates, i.e., the possibility of the structure interaction with floe ice enhances while operating in cold regions. For example, the Great Lakes are the most promising locations for the OWTs in the United States. The lakes are often substantially ice covered for the entire winter, and have wind and sea current driven ice floes at times [2]. An OWT operating under wind and ice conditions is shown in Figure 1. Ice loads should be taken as one of the important environmental impacts in addition to the aerodynamic loads. Therefore, it is necessary to predict the ice loads caused by the level ice-OWT interaction.

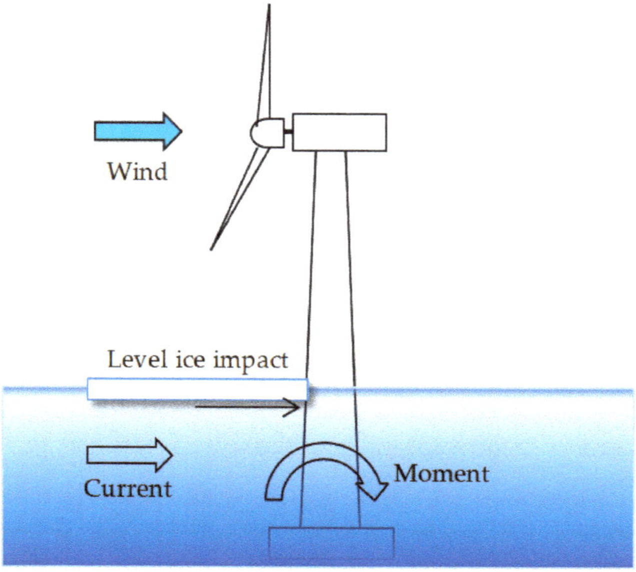

Figure 1. OWT exposed to wind field and in contact with ice.

The interaction between an OWT and level ice is a complex process. The magnitude and time variation of the ice loads depend strongly on the geometry of the wind turbine, the ice thickness, the ice strength, and their impact velocity. There are various ice failure modes observed in the ice-structure interaction. Typically, for level ice, bending, buckling, cracking/splitting, or crushing could take place, which is strongly governed by the shape of the structure at the water level [3,4]. The sloping shapes cause the level ice to fail by bending, whereas the vertical shapes induce the level ice to fail by crushing [5]. The level ice is weaker in bending than crushing. Therefore, the implementation of sloping shapes for OWTs can effectively reduce the magnitude of ice loads. Many model tests have verified the lower ice loads on a conical structure than on a cylindrical structure of similar size [6–8]. However, the application of sloping shapes will induce larger wave loads and enhance the foundation costs because of the additional material near the water level.

Many studies have been carried out to investigate the interaction between ice and cylindrical or sloped structures by model tests, full-scale tests and numerical simulations [9–14]. Yue et al. conducted full-scale tests on a cylindrical compliant monopod platform to investigate the dynamic ice forces and structure vibrations generated by crushing failure of the ice sheet [15]. The test results showed that three ice force modes take place in the loading speeds which make ice fail in ductile, ductile-brittle transition, and brittle range respectively. Kuutti et al. simulated ice crushing against a rigid vertical structure using cohesive surface methodology [16]. The simulated crushing forces agreed well with the experimental results. Lu et al. and Wang et al. carried out numerical simulations of interactions between level ice and sloping structure using the cohesive element method [17,18]. Zhou et al. proposed a numerical model to simulate the non-simultaneous crushing force acting on the cylindrical structures of wind turbines [19]. It was observed that the simulation results agree well with the measured data from the model tests in terms of the maximum ice force. Ranta et al. simulated ice rubble-structure interaction processes based on arbitrary Lagrangian-Eulerian finite element method [20]. However, there was a lack of validation on the characteristics of the obtained rubble pile geometries.

Some researchers focus on the predictions of the coupled dynamic loads and responses of an OWT [2,21]. Shi et al. studied the dynamic ice-structure interaction of a monopile-type OWT in drifting level ice in both parked and operating conditions by coupling a semi-empirical numerical model to the aero-hydro-servo-elastic simulation tool HAWC2 [22]. The effects of ice drifting speed and ice thickness were investigated by using the coupled dynamic analyses. It was found that the effect of the ice thickness on the response is significant, whereas the effect of drifting speed on the bending moment response in the fore-aft direction is negligible. Wells developed a simulation tool to study the effects of ice on both cylinder- and cone-shaped OWTs [23]. The simulation results indicated that the surface ice sheet loads can be much larger than the wind loads and could be the driving parameters of the OWT foundations design in areas where ice can occur. Heinonen and Rissanen conducted a feasibility study of the FAST simulation software to investigate the structural performance of OWTs [4]. They studied the ice interaction with vertically shaped structures at the water line and taking into account the coupling between the ice, wind, and structural response. However, there is a limitation in the ice model for describing a brittle crushing process.

For the load design, the ice crushing is the most important since it causes the biggest force and might induce severs steady-state vibrations as well [24]. When the cylindrical structures are interacting with drifting ice (of thickness 0.2 m and more), the ice crushing failure action can generate as high dynamic forces as 5 MN and are of critical concern for the structural designers [25]. Therefore, it is necessary to investigate the dynamic interaction between level ice and vertical structures where ice crushing failure takes place. Most of the present works established the ice forces from the existing ice models, in which the dynamic interaction process and the ice crushing failure could not be simulated.

This paper focuses on the numerical study to predict the ice crushing force acting on the cylindrical OWT foundation based on the nonlinear finite element method, in which the dynamic interaction process is simulated. The isotropic elasto-plastic material model is used for level ice to simulate ice crushing failure. The effects of ice mesh size and failure strain on the ice forces are investigated. Model tests on the interaction between level ice and nearly vertical wind turbine tower are used to calibrate and validate the numerical simulation results. Four impact cases are considered. The comparisons between the simulated and measured results including the maximum, mean, standard deviation, and time series of the ice forces are made. In addition, the studies on the effects of the impact speed and the size of wind turbine tower on the comparison are carried out.

2. Experimental Data

This section presents the experimental data used to calibrate and verify the numerical simulation results. The tests were conducted by Wu et al. at [26] the ice Basin of Tianjin University. The interaction between level ice and wind turbine tower was considered. The experimental scale was chosen to be 1:20. The force in full scale F_f is calculated by the following equation:

$$F_f = F_m \lambda^3 \tag{1}$$

The test represents the impacts between a 0.4 m thickness level ice and the monopile foundations of a 3-MW and a 4-MW wind turbine towers at speeds varying from 0.05 to 1.2 m/s in full scale. The target thickness of the level ice is 0.4 m in full scale. The bending and crushing strength of the level ice are expected to be 0.6 MPa and 2.06 MPa in full scale, respectively. A total of 12 impact tests were conducted. Tests #304, #306, #404, and #406 are selected for the analysis because the brittle ice crushing failure took place in these tests and the time histories of ice forces for these tests were available. The specific test matrix and the ice properties are given in Table 1.

Table 1. Test matrix and measured ice conditions (in full scale).

Test	Wind Turbine Tower	Ice Thickness (m)	Bending Strength (kPa)	Crushing Strength (kPa)	Ice Drifting Speed (m/s)
#304	3 MW	0.4	572	1980	0.6
#306	3 MW	0.4	664	2122	1.2
#404	4 MW	0.4	572	1980	0.6
#406	4 MW	0.4	664	2122	1.2

Figure 2 shows the geometry of the 3-MW and 4-MW monopile wind turbine towers. The foundations of the wind turbine towers are nearly vertical structures. The diameter of the 3-MW and 4-MW monopile wind turbine towers at waterline is 5.30 m and 5.83 m in full scale, respectively, and their slope angle is 87.2 degrees and 88.3 degrees, respectively.

An ice force experiment scenario for a 3 MW model test is shown in Figure 3. A force transducer measured the ice loads using a data acquisition system with a sampling frequency of 100 Hz.

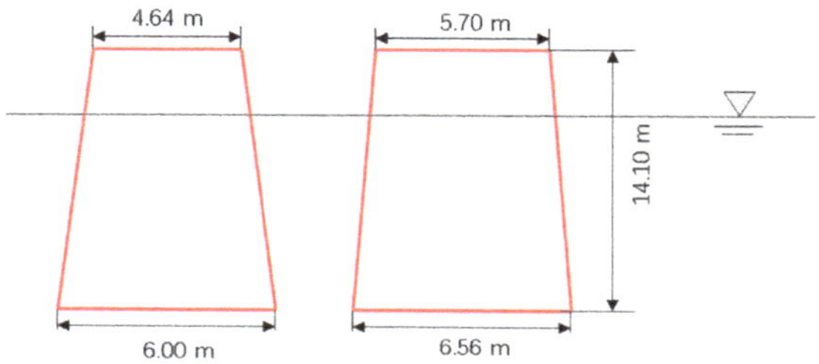

Figure 2. The geometry of the 3-MW and 4-MW wind turbine towers in full scale.

Figure 3. Photograph of a 3-MW model test from Zhou et al. [19].

3. Numerical Analysis

This section details the finite element modeling, the material models, and major results. All simulations were run on an 8 CPU workstation with Inter 3.60 GHz processors and 16.0 GB of RAM. The software used was LS-DYNA version R700 with double precision. LS-DYNA software has a number of contact algorithms and a large suite of material types that can be chosen for the interacting structures. It has been widely used to simulate ice-structure collisions. Patran software was used for the modeling and generation of meshes for the study.

3.1. Model Description

Figure 4 shows the numerical domain of the simulations. The dimensions of the level ice are 55 m × 55 m × 0.4 m. The ice model extent is sufficient to minimize the effect of boundary conditions. The dimensions of the wind turbine towers in the numerical models are the same as the experimental models.

Figure 4. Numerical model of ice-structure interaction.

The ice is modeled using eight-node solid elements. The wind turbine towers are discretized using four-node Belyscho-Tsay shell elements. The mesh size for the level ice is approximately 0.6 m × 0.6 m × 0.4 m, in which there is only one layer of meshes in the vertical direction. The wind turbine towers are meshed with an element size of 0.4 m. The number of elements for the ice and the wind turbine towers are 151250 and 1645, respectively.

To avoid the initial penetration and numerical instabilities, the translational velocity of the wind turbine tower at water plane ramps up from 0.0 m/s to 1.2 m/s before the impact occurs. After it reaches 1.2 m/s, the velocity is kept to be constant throughout the rest of the simulations. This can be achieved by using the LS-DYNA command "boundary prescribed motion rigid" with define curve. LS-DYNA offers a large number of contact types. The contact between the level ice and the wind turbine tower is implemented through the contact-eroding-surface-to-surface formulation, which is used with the segment-based contact option (soft = 2) in LS-DYNA. This eroding contact type contains logic which allows the contact surface to be updated to consider the ice element deletion [27]. The ice is defined as "slave "segment and the wind turbine tower is defined as "master" segment, a search for penetration of a "slave" node through the "master" segment is made every time step. When a penetration is found, a contact force proportional to the penetration depth is applied to resist and ultimately eliminate the penetration. The contact force is contained in the "rcforc" file produced by using a database-rcforc command. In order to consider the self-contact of the ice component, the contact-eroding-single-surface contact type which is the most widely used contact options in LS-DYNA is applied for the ice model. Both static and dynamic coefficients of friction are set to 0.15 at all the contacts, which is a reasonable assumption for the friction between the ice and the steel surfaces.

3.2. Material Models

For finite element analyses of ice-structure interactions, the constitutive material model for the ice is a critical factor to accurately predict maximum ice forces [28]. Wang et al. proposed an ice model for the interaction between sloping marine structure and level ice by using the cohesive element model [18]. In their model, the isotropic elasto-plastic linear softening constitutive model proposed by Hilding et al. was introduced to present the microscopic crushing of the ice sheet, while the bending failure of ice sheet was caused by the failure of cohesive elements [29]. In our case, the slope angle of the wind turbine towers is close to 90 degree. The ice crushing is in the dominant failure mode during the interaction between the level ice and the wind turbine towers. Therefore, the isotropic elasto-plastic material model is used for the level ice in this paper.

Figure 5 shows the relationship between the yield stress and effective plastic strain for the ice model. The effective plastic strain is defined as:

$$\varepsilon^p = \int_0^t \left(\frac{2}{3} D_{ij}^p D_{ij}^p\right)^{1/2} dt \tag{2}$$

The ice performance is assumed to have three stage states: The ice material is elastic before reaching the crushing initial point; after the first crack, the ice material shows a linear softening behavior; when the ice is totally crushed, it behaves as a viscous fluid. To describe the ice behavior, the "mat-piecewise-linear-plasticity" material type from LS-DYNA's suite of material types is used here, in which an elasto-plastic material with the yield stress versus strain curve and failure based on a plastic strain can be defined. For the wind turbine towers, the rigid material model is used, in which the deformation of the structure during interaction is not considered. The input material parameters to both the level ice and the wind turbine towers models are given in Table 2.

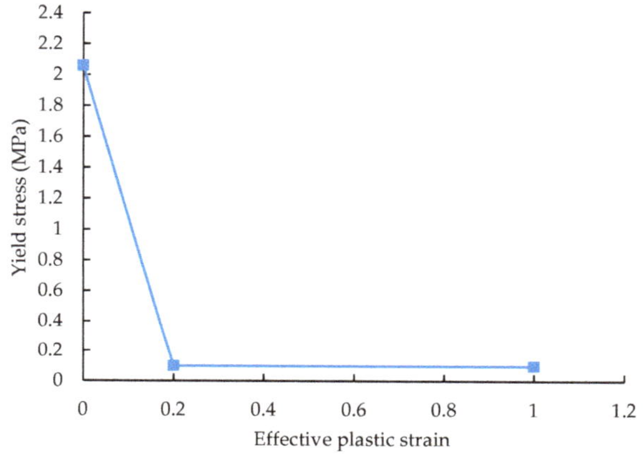

Figure 5. Hardening curve for ice material model.

Table 2. Material parameters used in the simulations.

Items	Level Ice	Wind Turbine Towers
Density (kg/m^3)	900	7850
Young's modulus (GPa)	2	210
Poisson ratio (-)	0.3	0.3
Yield stress (MPa)	2.06	-
Failure strain (-)	0.2	-

3.3. *Effect of Ice Mesh Size*

To investigate the effect of the ice mesh size on the ice force, four meshes with characteristic element lengths of 0.2 m, 0.4 m, 0.6 m, and 0.8 m are considered, and corresponding size ratio (mesh size/ ice thickness) is 0.5, 1, 1.5, and 2, respectively. It is noted that there are two layers of meshes in the vertical direction for mesh size of 0.2 m, and only one layer for other mesh size (shown in Figure 6). The ice failure strain of 0.2 is used in all simulations. The other parameters are equal to the basic values according to the setup of test #306.

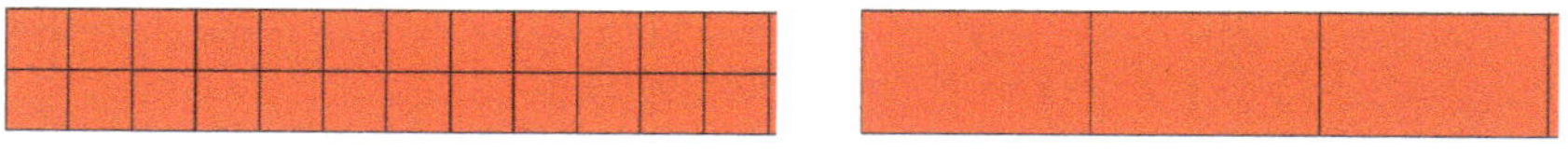

(a) (b)

Figure 6. Side view of ice model: (**a**) mesh size of 0.2 m (**b**) mesh size of 0.8 m.

Figure 7 shows the horizontal ice force histories for different ice mesh sizes. It is found that the ice mesh size has a significant effect on both the fluctuated frequency and peak forces. With the refinement of mesh, the frequency increases, while the peak forces decrease. The mean and standard deviation of force are calculated by the following equations:

$$F_{mean} = \frac{1}{N}\sum_{i}^{N} F_i \tag{3}$$

$$\sigma = \sqrt{\frac{1}{N}\sum_{i}^{N} (F_i - F_{mean})^2} \tag{4}$$

The comparison of the mean, standard deviation, and maximum forces are tabulated in Table 3. The simulated maximum force varies from 1.88 MN to 4.34 MN. Figure 8 shows the mean, standard deviation, and maximum forces varying with the mesh size. It is shown that the simulation with coarse mesh yields higher standard deviation and maximum force. Overall, both the standard deviation and maximum forces present an approximately linear relationship with the mesh size. However, for the mean force, the simulated values with different mesh sizes are around 0.86. The simulated results indicate that the mesh size has a significant effect on both the standard deviation and the maximum loads, while it has a slight effect on the mean load.

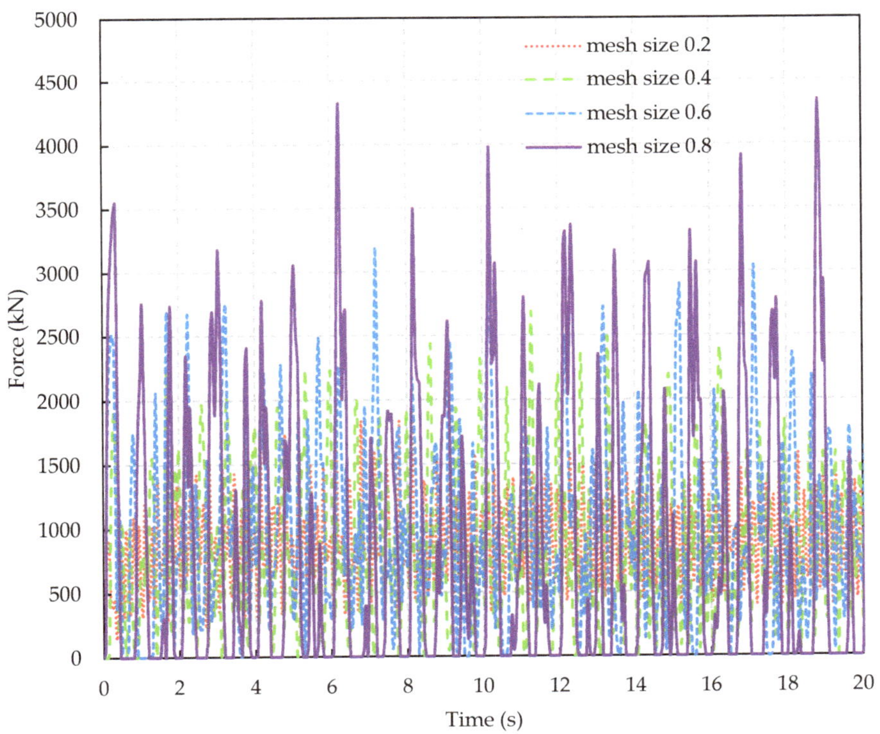

Figure 7. Ice force histories from the simulations with different mesh sizes.

Table 3. Comparison of the ice forces for the simulations with different mesh sizes.

Mesh Size (m)	0.20	0.40	0.60	0.80
Mean (MN)	0.88	0.83	0.98	0.87
Std. (MN)	0.35	0.59	0.70	1.11
Maximum (MN)	1.88	2.65	3.20	4.34

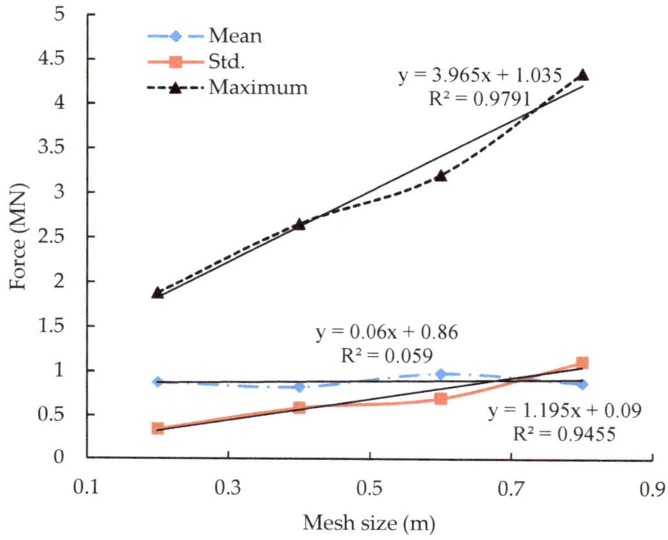

Figure 8. The mean, std. and maximum forces varying with the mesh size.

3.4. Effect of Ice Failure Strain

To investigate the effect from the ice failure strain, numerical simulations of the interaction between the level ice and the 3 MW wind turbine tower with different failure strain coefficients are carried out. The values of the ice failure strain in different simulations are set as 0.15, 0.2, 0.25, and 0.3, respectively. The other parameters are kept constant and equal to the basic values according to the setup of test #306 where the drift speed is 1.2 m/s in full scale.

Figure 9 shows the comparison of the horizontal ice force histories for various failure strains. It is observed that the fluctuated frequencies in the four curves are similar. The peak load increases with increasing failure strain. The mean, standard deviation, and maximum values are listed in Table 4 and these values varying with failure strain is shown in Figure 10. It is seen that the mean, standard deviation, and maximum forces present a linear increasing tendency with the larger failure strain. The linear curves which are fitted to the simulated mean, standard deviation, and maximum data are also presented in the figure, in which the slope of the curves is 4.48, 1.76, and 11.68, respectively. The simulated results indicate that the failure strain plays an important role in the simulated ice forces.

Table 4. Comparison of the ice forces for the simulations with different failure strains.

Failure Strain	0.15	0.20	0.25	0.30
Mean (MN)	0.70	0.98	1.21	1.37
Std. (MN)	0.59	0.70	0.77	0.86
Maximum (MN)	2.38	3.20	3.67	4.17

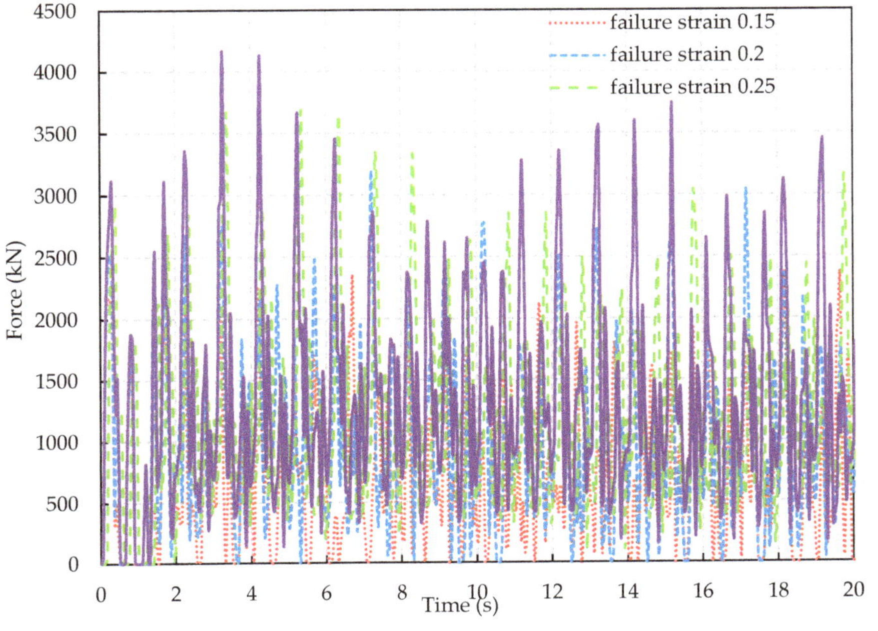

Figure 9. Ice force histories from the simulations with different failure strains.

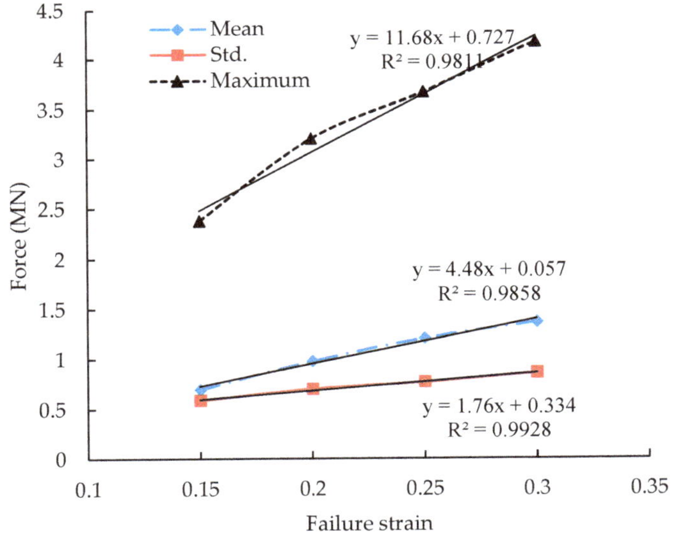

Figure 10. The mean, standard deviation, and maximum forces varying with the failure strain.

In summary, both the ice failure strain and the ice mesh size are crucial to ice force, including the mean, standard deviation, and maximum values. In addition, the fluctuated frequency is lower in the simulation with coarse mesh. Therefore, the failure strain should be determined from the numerical simulation with a given mesh size and the parameters should be calibrated using available experimental data.

4. Comparison of the Numerical Simulations and Test Results

This section presents the comparisons of the horizontal ice force histories, maximum, mean, and standard deviation values between the simulated and measured results for tests #306, #304, #404, and #406. Four groups of meshes are considered.

4.1. Comparison of Test #306

According to the results from Sections 3.3 and 3.4, the selections of ice failure strain for different ice mesh sizes are justified by a trial and error procedure which yields the better results for the maximum load, i.e., the simulated maximum force for the interaction between the level ice and the 3-MW wind turbine tower is in good agreement with the experimental measurement for test #306.

The ice failure strains of 0.43, 0.29, 0.2, and 0.12 are determined for using in the numerical simulations with the mesh size of 0.2 m, 0.4 m, 0.6 m, and 0.8 m, respectively. The relationship between the failure strain and the size ratio is shown in Figure 11. It is observed that the failure strain decreases with increasing size ratio. In this figure, an exponential curve $y = Ce^{Ax}$, where $C = 0.67$ and $A = -0.84$, is fitted to the simulation data. It is shown that the difference between the two curves is small, in which the coefficient of determination R^2 is equal to 0.99. The results indicate that $y = 0.67e^{-0.84x}$ can be adopted to describe the relationship between failure strain and size ratio for this case.

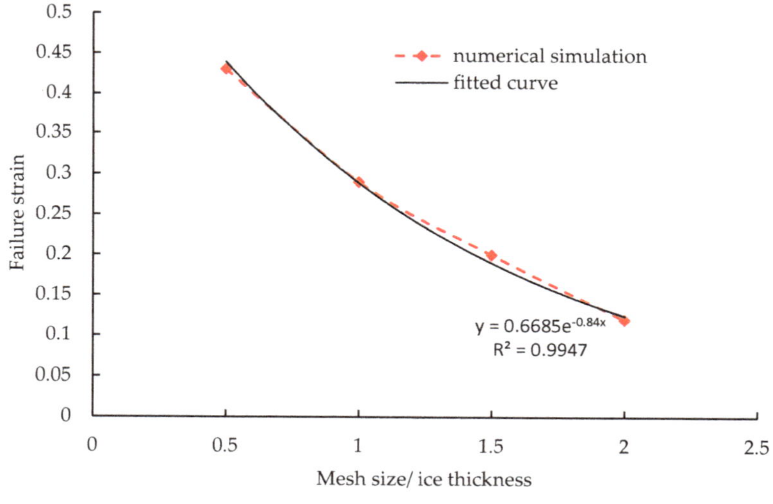

Figure 11. Relationship between failure strain and size ratio.

Figures 12–15 show the comparison of horizontal ice force histories between the simulations and measurement for test #306. In general, both the simulated and measured ice forces present strong nonlinear and vibrating behavior, and their trends are similar. It is observed that all simulations capture the maximum force well. Overall, the simulation with mesh size of 0.2 m gives better results: most of the peak and valley values are around 2.8 MN and 0.8 MN, respectively (see the dash line in Figure 12), which are in good agreement with the model test. However, the valley values in the other simulations are much smaller than the measurement, especially in the simulation with mesh size of 0.8 m. It can be seen that the zero forces obtained from the simulation with mesh size of 0.8 m are much more than those obtained from the other simulations and the model test (see Figure 15). This is mainly because the accumulation and sliding forces from the interaction between the wind turbine tower and the ice fragments are not considered in the simulations where the failed ice elements are removed.

With increasing mesh size of ice, the simulated ice breaking length increases, and consequently the gap between the wind turbine tower and the unbroken ice sheet will increase.

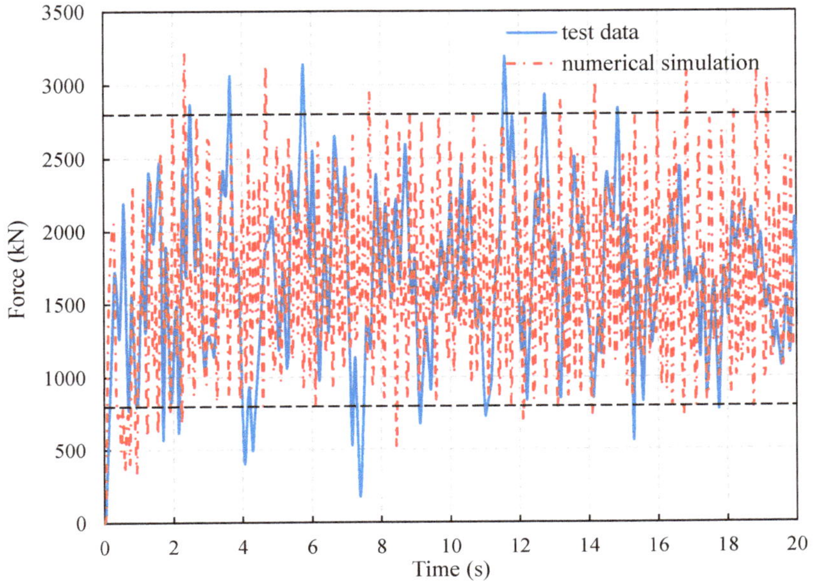

Figure 12. Ice force histories from the simulation with mesh size of 0.2 m and measurement for test #306.

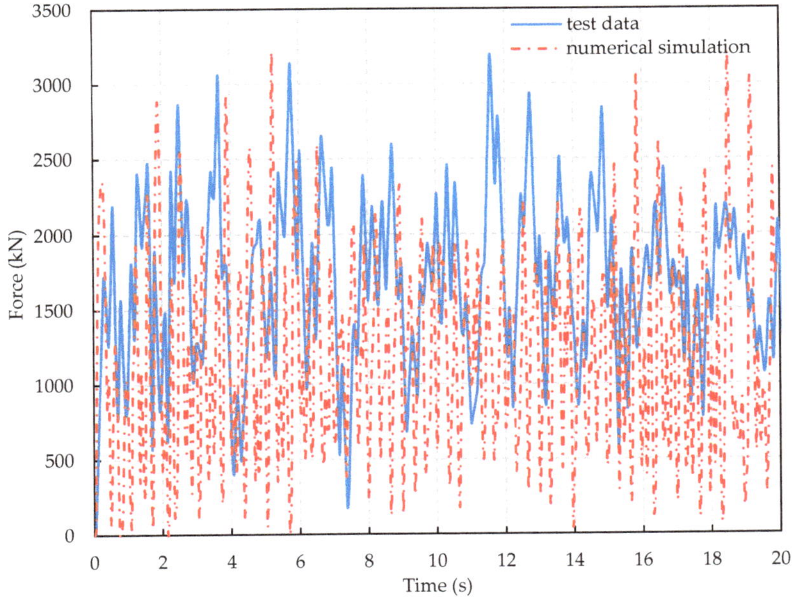

Figure 13. Ice force histories from the simulation with mesh size of 0.4 m and measurement for test #306.

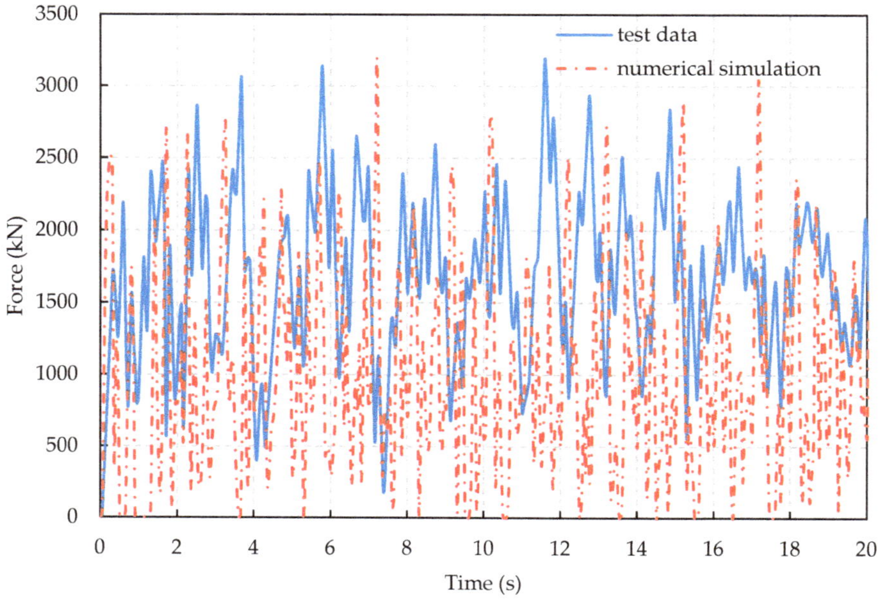

Figure 14. Ice force histories from the simulation with mesh size of 0.6 m and measurement for test #306.

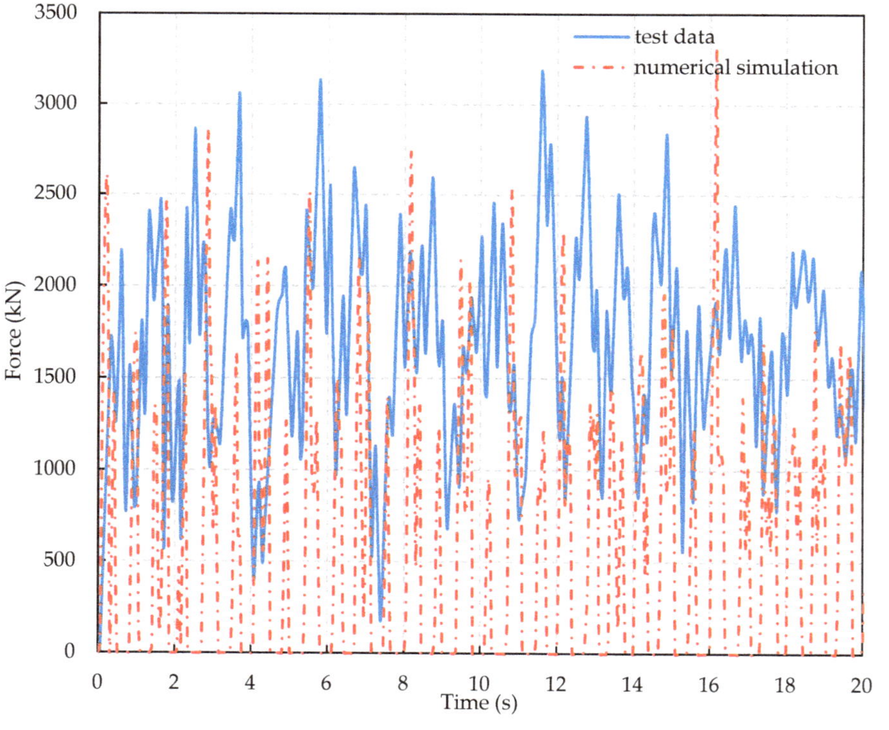

Figure 15. Ice force histories from the simulation with mesh size of 0.8 m and measurement for test #306.

To have a better understanding of the icebreaking process, Figures 16 and 17 give the partial magnification snapshots of the ice sheet at t = 10 s in the simulation with mesh sizes of 0.2 m and 0.8 m. It is obviously shown that there is a big gap between the wind turbine tower and the unbroken ice sheet in the simulation with mesh size of 0.8 m, which results in many zero forces. For the simulation with mesh size of 0.2 m, full contact between the structure and the ice can be found, which means that the interaction between the structure and the level ice is continuous. This phenomenon can explain why the mesh size affects the forces. Figure 18 shows an ice crushing scenario from test #306. It is seen that there are many small ice pieces accumulating in front of the wind turbine tower during the interaction. However, there is no interaction between the wind turbine tower and the broken ice pieces as failed ice elements are removed in the simulations. This limitation of element deletion in the numerical simulations results in the differences of the ice forces between the simulated and measured results.

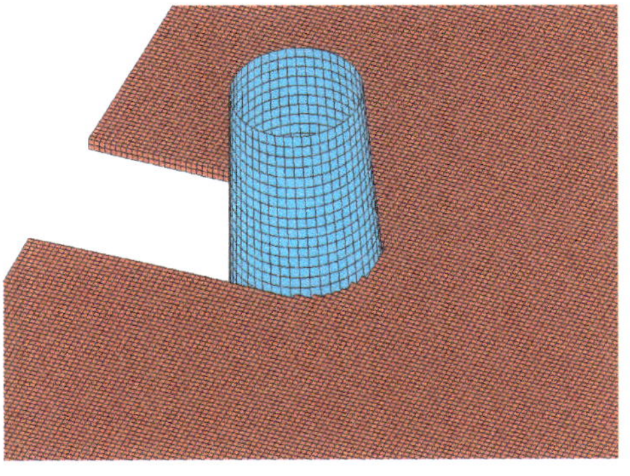

Figure 16. An image extracted from the simulation with mesh size of 0.2 m at t = 10 s.

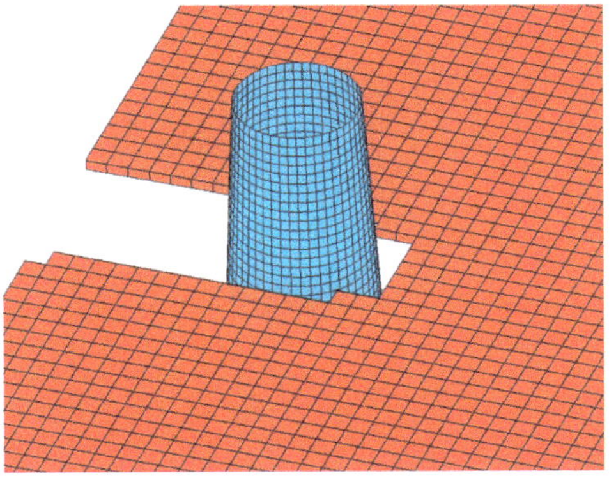

Figure 17. An image extracted from the simulation with mesh size of 0.8 m at t = 10 s.

Figure 18. An image extracted from test #306.

The simulated results including the mean, standard deviation, and maximum loads for different mesh sizes are presented in Table 5. In addition, the corresponding model test data are given for comparison. The comparison shows that the discrepancy of the maximum ice force between the simulated and measured results is small for all mesh sizes. The largest discrepancy is 2.82% for mesh size of 0.8 m, and the smallest is only 0.3% for mesh size of 0.6 m. However, the mean load decreases with increasing mesh size. It is also found that the simulation with mesh size of 0.2 m provides better predictions of mean load and standard deviation than the other simulations, i.e., there is a good agreement between the simulation with mesh size of 0.2 m and the experiment. The discrepancies of the mean load and the standard deviation are 3.1% and 3.3%, respectively. The simulations when the mesh size is larger than 0.2 m underestimate the mean load, while overestimating the standard deviation. This is due to more zero forces in the simulations with larger mesh size.

Table 5. Comparison between the simulated and measured results for test #306.

Items	Numerical Simulations (MN)				Model Test (MN)
	Mesh Size 0.2 m	Mesh Size 0.4 m	Mesh Size 0.6 m	Mesh Size 0.8 m	
Maximum	3.21	3.21	3.20	3.28	3.19
Mean	1.64	1.16	0.98	0.55	1.59
Std.	0.62	0.67	0.70	0.70	0.60

Figures 19–22 show the spectra of the ice force from the measurement and simulations with different mesh sizes for test #306. It is observed that the main frequency in the measurement is 1.05 Hz, compared to 6.05 Hz in the simulation with mesh size of 0.2 m, 3.05 Hz in the simulation with mesh size of 0.4 m, 2.05 Hz in the simulation with mesh size of 0.6 m, and 1.55 Hz in the simulation with mesh size of 0.8 m, which indicates that the main frequency in the simulation with larger mesh size is lower and closer to the experimental data. It is noted that there is a good agreement of a peak between the simulation with mesh size of 0.6 m and the measurement at the frequency at 1.05 Hz.

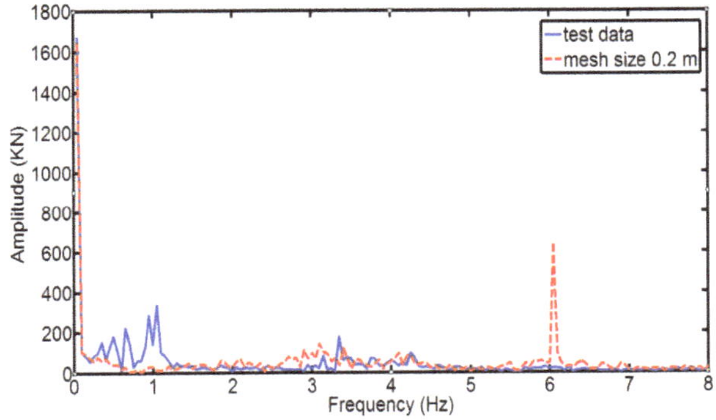

Figure 19. Spectrum of ice force from the simulation with mesh size of 0.2 m and measurement for test #306.

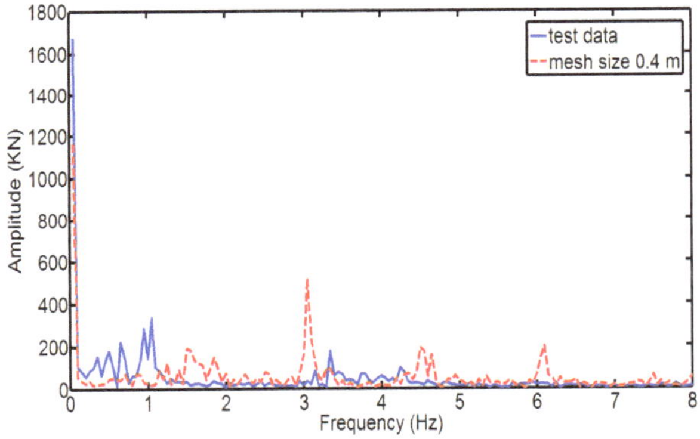

Figure 20. Spectrum of ice force from the simulation with mesh size of 0.4 m and measurement for test #306.

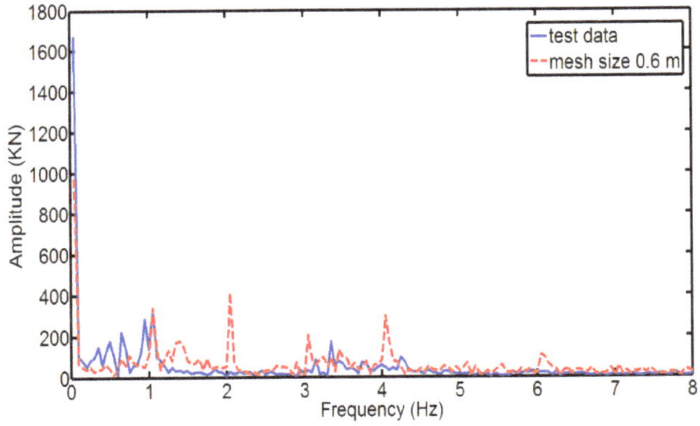

Figure 21. Spectrum of ice force from the simulation with mesh size of 0.6 m and measurement for test #306.

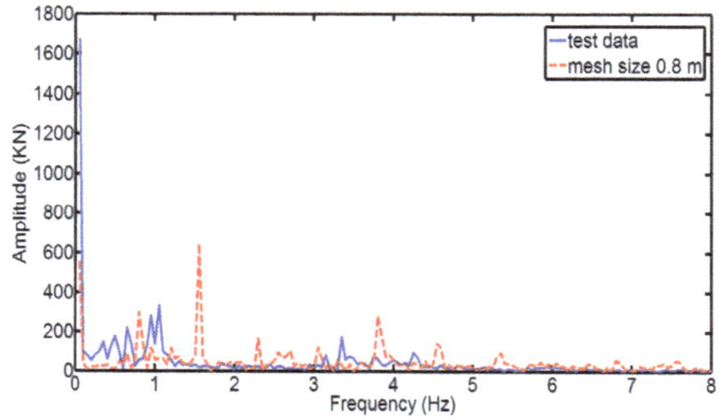

Figure 22. Spectrum of ice force from the simulation with mesh size of 0.8 m and measurement for test #306.

In summary, the simulations when the relationship between failure strain and size ratio is $y = 0.67e^{-0.84x}$ can provide an accurate prediction of maximum ice force for test #306. When comparing the ice force histories between the simulated and measured results, more zero forces are found in the simulations when the mesh size is larger than 0.2 m. Besides, the simulations overestimate the main frequency than the model test.

4.2. Comparison of Test #304

It is interesting to investigate the effect of the impact speed on the comparison between the simulated and measured results. Therefore, test #304 is simulated, in which the impact speed is 0.6 m/s. The other parameters and mesh size are the same as they are in Section 4.1.

The time series of the horizontal ice forces from both the simulations and the measurements for test #304 are given in Figures A1–A4. The comparisons show similar ice force characteristics with those for test #306. The maximum, mean, and standard deviation values derived from the simulated and measured ice forces are presented in Table 6. The comparison shows that the simulation with mesh size of 0.6 m provides a more accurate prediction of maximum ice force than the other simulations. The discrepancy between all simulated and measured maximum force ranges from 0.6% to 6.9%. In addition, the mean ice force decreases with increasing mesh size, and the mean value obtained from the simulation with mesh size of 0.4 m is closest to the measured data. All simulations overestimate the standard deviation.

It is concluded that there is a good agreement between all simulated and measured results with regard to the maximum ice force, which is consistent with the conclusion for test #306. The results indicate that the impact speed has little effect on the comparison of maximum ice forces between the simulations with different mesh sizes and the model tests.

Table 6. Comparison between the simulated and measured results for test #304.

Items	Numerical Simulations (MN)				Model Test (MN)
	Mesh Size 0.2 m	Mesh Size 0.4 m	Mesh Size 0.6 m	Mesh Size 0.8 m	
Maximum	3.25	3.28	3.47	3.37	3.49
Mean	1.63	1.13	0.94	0.57	1.15
Std.	0.66	0.72	0.75	0.73	0.60

4.3. Comparison of the Tests #404 and 406

To investigate the effects of the dimension of the wind turbine tower on the comparison between the simulated and measured results, test #404 and test #406 are also simulated, in which a 4 MW wind turbine tower is used. The drift speeds in test #404 and test #406 are 0.6 m/s and 1.2 m/s, respectively. The other parameters and mesh size are the same as they are in Section 4.1.

The maximum, mean, and standard deviation values obtained from both the simulated and measured ice forces for test #404 are presented in Table 7. It is observed that the maximum ice force calculated from the simulation with mesh size of 0.6 m fits well with the measured result in which the discrepancy is 2.1%. The discrepancy between the other simulations and the measured results ranges from 10.8% to 21.2%. For the mean force, the simulation with mesh size of 0.2 m is closest to the measured results. In addition, the standard deviation is higher than the model test data when the mesh size is larger than 0.2 m.

The comparisons for test #406 are shown in Table 8. Similar results with test #404 are found. The maximum ice force obtained from the simulation with mesh size of 0.6 m and the mean force obtained from the simulation with mesh size of 0.2 m are closest to the measured results, in which the discrepancies are 9.3% and 1.8%, respectively. The discrepancy of maximum force between the other simulations and the measured results ranges from 13.3% to 23.7%.

Table 7. Comparison between the simulated and measured results for test #404.

Items	Numerical Simulations (MN)				Model Test (MN)
	Mesh Size 0.2 m	Mesh Size 0.4 m	Mesh Size 0.6 m	Mesh Size 0.8 m	
Maximum	3.35	4.71	4.16	4.85	4.25
Mean	1.73	1.35	0.85	0.75	1.58
Std.	0.49	1.01	1.02	1.00	0.85

Table 8. Comparison between the simulated and measured results for test #406.

Items	Numerical Simulations (MN)				Model Test (MN)
	Mesh Size 0.2 m	Mesh Size 0.4 m	Mesh Size 0.6 m	Mesh Size 0.8 m	
Maximum	3.26	4.61	4.11	4.65	3.76
Mean	1.74	1.37	0.84	0.76	1.71
Std.	0.45	0.97	0.96	0.93	0.72

5. Discussion and Conclusions

In this paper, numerical simulations of the interaction between level ice and wind turbine tower have been performed using software LS-DYNA. The study confirms that both the mesh size and the failure strain of the ice model play a significant role in the simulated ice forces. With the refinement of mesh size, the simulated maximum ice force decreases, while the fluctuated frequency of ice force increases. The mesh size influences both the maximum load and the load frequency greatly. This finding is similar to those in Wang et al. and Lu et al. for simulating ice-sloping structure interactions [17,18]. It is also found that the mean, standard deviation, and maximum values derived from the simulated ice forces increase with increasing failure strain. The failure strain of the ice model is not a strictly material property but rather a numerical remedy to excessive mesh distortions. Therefore, its application to the simulation of a physical phenomenon requires the calibration with experimental results.

In our study, the measured maximum ice force derived from test #306 is used to calibrate the failure strain of the ice model with different mesh sizes. The relationship between failure strain and size ratio is obtained, i.e., $y = 0.67e^{-0.84x}$. It is found that a larger failure strain should be applied for the simulations with smaller mesh size to achieve an accurate prediction of maximum ice force. This is due to the combined effect of mesh size and failure strain on the simulated ice force. In order to investigate the effect of the impact speed and the wind turbine tower diameter, the same numerical models are applied to simulate the interaction for tests #304, #404, and #406.

It is found that a mesh size (0.6 m) that is 1.5 times the ice thickness (0.4 m) predicts maximum ice force with reasonable accuracy for all tests (see Figure A5), in which the discrepancy between the simulations and the model tests ranges from 0.3% to 9.3%. The size ratio (mesh size/ice thickness) is similar to that in Wang et al. for simulating the interaction between sloping marine structure and the level ice [18].

The other simulations can provide accurate predictions of maximum force for test #304 and #306 (see Figure A5). It indicates that the impact speed has little effect on the comparison between the numerical simulations with different mesh sizes and the model tests. These numerical models can be used to study the effect of speed for the same impact objects with regard to maximum force.

The discrepancy of maximum ice force between the simulated and measured results for test #404 and #406 ranges from 2.1% to 23.7%, which is larger than that for test #304 and #306 (i.e., 0.3% to 6.9%). It indicates that the predictive accuracy may decrease when the impacted structure is changed.

It is also found that the simulated maximum forces for tests #404 and 406 are higher than those for tests #304 and #306 (see Figure A5). This is because the diameter of the wind turbine tower at the mean waterline in tests #404 and #406 are relatively larger. Besides, when comparing the simulated maximum ice forces for tests #304 and #306, or tests #404 and #406, it is seen that slower impact speed results in a larger maximum force. These are confirmed by the model test results.

There exists significantly discrepancy of mean load and standard deviation between the simulated and measured results (see Figures A6 and A7). This is mainly caused by the limitations of the ice model. As an element deletion technique is used to remove failed ice elements from the calculation, numerical simulation of ice crushing generates zero contact that is created upon the deletion of elements. In addition, the rotation, accumulation and sliding forces that are contributed by the crushed ice could not be considered using this ice model. These limitations of the ice model will be addressed in the future work.

It should be noted that the model tests with the experimental scale of 1:20 were used for the comparison with the numerical simulations. The effect of the experimental scale on the comparison will be investigated by using more physical tests including model and full scale tests in the future work.

Author Contributions: Conceptualization, M.S.; methodology, M.S.; software, M.S.; validation, M.S., L.Z., and W.S.; formal analysis, M.S.; investigation, M.S.; resources, W.S. and Z.R.; data curation, W.S.; writing—original draft preparation, M.S.; writing—review and editing, W.S., Z.R.; visualization, M.S.; supervision, L.Z.; project administration, L.Z.; funding acquisition, L.Z., W.S.

Funding: This research was funded by the National Natural Science Foundation of China, grant number 51809124, 51911530156, 51709039; Natural Science Foundation of Jiangsu Province of China, grant number BK20170576; Natural Science Foundation of the Higher Education Institutions of Jiangsu Province of China, grant number 17KJB580006; State Key Laboratory of Ocean Engineering (Shanghai Jiao Tong University), grant number 1704 and 1807.

Acknowledgments: The authors would like to thank the Jiangsu University of Science and Technology (JUST).

Conflicts of Interest: The authors declare no conflict of interest.

Nomenclature

OWT	Offshore Wind Turbine
HAWC2	Horizontal Axis Wind turbine simulation Code 2nd generation
FAST	Fatigue Aerodynamics Structures and Turbulence
Std.	Standard Deviation
ε^p	Effective plastic strain
t	Time
D^p_{ij}	Plastic component of the rate of deformation tensor
F_f	Ice force in full scale
F_m	Ice force in model scale
λ	Experimental scale
F_{mean}	Mean force
F_i	Ice force at each time
N	Total number of output force
σ	Standard deviation of force

Appendix A. The Time Series of the Horizontal Ice Forces from Both the Simulations and Measurement for Test #304

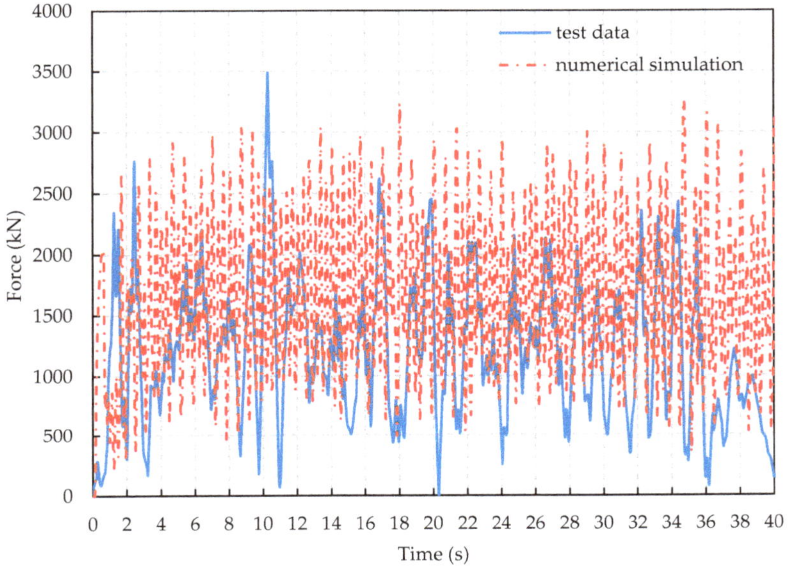

Figure A1. Ice force histories from the simulation with mesh size of 0.2 m and measurement for test #304.

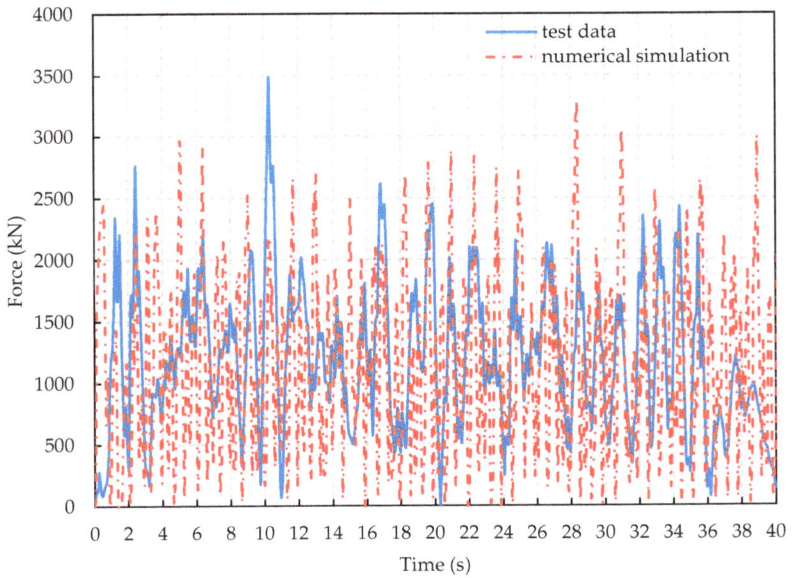

Figure A2. Ice force histories from the simulation with mesh size of 0.4 m and measurement for test #304.

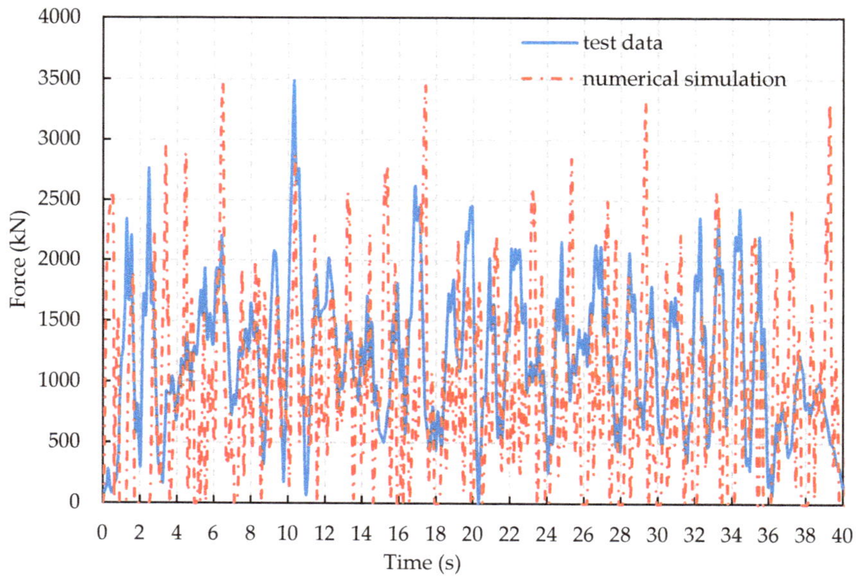

Figure A3. Ice force histories from the simulation with mesh size of 0.6 m and measurement for test #304.

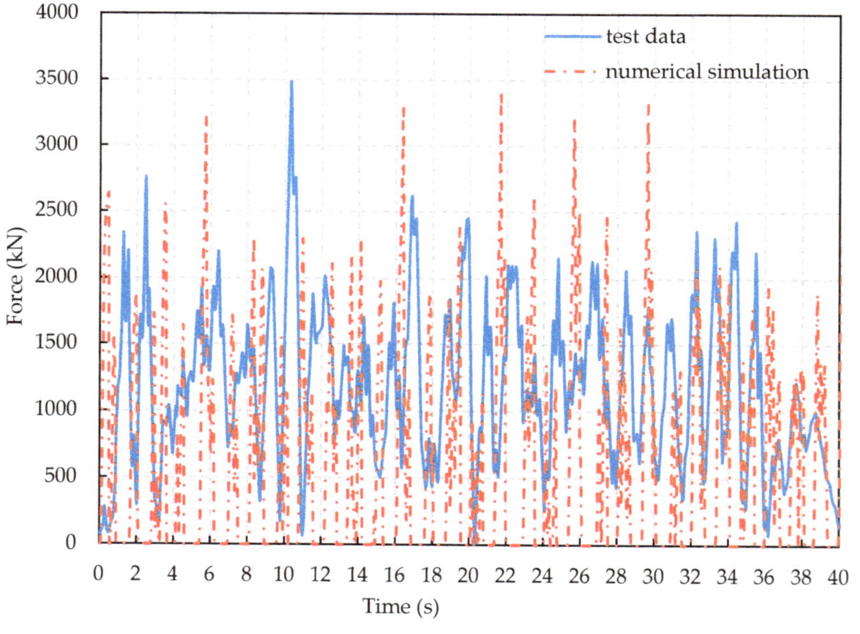

Figure A4. Ice force histories from the simulation with mesh size of 0.8 m and measurement for test #304.

Appendix B. The Comparison of the Maximum Force, Mean Force, and Standard Deviation between the Simulated and Measured Results for All Tests

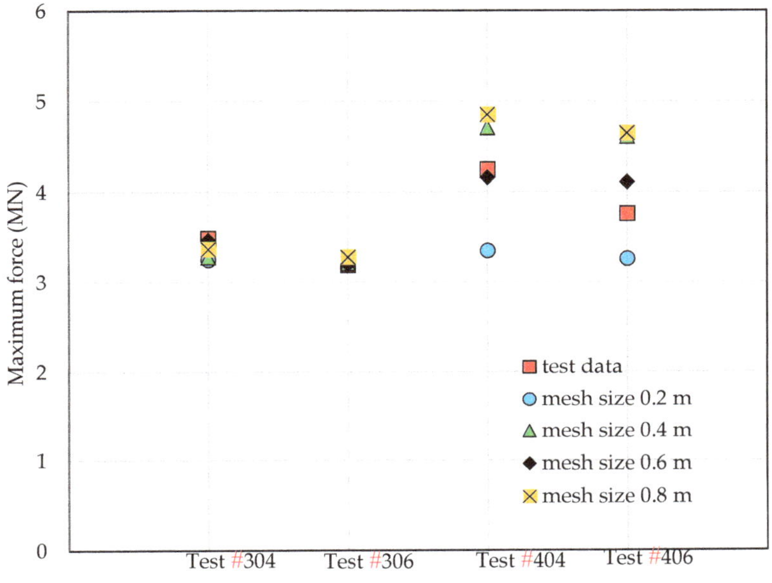

Figure A5. Comparison of the maximum ice force between the simulated and measured results for all tests.

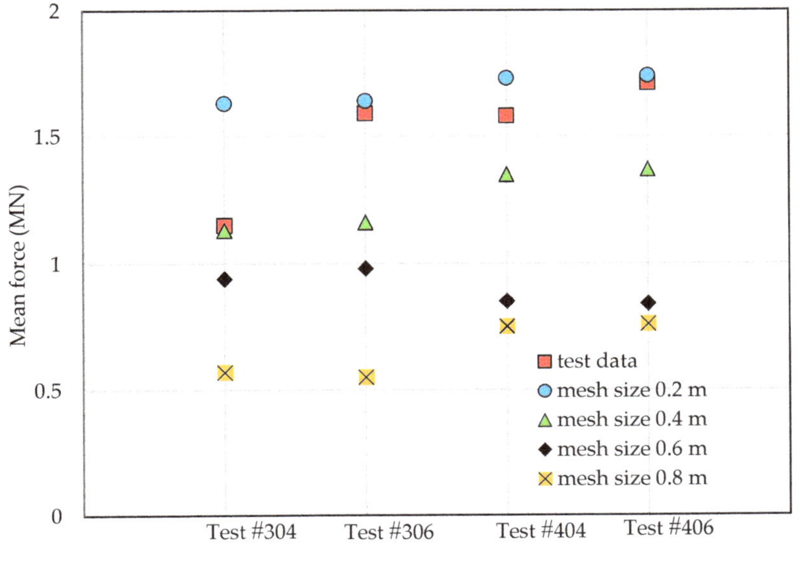

Figure A6. Comparison of the mean ice force between the simulated and measured results for all tests.

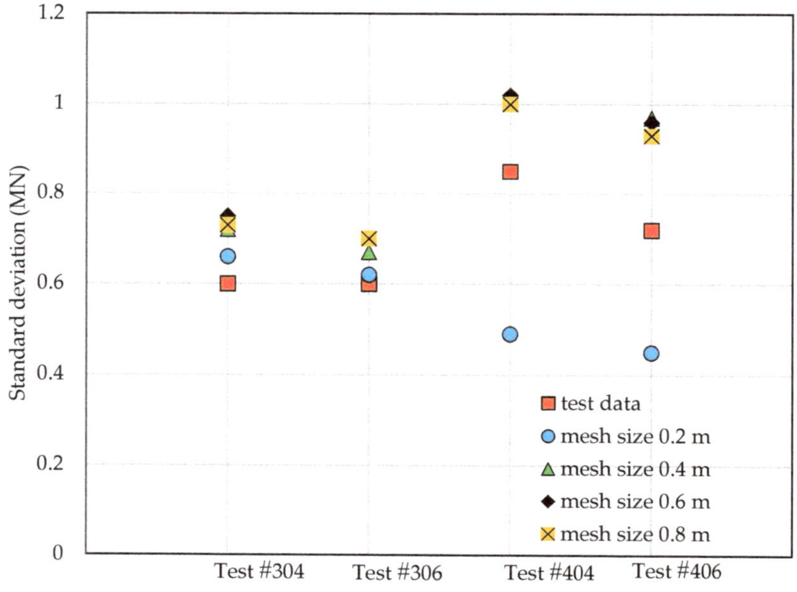

Figure A7. Comparison of the standard deviation between the simulated and measured results for all tests.

References

1. Ren, Z.; Skjetne, R.; Jiang, Z.; Gao, Z.; Verma, A.S. Integrated GNSS/IMU hub motion estimator for offshore wind turbine blade installation. *Mech. Syst. Signal Process.* **2019**, *123*, 222–243. [CrossRef]
2. Yu, B.; Karr, D.G.; Song, H.; Sirnivas, S. A surface ice module for wind turbine dynamic response simulation using FAST. In Proceedings of the 32nd International Conference on Ocean, Offshore and Arctic Engineering (OMAE), Nantes, France, 9–14 June 2013.
3. Timco, G.W. Indentation and penetration of edge-loaded freshwater ice sheets in the brittle range. *J. Offshore Mech. Arct. Eng.* **1987**, *109*, 287–294. [CrossRef]
4. Heinonen, J.; Rissanen, S. Coupled-crushing analysis of a sea ice-wind turbine interaction-feasibility study of FAST simulation software. *Ships Offshore Struct.* **2017**, *12*, 1–8. [CrossRef]
5. Karna, T.; Jochmann, P. Field observations on ice failure modes. In Proceedings of the 17th International Conference on Port and Ocean Engineering Under Arctic Conditions, Trondheim, Norway, 16–19 June 2003.
6. Hirayama, K.; Obara, I. Ice forces on inclined structures. In Proceedings of the Fifth International Conference on Offshore Mechanic and Arctic Engineering, Tokyo, Japan, 13–18 April 1986; pp. 515–520.
7. Sodhi, D.S.; Morris, C.E.; Cox, G.F.N. Dynamic analysis of failure modes on ice sheets encountering sloping structures. In Proceedings of the Sixth International Conference on Offshore Mechanics and Arctic Engineering, Houston, TX, USA, 1 March 1987; Volume IV, pp. 281–284.
8. Irani, M.B.; Timco, G. Ice load on a multifaceted conical structure. *Int. J. Offshore Polar. Eng.* **1993**, *3*, 313–321.
9. Barker, A.; Timco, G.; Gravesen, H.; Vølund, P. Ice loading on Danish wind turbines: Part 1: Dynamic model tests. *Cold Reg. Sci. Technol.* **2005**, *41*, 1–23. [CrossRef]
10. Gravesen, H.; Sørensen, S.L.; Vølund, P.; Barker, A.; Timco, G. Ice loading on Danish wind turbines: Part 2. Analyses of dynamic model test results. *Cold Reg. Sci. Technol.* **2005**, *41*, 25–47. [CrossRef]
11. Yue, Q.J.; Qu, Y.; Bi, X.J.; Tuomo, K. Ice force spectrum on narrow conical structures. *Cold Reg. Sci. Technol.* **2007**, *49*, 161–169. [CrossRef]
12. Gürtner, A.; Bjerkas, M.; Kuhnlein, W.; Jochmann, P.; Konuk, I. Numerical simulation of ice action to a lighthouse. In Proceedings of the 28th International Conference on Ocean, Offshore and Arctic Engineering (OMAE), Honolulu, HI, USA, 31 May–5 June 2009.

13. Määttänen, M.; Marjavaara, P.; Saarinen, S.; Laakso, M. Ice crushing tests with variable structural flexibility. *Cold Reg. Sci. Technol.* **2011**, *67*, 120–128. [CrossRef]
14. Tian, Y.; Huang, Y. The dynamic ice loads on conical structures. *Ocean Eng.* **2013**, *59*, 37–46. [CrossRef]
15. Yue, Q.J.; Guo, F.; Kärnä, T. Dynamic ice forces of slender vertical structures due to ice crushing. *Cold Reg. Sci. Technol.* **2009**, *56*, 77–83. [CrossRef]
16. Kuutti, J.; Kolari, K.; Marjavaara, P. Simulation of ice crushing experiments with cohesive surface methodology. *Cold Reg. Sci. Technol.* **2013**, *92*, 17–28. [CrossRef]
17. Lu, W.; Lubbad, R.; Loset, S. Simulating ice-sloping structure interactions with the cohesive element method. *J. Offshore Mech. Arct. Eng.* **2014**, *136*, 1–16. [CrossRef]
18. Wang, F.; Zou, Z.; Zhou, L.; Ren, Y.; Wang, S. A simulation study on the interaction between sloping marine structure and level ice based on cohesive element model. *Cold Reg. Sci. Technol.* **2018**, *149*, 1–15. [CrossRef]
19. Zhou, L.; Ding, S.; Song, M.; Gao, J.; Shi, W. A Simulation of Non-Simultaneous Ice Crushing Force for Wind Turbine Towers with Large Slopes. *Energies* **2019**, *12*, 2608. [CrossRef]
20. Ranta, J.; Santaoja, K.; Tuhkuri, J. Simulation of ice rubble failure against a conical structure with arbitrary Lagrangian-Eulerian finite element method. In Proceedings of the 20th IAHR International Symposium on Ice, At Lahti, Finland, 14–18 June 2010.
21. Ren, Z.; Skjetne, R.; Gao, Z. A crane overload protection controller for blade lifting operation based on model predictive control. *Energies* **2019**, *12*, 50. [CrossRef]
22. Shi, W.; Tan, X.; Gao, Z.; Moan, T. Numerical study of ice-induced loads and responses of a monopile-type offshore wind turbine in parked and operating conditions. *Cold Reg. Sci. Technol.* **2016**, *123*, 121–139. [CrossRef]
23. Wells, E. An Assessment of Surface Ice Sheet Loads and Their Effects on an Offshore Wind Turbine Structure. Ph.D. Thesis, The University of Toledo Digital Repository, Toledo, OH, USA, 2012.
24. Sanderson, T.J.O. *Ice Mechanics: Risks to Offshore Structures*; Graham and Trotman: London, UK, 1988.
25. Mróz, A.; Holnickiszulc, J.; Kärnä, T. Mitigation of ice loading on off-shore wind turbines: Feasibility study of a semi-active solution. In Proceedings of the II Eccomas Thematic Conference on Smart Structures and Materials, Lisbon, Portugal, 18–21 July 2005.
26. Wu, H.; Huang, Y.; Li, W. Experimental study on the ice load of large-diameter monopile wind turbine foundations. *Ocean. Eng.* **2018**, *36*, 83–91.
27. Livermore Soft Technology Corporation. *LS-DYNA User's Manual*; Livermore Soft Technology Corporation: Troy, MI, USA, 2011.
28. Derradji-Aouat, A. Explicit FEA and constitutive modelling of damage and fracture in polycrystalline ice-simulations of ice loads on structures. In Proceedings of the 18th International Conference on Port and Ocean Engineering Under Arctic Conditions, Potsdam, NY, USA, 26–30 June 2005.
29. Hilding, D.; Forsberg, J.; Gürtner, A. Simulation of ice action loads on offshore structures. In Proceedings of the 8th European LS-DYNA Users Conference, Strasbourg, France, 23–24 May 2011; pp. 1–12.

Journal of
Marine Science and Engineering

MDPI

Article

Influence of Central Platform on Hydrodynamic Performance of Semi-Submerged Multi-Buoy Wave Energy Converter

Yuan Hu [1], Shaohui Yang [1,2,3,*], Hongzhou He [1,2,3] and Hu Chen [1,2,3]

1 College of Mechanical and Energy Engineering, Jimei University, Xiamen 361021, China; huyuan_1995@outlook.com (Y.H.); hhe99@jmu.edu.cn (H.H.); chenhu2000@jmu.edu.cn (H.C.)
2 Key Laboratory of Energy Cleaning Utilization and Development of Fujian Province, Xiamen 361021, China
3 Key Laboratory of Ocean Renewable Energy Equipment, Xiamen 361021, China
* Correspondence: shaohuiyang@jmu.edu.cn

Received: 27 November 2019; Accepted: 12 December 2019; Published: 23 December 2019

Abstract: The influence of the central platform on hydrodynamic performance of a wave energy converter (WEC) has remained elusive. To approach this dearth of relevant theoretical research, this paper presents a semi-submerged multi-buoy WEC and the results of the numerical analysis at different dimension parameters of the central platform of the WEC. The WEC consists of three oscillating buoys hinged with a central platform through multiple actuating arms. Numerical analysis revealed that there exists a relationship between the hydrodynamic performance of device and the geometry of the central platform. At the given wave condition, different central platform size would obviously affect the hydrodynamic performance and wave energy capture width ratio of the semi-submerged multi-buoy WEC. Additionally, appropriately increasing central platform draft would help to improve the wave energy capture capability of the oscillating buoys.

Keywords: WEC; oscillating buoy; hydrodynamic performance; capture width ratio; central platform

1. Introduction

Due to the urgency to act against climate change caused by the growing carbon dioxide in the atmosphere, renewable energy has received extensive attention around the world [1]. Wave energy is the integration of kinetic and potential energy of a water point relative to displacement of hydrostatic surface [2]. Compared to traditional energy sources, it produces fewer waste products such as chemical pollution and carbon dioxide, which means the negative influence upon the environment is minimal. The research on wave energy conversion started in the 1970s when the oil crises provoked the exploitation of a wide variety of renewable energy sources. An advantage of wave energy technology is the wide diversity of design concepts. Different technologies use different solutions to harness energy from sea waves and are suited to operate at different water depths and locations [3]. A multi-buoy wave energy converter (WEC) primarily consists of three parts: a central platform, multiple oscillating bodies, and multiple actuating arms [4]. Additionally, the multi-buoy WEC can be classified into three types: stationary type, floating type, and semi-submerged type.

First proposed in 2000 by Niels and Keld Hansen, the stationary type WEC, Wavestar has become one of the world-leading wave energy technologies [5], which generates electricity through the oscillatory motion of hemispherical floats which in turn drive hydraulic power take-off systems. Kramer, who has already investigated the effect of a hydraulic power take-off system on device efficiency, found that power conversion efficiencies could improve by 70% by using the hydraulic system proposed in the paper [6]. Additionally, Kramer also presented a reliability-based structural optimization of the Wavestar WEC, which focused on improving the reliability of the device [7], which presented how structural designs of WECs can be optimized by using a reliability-based approach, and the ultimate goal is also to improve the energy efficiency and life of the device. However, the Wavestar WEC is fixed on the seabed which results in the water depth in the installation area being limited.

For the floating type WEC, our research group already has built two multi-buoy WEC, which are called "JMU I" and "JMU II", respectively. Both of them have a floating structure, but the former uses a mechanical energy conversion system, while the latter is a hydraulic system [8]. In [9] the authors investigated influence of buoy placement and platform wedge angle on hydrodynamic characteristics of the "JMU II", and the results show that different buoy placements and platform wedge angles would affect the amplitude of buoys. However, the investigation of the platform in the paper was limited to the platform wedge angle. Additionally, for the floating type WEC, the central platform has a shock amplitude of oscillation, which means it would indirectly affect buoy oscillation and also reduce the efficiency of wave energy collection.

For the semi-submerged multi-buoy WEC, the Guangzhou Institute of Energy Research of the Chinese Academy of Sciences successfully launched the two Eagle-type wave power generation equipment "Eagle I" and "Wanshan" in the waters of Wanshan Island, Zhuhai City, which is a semi-submerged barge and wave energy capture equipment [10,11]. The combination body can be parked like a ship, or can dive to a designated position to become a wave energy generating device. The "eagle" floating body used in the device is an improved type of "nodding duck" WEC. In [12], the frequency domain motion simulation of the "Eagle I" WEC under different wave elements and different PTO (power take-off) damping is presented, and the optimal load damping is obtained, which provides the theoretical basis for load design of prototype device. Additionally, the authors of [13] introduced the sea trial data of "Eagle I", and the experimental results verified the feasibility of the WEC. The "Wanshan" WEC is an improved model of the former, and its installed power could reach 100 kW which is 10 times of the former model. In [14] the authors investigated the hydrodynamic characteristics of a sharp eagle WEC, which could be considered that the buoy of the device is almost as same as "Wanshan", and the result shows that eagle head power prime mover has a strong ability to capture wave energy. In [15] the authors systematically introduced the structure, working principle, and experimental data of the sea trial of the device. It can be seen that the buoys of the device have been optimized, but the influence of the central platform on wave energy conversion efficiency of the device has not been mentioned.

To date, most of the studies are limited to the influence of the arrangement, numbers, and structure of the buoy, and the energy load damping on the efficiency of the device. As far as we know, the influence of the central platform on hydrodynamic performance of WEC have remained elusive. Hence, this paper presents a semi-submerged multi-buoy WEC, which uses the relative motion between buoys and central platform to drive the hydraulic system, such that the WEC could transfer wave energy to hydraulic power and be further utilized. Compared with other WECs, the presented WEC is proven to have higher sea condition adaptability due to the semi-submerged structure and energy output is more stable by using the hydraulic system. The authors of [16] have already analyzed the influence of buoys on hydrodynamic performance of the WEC, which shows the geometry of buoys would affect the wave energy capture width ratio of the WEC. The emphasis of this paper is to analyze the influence of the central platform on hydrodynamic performance and wave energy capture ratio of the WEC, which may provide a new strategy to increase wave energy conversion efficiency of multi-buoy WEC.

2. Materials and Methods

2.1. Methods

The dynamic equation of motion in the frequency domain can be written as [17]:

$$\begin{bmatrix} -\omega^2(M_a+M)-i\omega(C_d+C_{PTO})+K_S+K_{PTO} & A_J^T \\ A_J & 0 \end{bmatrix}\begin{pmatrix} X \\ F_J \end{pmatrix}=\begin{pmatrix} F_e \\ 0 \end{pmatrix}, \quad (1)$$

where F_e and X are the frequency dependent complex amplitude of wave force array and complex amplitude of corresponding displacements array, respectively; M is the structure mass matrix for the central platform; M_a and C_d are the hydrodynamic added mass matrix and the hydrodynamic damping matrix, respectively; K_S is the hydrostatic restoring matrix; C_{PTO} and K_{PTO} are the PTO (power take-off) system damping and stiffness matrices, respectively; A_J is the displacement constraint matrix; F_J is the joint force vector; and F_e , M_a, and C_d can be obtained by solving the wave diffraction and radiation problem by using a boundary element method (ANSYS-AQWA).

To simplify the numerical simulation, a time domain analysis is carried out with a linear PTO. The basic equation of structural motion in the time domain can be written as [18]:

$$(M + M_\infty)\ddot{x}(t) + \int_{-\infty}^{t} C(t-\tau)\dot{x}(\tau)d\tau + K_s x(t) + F_{joint}(t) + F_{PTO}(t) = F_{exc}(t), \tag{2}$$

where M_∞ is the hydrodynamic added mass matrix for $\omega \to \infty$; $x(t)$, $\dot{x}(t)$ and $\ddot{x}(t)$ are the generalized displacement, velocity and acceleration vectors in the time domain, respectively; $C(t)$ is the retardation function matrix; $F_{joint}(t)$ is the resistant force vector due to the joint connection; $F_{PTO}(t)$ is the resistant force vector due to the PTO system, which is the hydraulic system damping in this paper and in order to simplify the numerical simulation, the $F_{PTO}(t)$ is constant.

2.2. Model of the WEC

A three-dimensional model of the full-size device was established using SolidWorks. The semi-submerged multi-buoy WEC proposed in the paper is composed of three oscillating buoys, a central platform, three actuating arms, and hydraulic systems at the joint as illustrated in Figure 1, which was built according to the original proportion of the solid model. The angle between the buoys is 120°, and Buoy 1 is located at the bow of the floating platform, Buoy 2 and Buoy 3 are symmetrically distributed on both sides of the central platform. The central platform can be divided into three layers: damping layer, support layer, and floating layer, as shown in Figure 2, which also presents the dimension of central platform: where H is the height of the damping layer, L is the length of the damping layer, and W is width of the damping layer. In addition, the initial parameters of the damping layer and buoys are presented in Table 1. The device has the capability of autonomous navigation with the buoys at the top of the central platform. Additionally, when it has arrived at the specified sea area, the central platform ballast tank injects seawater to sink the hull, after that buoys would descend to the surface of the sea. As the wave passed down the WEC, the oscillating buoys and central platform make a relative motion around the joint. This wave induced relative motion of the rafts is resisted by hydraulic systems. The hydraulic cylinders pump the high-pressure fluid via smooth gas accumulators to the hydraulic motor, which could be further utilized. This paper mainly investigates the hydrodynamic performance of the WEC under different geometry parameters of the damping layer and the draught depth of the central platform.

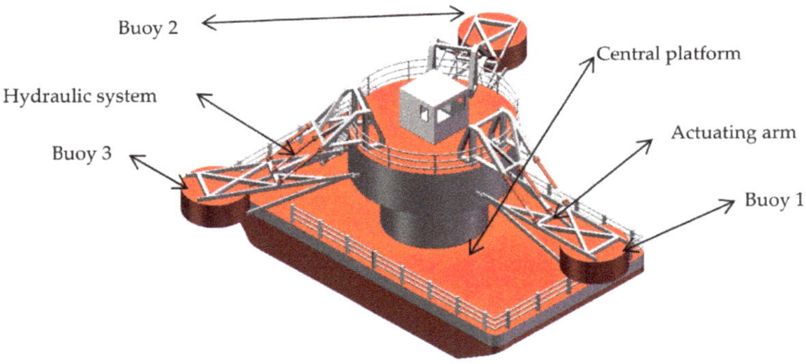

Figure 1. Semi-submerged multi-buoy wave energy converter (WEC).

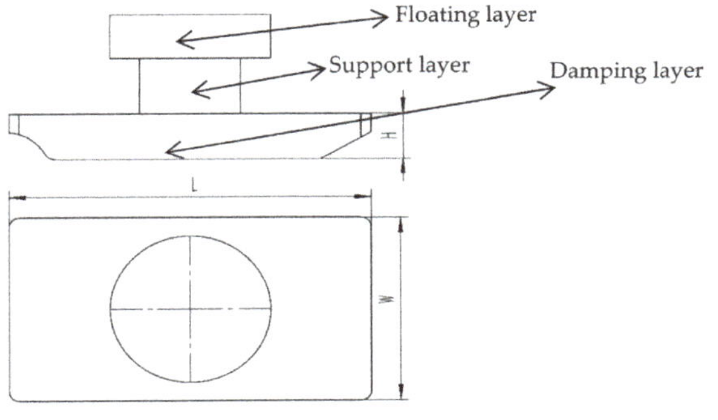

Figure 2. The dimensions of the central platform.

Table 1. Initial parameters of the damping layer and buoys.

Designation	Parameters
Molded length	18 m
Molded width	10 m
Molded height	2.5 m
Diameter of buoy	3.2 m
Height of buoy	2 m
Draft depth of the device	5.1 m

In the numerical calculation, we used the ANSYS-AQWA for simulation, which is based on potential flow theory and boundary elements method. Additionally, before running the simulation, the outside surface of the WEC model is divided into meshes, as shown in Figure 3, and the sea state parameters are set to water depth of 15 m and sea area of 100 m^2. The maximum grid size of the central platform and buoys are 0.6 and 0.2 m, respectively. For the sea state parameters setting of the time domain calculation, which refer to the sea conditions of the sea trial experiment, the wave height is 1 m, the period is 3 s, and the direction is pointed by the bow to the stern. The hinge damping of the buoys and central platform is set to 3000 (N*m/(°/s)), the additional damping parameter is set to 3500 (N/(m/s)), and the additional damping is intended to simulate the damping force of the hydraulic system of the device. The simulation time is set as 120 s, and the time step is 0.1 s.

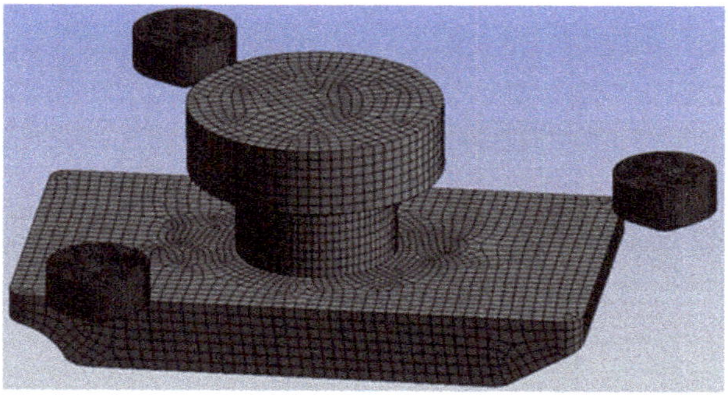

Figure 3. Grid model of semi-submerged multi-buoy WEC.

3. Results of the Numerical Analysis

In this section, we investigate the hydrodynamic performance of the device in different damping layer aspect ratios, cross-sectional areas, heights, and drafts of the central platform. In addition, the hydrodynamic performance of the central platform and the instantaneous power curves of the buoys are shown in each section. Since Buoy 2 and Buoy 3 are symmetrical at the platform in the wave direction, both of them have the same hydrodynamic performance. Therefore, only the simulation results of Buoy 2 are presented in this paper.

3.1. Numerical Results at Different Aspect Ratio of the Damping Layer

3.1.1. Hydrodynamic Performance of the Central Platform

The aspect ratio of the damping layer could be obtained by dividing the damping layer lengths (along the wave direction) L by its width (at a 90° angle to the direction of the wave) W, which can use B to represent:

$$B = L/W. \tag{3}$$

The influence of a different aspect ratio on hydrodynamic performance is examined with five different B values, ranging from 1 to 2.69. Additionally, the values of the B in each computation is presented in Table 2. The other parameters used are the initial parameters in this simulation.

Table 2. The values of L and W in each computation.

B	L/W (m)
1	13.4/13.4
1.25	15/12
1.8	18/10
2.22	20/9
2.69	22/8.18

Figure 4 shows the hydrodynamic performance curves of the central platform, it can be seen that wave force, diffraction force, and radiation damping all decrease with the increase of aspect ratio of the damping layer. For the wave force and the diffraction force, the maximum value occurs when the wave frequency is around 1.1 rad/s; for the radiation damping, the maximum value occurs when the wave frequency is around 1.3 rad/s; and for added mass, the maximum value occurs when the wave frequency is around 1 rad/s; which means that when the aspect ratio changes, the wave frequency at the peak of the curves are not changed. The added mass is decreased with the aspect ratio increases, except that the wave frequency is in the range of 1.2–2 rad/s.

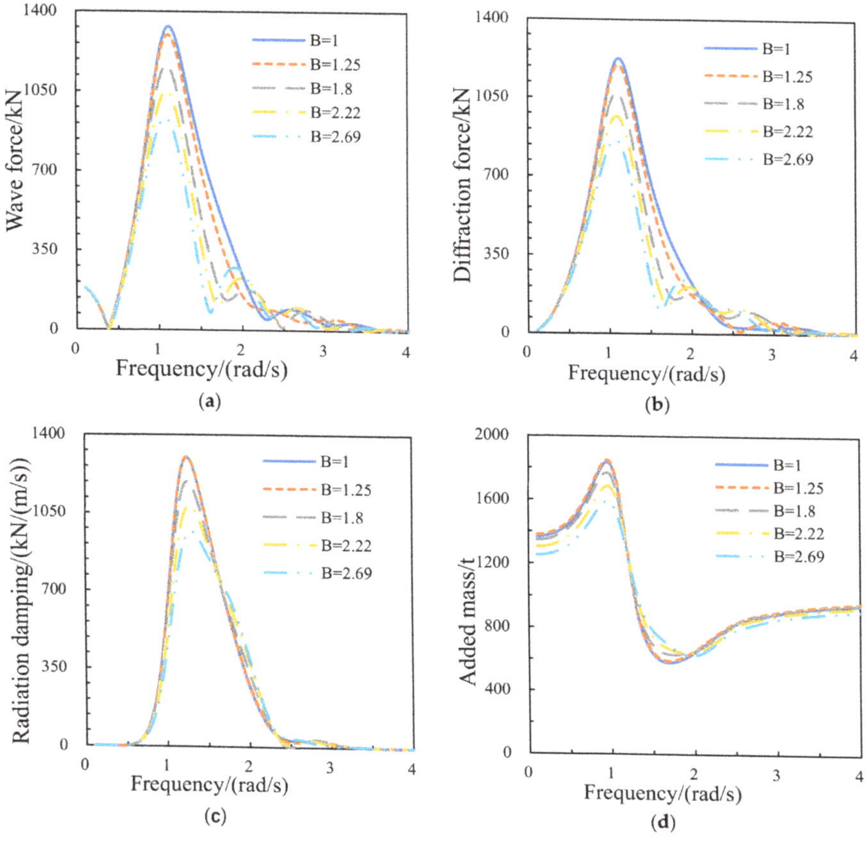

Figure 4. Four hydrodynamic parameter curves: (**a**) wave force, (**b**) diffraction force, (**c**) radiation damping, (**d**) added mass.

Figure 5 shows the oscillation curves of the central platform, which indicated that when the aspect ratio of damping layer is 1.25, the central platform has the smallest oscillation amplitude. However, the difference in oscillation amplitude of the central platform is small under different aspect ratios, which means the aspect ratio of the damping layer has little effect on the platform oscillation.

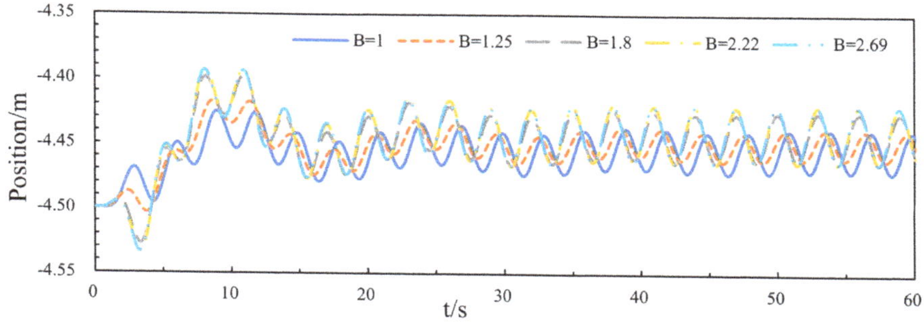

Figure 5. Oscillation curve of the central platform.

3.1.2. Wave Energy Capture Width Ratio of the Device

To better reflect the hydrodynamic performance of a WEC, capture width ratio is mentioned, which is a measure of the hydrodynamic efficiency [19]. It is obtained by dividing the absorbed wave power P by wave resource J and characteristic dimension D:

$$\eta = P/JD, \tag{4}$$

where P is the absorbed wave power, which could be obtained by the instantaneous power curves of the buoys; and for the device presented in the paper, D is the diameter of buoy. As for J, it can be shown as [20]:

$$J = (\rho g^2)/32\pi H^2 T, \tag{5}$$

where ρ is the density of sea water which is 1025 kg/m^3; the gravity g is 9.81 m/s^2; H and T are the height of wave and period of wave, respectively.

Figures 6 and 7 show the instantaneous power curves of Buoy 1 and Buoy 2, respectively. It can be seen that the maximum value of instantaneous power of Buoy 1 is 8.6 kW when the aspect ratio is 1 and the minimum value occurs when the aspect ratio is 2.22, which is about 7 kW. For Buoy 2, the maximum value is 8.2 kW when the aspect ratio is 1 and the minimum value is 6.8 kW when the aspect ratio is 2.22. Figure 8 is the curve of capture width ratio with aspect ratio, which shows the capture width ratio of the device generally decreases as the aspect ratio of the damping layer increases. In order to better show its regularity, two sets of simulation data were added in the process of drawing the curve, which are $B = 3$ and $B = 0.45$, respectively (L/W are 23.2/7.7 and 9/20, respectively). At a lower aspect ratio, the capture width ratio of Buoy 1 is better than Buoy 2, and the difference is narrowed as the aspect ratio increases. The capture width ratio of all buoys refers to the average wave energy capture width ratio of three buoys. However, when the aspect ratio of the damping layer is further increased, the wave energy flowing through Buoy 1 is consumed too much by the damping layer, resulting in a rapid decrease in the capture width ratio.

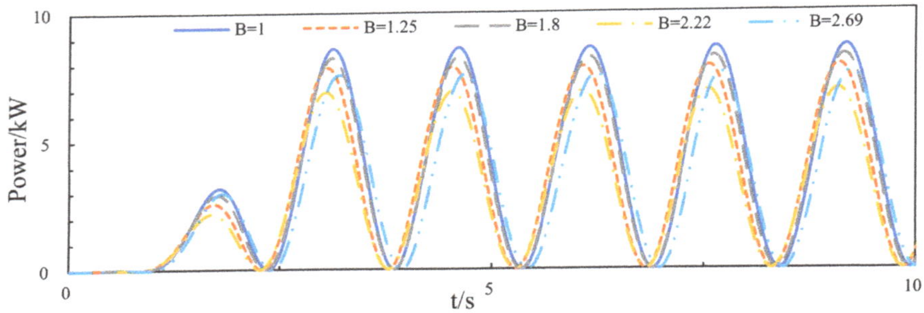

Figure 6. The instantaneous power curves of Buoy 1.

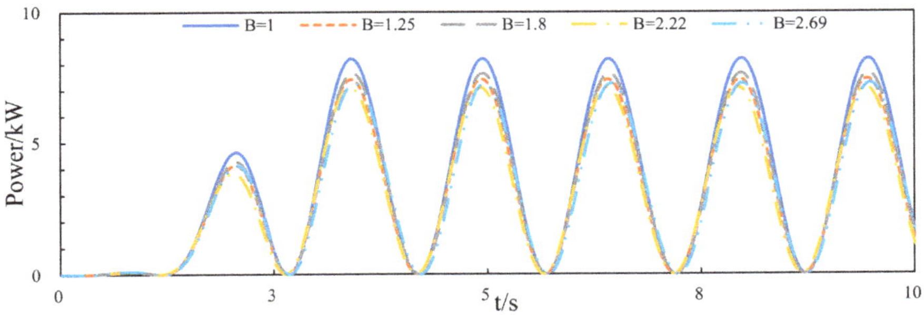

Figure 7. The instantaneous power curves of Buoy 2.

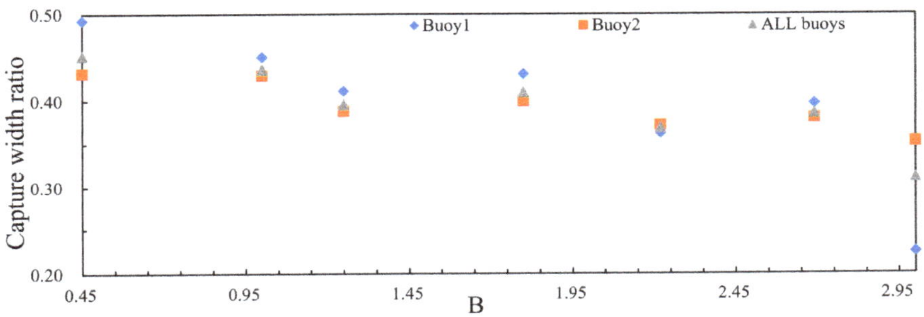

Figure 8. Curve of capture width ratio with aspect ratio.

The above results show that appropriately reducing the aspect ratio of the damping layer of the central platform is beneficial to increasing the wave energy capture width ratio of the device, which could improve the wave energy collection efficiency.

3.2. Numerical Results at Different Areas of the Damping Layer

3.2.1. Hydrodynamic Performance of the Central Platform

The area of damping layer refers to L multiplied by W, which can use S to represent:

$$S = LW. \tag{6}$$

The influence of different areas on hydrodynamic performance is examined with four different S ranging from 108 to 320 m^2. As we considered, the wave energy collection efficiency is better when the aspect ratio of the damping layer is 0.45. Hence, in this simulation when the S varies, B remains at 0.45. Additionally, the values of S in each computation is presented in Table 3. The other parameters use the initial parameters in this simulation.

Table 3. The values of L and W in each computation.

S (m^2)	LW (m)
108	7 × 5.56
180	9 × 20
222	10 × 22.2
320	12 × 26.67

Figure 9 shows the hydrodynamic performance curves of the central platform, and it can be seen that wave force, diffraction force, radiation damping, and added mass all increase with the increase in area of the damping layer. For the wave force, the diffraction force, the radiation damping, and added mass, the range of wave frequency corresponding to the peak value of the curve is between 0.95–1.45, 0.95–1.34, 1.03–1.5, and 0.83–1.14 rad/s, respectively, which means the wave frequency at the peak of the curves decreases slightly with the increase in the area of the damping layer.

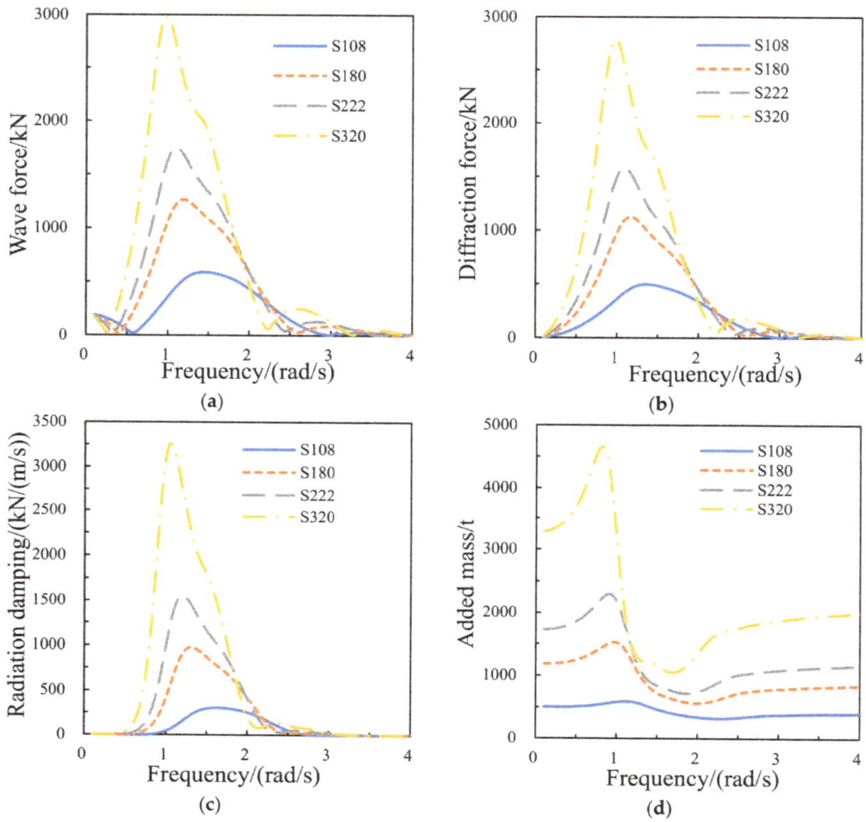

Figure 9. Four hydrodynamic parameter curves. (**a**) Wave force, (**b**) diffraction force, (**c**) radiation damping, (**d**) added mass.

Figure 10 shows the oscillation curves of the central platform, which indicated that when the area of damping layer is 222 m^2, the central platform has the smallest amplitude of oscillation, and the difference in oscillation amplitude of the central platform is obvious under different areas of the damping layer, which means the area of the damping layer has great effect on the platform oscillation. Additionally, it can be seen that when the area of the damping layer is 320 m^2, the oscillation of the central platform is not transient, which is due to large area and low aspect ratio of the damping layer, resulting in excessive platform roll. In addition, the curve is based on the mass point of the central platform, and the excessive platform roll would result in volatility of the oscillation of the central platform.

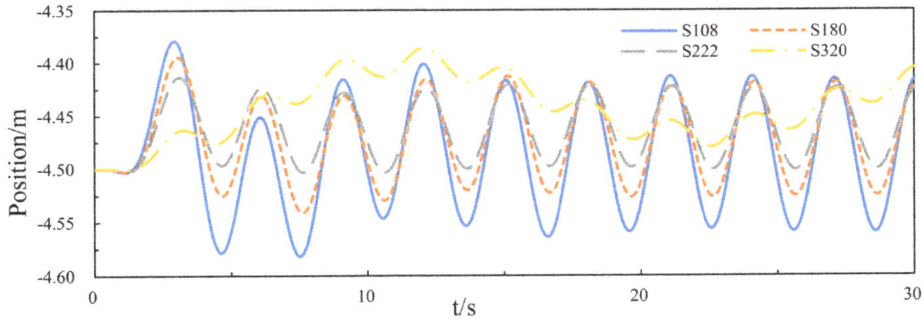

Figure 10. Oscillation curve of the central platform.

3.2.2. Wave Energy Capture Width Ratio of the Device

Figures 11 and 12 show the instantaneous power curves of Buoy 1 and Buoy 2, respectively. It can be seen that the maximum value of instantaneous power of Buoy 1 is 9.1 kW when the area of damping layer is 320 m^2 and the minimum value occurs when the area is 108 m^2, which is about 6.8 kW. For Buoy 2, the maximum value is 12 kW when the area is 320 m^2 and the minimum value is 6.3 kW when the area is 108 m^2. Figure 13 is the curve of capture width ratio with the area of the damping layer, which shows the capture width ratio of the device generally increases as the area of the damping layer increases. In order to better show its regularity, one set of simulation data was added in the process of drawing the curve, which is S = 268.8 m^2 (L*W is 11*24.4). When the area of damping layer is smaller than 222 m^2, the capture width ratio of Buoy 1 is better than Buoy 2, and the difference is narrowed as the area increases. However, when the area of the damping layer is further increased, the wave energy capture ratio of Buoy 1 reduced significantly while Buoy 2 increased gradually and finally stabilized.

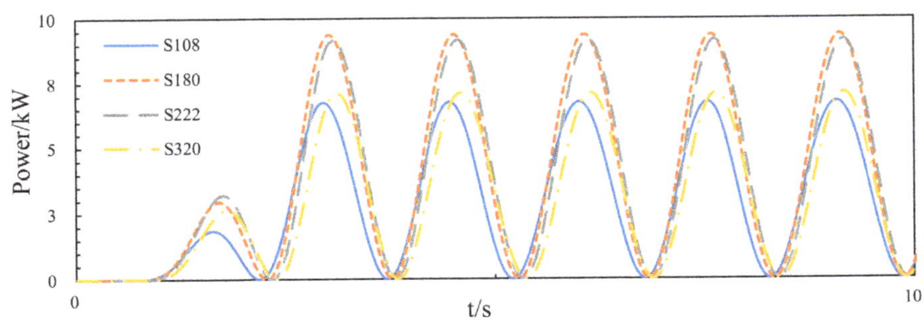

Figure 11. The instantaneous power curves of Buoy 1.

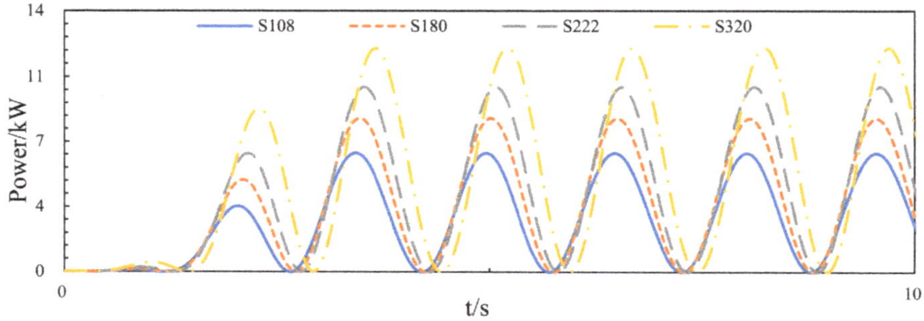

Figure 12. The instantaneous power curves of Buoy 2.

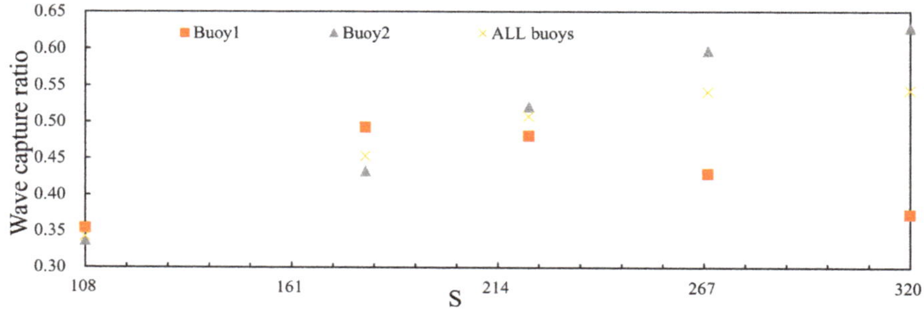

Figure 13. Curve of capture width ratio with area.

The above results show that appropriately increasing the area of the damping layer of the central platform is beneficial to increasing the wave energy capture width ratio of the device, and could improve the wave energy collection efficiency.

3.3. Numerical Results at Different Heights of the Damping Layer

3.3.1. Hydrodynamic Performance of the Central Platform

The height of damping layer refers to H, which has already been presented in Figure 2. The influence of different heights on hydrodynamic performance was examined with four different H ranging from 2 to 3.5 m. According to the simulation results above, the WEC has better efficiency when the aspect ratio and area of the damping layer are 0.45 and 320 m^2, respectively. Hence, in this simulation, the aspect ratio and area of the damping layer are 0.45 and 320 m^2, respectively.

Figure 14 shows the hydrodynamic performance curves of the central platform, it can be seen that wave force, diffraction force, radiation damping, and added mass all increase with the increasing height of the damping layer. For the wave force, the diffraction force, the radiation damping, and added mass, the range of wave frequency corresponding to the peak value of the curve is between 0.82–1.05, 0.86–1.07, 0.86–1.1, and 0.74–0.86 rad/s, respectively, which means the wave frequency at the peak of the curves decrease slightly with the increasing height of the damping layer.

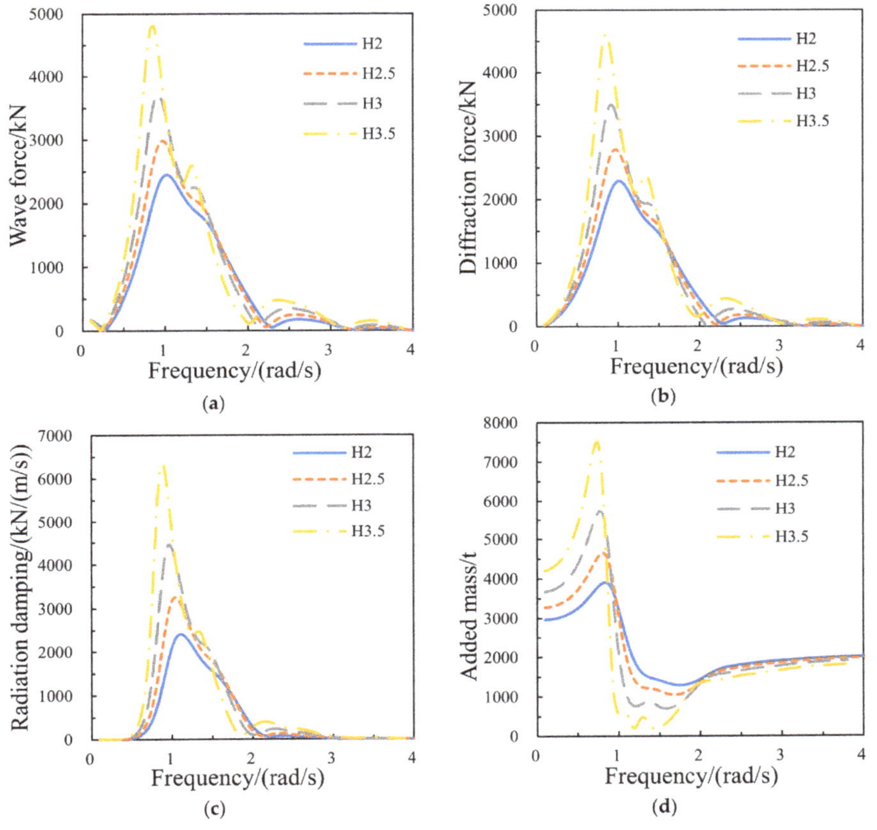

Figure 14. Four hydrodynamic parameter curves. (**a**) Wave force, (**b**) diffraction force, (**c**) radiation damping, (**d**) added mass.

Figure 15 shows the oscillation curves of the central platform, which indicated that when the height of damping layer is 2 m, the central platform has the smallest amplitude of oscillation, and the difference in oscillation amplitude of the central platform is obvious under different heights of damping layer, which means the height of the damping layer has a large effect on the platform oscillation. We considered that when the height of the damping layer increases, the damping layer would be subjected to stronger wave forces owing to increasing the area of contact with the wave, which results in stronger oscillation of the central platform.

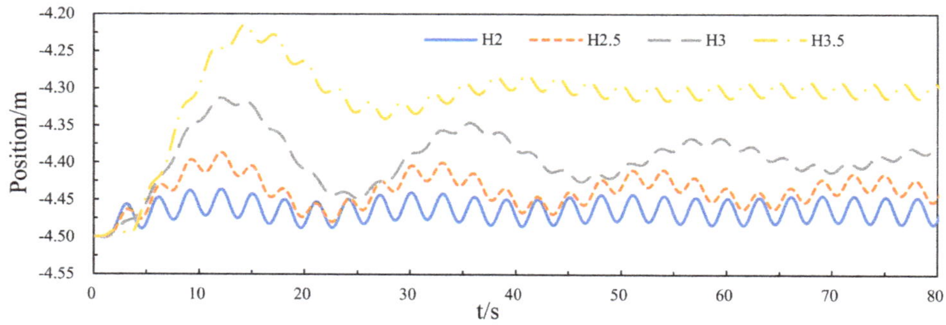

Figure 15. Oscillation curve of the central platform.

3.3.2. Wave Energy Capture Width Ratio of the Device

Figures 16 and 17 show the instantaneous power curves of Buoy 1 and Buoy 2, respectively. It can be seen that Buoy 1 and Buoy 2 have completely opposite power characteristics, by which the maximum value of instantaneous power of Buoy 1 is 7.7 kW when the height of damping layer is 2 m and the minimum value occurs when the height is 3.5 m, which is about 2.5 kW. For Buoy 2, the maximum value is 15.1 kW when the height is 3.5 m and the minimum value is 10 kW when the height is 2 m. Figure 18 is the curve of capture width ratio with height of damping layer, which shows the capture width ratio of the device is slightly increased as the height of damping layer increases. However, Buoy 1 and Buoy 2 have completely different characteristics, which means the wave energy capture ratio of Buoy 1 decreases as the height of the damping layer increases, while Buoy 2 increases as the height of the damping layer increases.

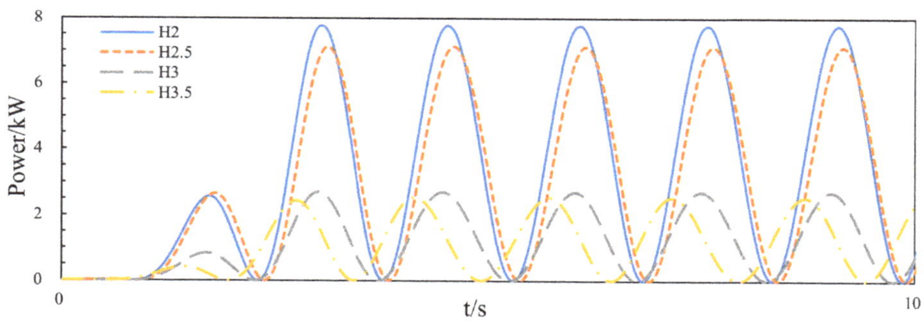

Figure 16. The instantaneous power curves of Buoy 1.

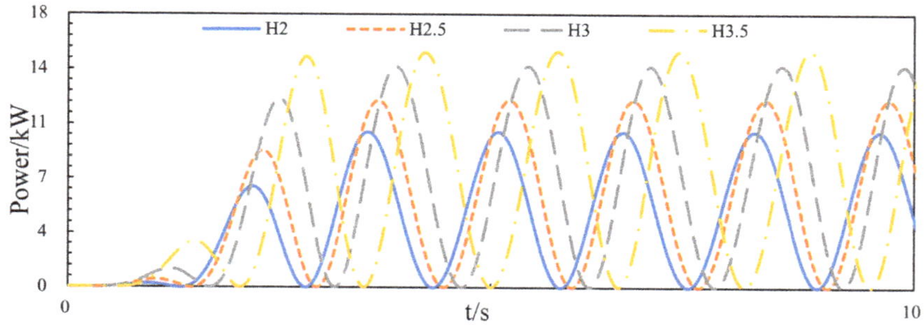

Figure 17. The instantaneous power curves of Buoy 2.

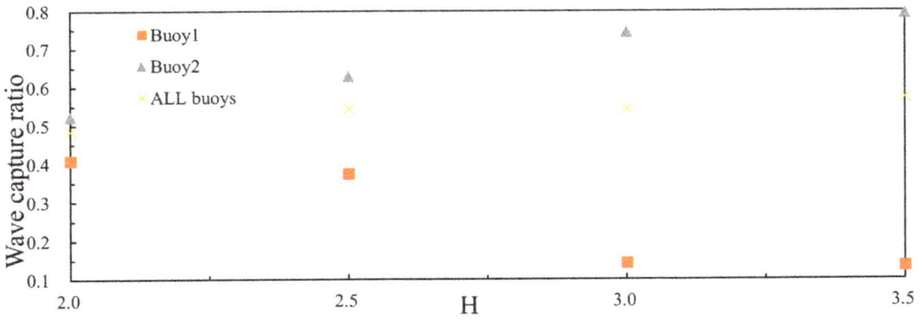

Figure 18. Curve of capture width ratio with height.

The above results show that increasing the height of the damping layer of the central platform only has little benefit to increasing the wave energy capture width ratio of the device, but it has a huge impact on the buoys in different positions.

3.4. Numerical Results at Different Drafts of the Damping Layer

3.4.1. Hydrodynamic Performance of the Central Platform

The draft of the central platform indicates the depth of the central platform below the sea surface, which could be represented by h. The influence of different drafts on hydrodynamic performance was examined with four different h ranging from 4.1 to 5.6 m. According to the simulation results above, we already found a better geometric parameter of the damping layer. Hence, the geometric parameter of the damping layer remains unchanged, which means the aspect ratio of the damping layer, the area of the damping layer, and the height of the damping layer are 0.45, 320 m^2, and 3.5 m, respectively.

Figure 19 shows the hydrodynamic performance curves of the central platform, it can be seen that wave force, diffraction force, radiation damping, and added mass all decrease with the increasing draft of the central platform. For the wave force and the diffraction force, the range of wave frequency corresponding to the peak value of the curve is between 0.63 and 1 rad/s; for the radiation damping, the range of wave frequency corresponding to the peak value of the curve is between 0.67 and 1.1 rad/s; and for the added mass, the range of wave frequency corresponding to the peak value of the curve is between 0.63 and 0.9 rad/s, which means the wave frequency at the peak of the curves increases as the draft of the central platform increases.

Figure 20 shows the oscillation curves of the central platform, which indicated that when the draft of central platform is 5.1 m, the central platform has the smallest amplitude of oscillation, and the difference in oscillation amplitude of the central platform is great under different drafts of the central platform. This means the draft of the damping layer has an effect on the platform oscillation. We considered that the oscillation amplitude of the platform can be reduced by increasing the platform draft appropriately, then it would indirectly promote the relative oscillation between the buoys and central platform. However, as the draft of the platform increased, the shape of the draft part of the platform would change (a part of the floating layer would be below the waterline), which may lead to changes in the hydrodynamic performance of the platform, and would affect the oscillation of the platform.

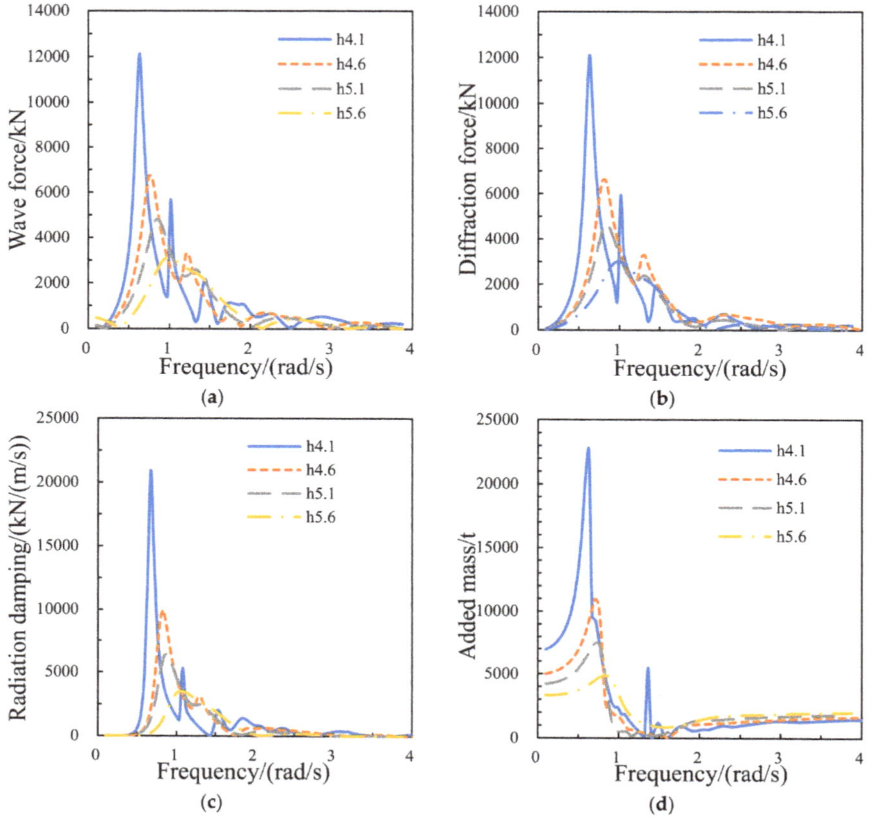

Figure 19. Four hydrodynamic parameter curves. (**a**) Wave force, (**b**) diffraction force, (**c**) radiation damping, (**d**) added mass.

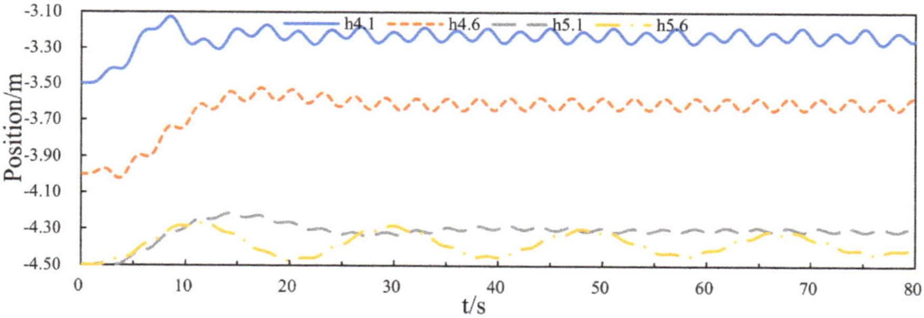

Figure 20. Oscillation curve of the central platform.

3.4.2. Wave Energy Capture Width Ratio of the Device

Figures 21 and 22 show the instantaneous power curves of Buoy 1 and Buoy 2, respectively. It can be seen that the maximum value of instantaneous power of Buoy 1 is 11.2 kW when the draft of the central platform is 4.6 m and the minimum value occurs when the height is 5.1 m, which is about 2.5 kW. For Buoy 2, the maximum value is 19.7 kW when the draft of central platform is 4.6 m and the minimum value is 11.8 kW when the draft is 4.1 m. Figure 23 is the curve of capture width ratio with height of damping layer, which shows the capture width ratio of the device is increased when the draft of central platform is between 4.1 and 4.6 m, slightly reduced as the draft further increases, and finally slightly increases. Additionally, the characteristics of Buoy 1 and Buoy 2 are consistent.

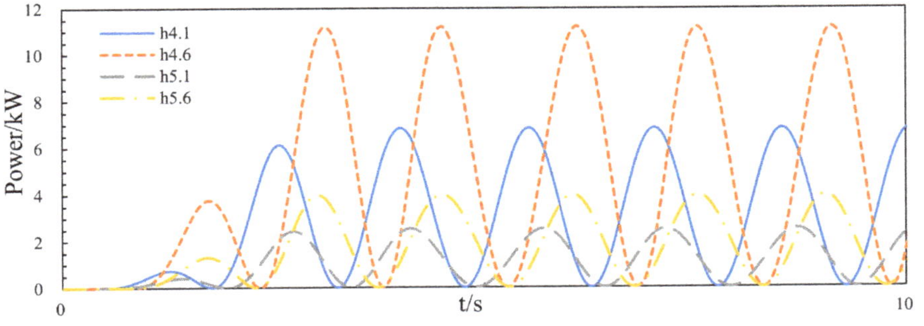

Figure 21. The instantaneous power curves of Buoy 1.

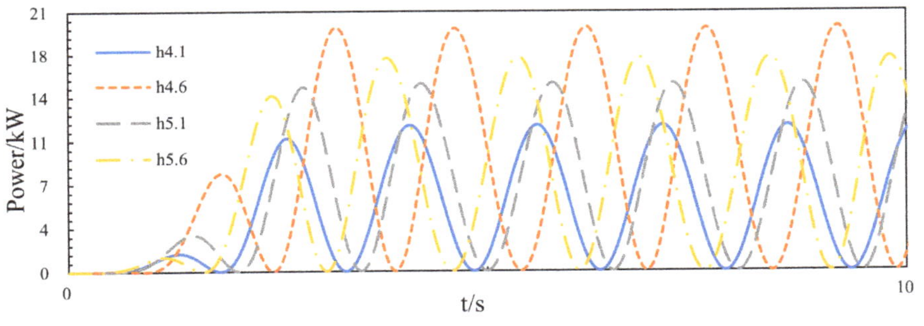

Figure 22. The instantaneous power curves of Buoy 2.

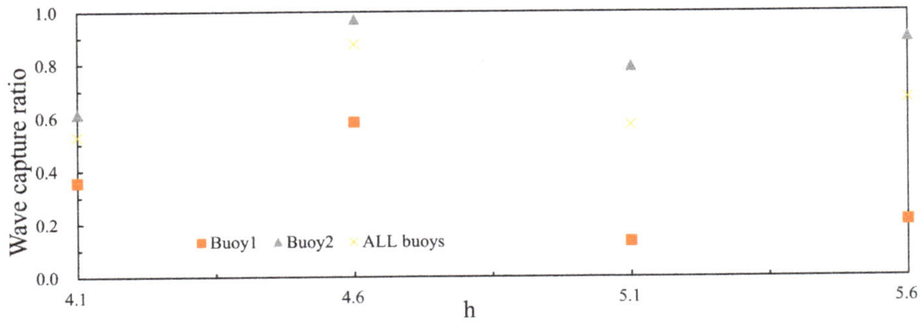

Figure 23. Curve of capture width ratio with draft.

The above results show that the relationship between the wave energy capture width ratio and the draft of the central platform is nonlinear. The maximum value of wave energy capture is about 0.87 when the draft of central platform is 4.6 m, which means when the depth of the device reaches 4.6 m, the wave energy collection efficiency of the device is the best.

4. Discussion

The numerical analysis results indicated that the geometry of the central platform does have a great effect on hydrodynamic performance of the WEC. Additionally, the wave energy acquisition efficiency of buoys at different positions is also greatly influenced by the platform geometry. The analysis of the aspect ratio of the damping layer of the platform is to determine whether the dam class (platform has a small aspect ratio) is beneficial to the wave energy absorption of the WEC or the ship class (platform has a large aspect ratio). The results show that when the damping layer has a small aspect ratio, the WEC has better wave energy acquisition efficiency. This may provide an idea for the future design of this type of WEC, which is to design the shape of the central platform as a dam. The effect of the damping layer area on the wave energy acquisition efficiency of the WEC is expected, since the energy collection principle of the WEC is based on the relative motion between the buoys and central platform. The larger platform has a smaller oscillation amplitude, which indirectly improves the relative motion between the buoys and central platform. Additionally, to properly increase the platform's draft is also to reduce the platform's oscillation amplitude, which is consistent with increasing the area of the damping layer.

The analysis of the height variation of the damping layer did have some interesting results. When the height of the damping layer changes, the wave energy capture ratio of the WEC remains nearly constant. However, there is a huge difference in the wave energy capture ratio of Buoy 1 and Buoy 2. Unfortunately, we have not found out the specific cause of this phenomenon, it can only be assumed that the change in the height of the damping layer results in the change of distribution of the wave field around the buoys. Additionally, the sea trial test of the full-sized device is planned to be carried out in 2020, which may help to validate the results of the numerical analysis.

Since the author's learning ability is limited, there is insufficiency in this article's profundity and breadth, which needs to be improved by the author's further study in the future. In addition, the purpose of this paper is to prove that the geometry of the central platform has great effect on hydrodynamic performance of the WEC instead of finding the optimal structure of the central platform in the full space of admissible parameters. Hence, the parameters of the central platform presented in the paper have limitations.

5. Conclusions

In summary, the hydrodynamic analysis of a semi-submerged multi-buoy WEC was carried out in the frequency and time domains. The effect of aspect ratio, area, and height of the damping layer and the draft of the central platform on the wave energy capture width ratio of the WEC were explored. The comparison between Buoy 1 and Buoy 2 was made to elucidate the difference in the wave energy capture performance of buoys on the same wave condition but at different positions. Based on the research above, the following conclusions can be drawn:

(1) There exists a relationship between the hydrodynamic performance of the WEC and the geometry of central platform.
(2) For a certain wave condition, there exists an optimal geometry of the central platform. At the wave condition mentioned in the paper, when the aspect ratio of the damping layer is 0.45, the area of the damping layer is 320 m^2 and the height of the damping layer is 3.5 m, the wave energy capture width ratio of the WEC is better, so that more wave energy can be extracted from the ocean.
(3) It is found that increasing the draft of the central platform is conducive to improving the wave energy capture width ratio of the WEC, and in this paper, the wave energy capture width ratio of the WEC is the largest when the draft of the central platform is 4.6 m.
(4) Further related research should be carried out in a physical prototype test and focused on fluid analysis of the central platform, to find out the influence of different shaped platforms on wave distribution.

Author Contributions: Conceptualization, S.Y.; methodology, S.Y.; software, Y.H.; validation, Y.H.; formal analysis, Y.H.; investigation, Y.H.; resources, S.Y., H.H. and H.C.; data curation, Y.H.; writing—original draft preparation, Y.H.; writing—review and editing, S.Y.; visualization, Y.H.; supervision, S.Y.; project administration, S.Y., H.H. and H.C.; funding acquisition, Y.H. All authors have read and agreed to the published version of the manuscript.

Funding: This research was funded by the National Natural Science Foundation of China (Grant No. 51779104), the Natural Science Foundation of Fujian Province, China (Grant No. 2016J01247), and the Foreign Cooperation Project of Department of Science and Technology of Fujian Provincial (Grant No. JA13170).

Acknowledgments: The authors would like to thank Shaohui Yang from Jimei University for his support on experiments and paper modification.

Conflicts of Interest: The authors declare no conflict of interest.

References

1. Wang, L.G.; Isberga, J. Review of control strategies for wave energy conversion systems and their validation, the wave-to-wire approach. *Renew. Sustain. Energy Rev.* **2018**, *81*, 366–379. [CrossRef]
2. Khana, N.; Kalaira, A.; Haiderc, N.A. Review of ocean tidal, wave and thermal energy technologies. *Renew. Sustain. Energy Rev.* **2017**, *72*, 590–604. [CrossRef]
3. Soukissian, T.H. Marine renewable energy in the Mediterranean Sea: Status and perspectives. *Energies* **2017**, *10*, 1512. [CrossRef]
4. Yang, S.H.; Wang, Y.Q.; He, H.Z.; Zhang, J.; Chen, H. Dynamic properties and energy conversion efficiency of a floating multi-body wave energy converter. *China Ocean Eng.* **2017**, *3*, 108–118. [CrossRef]
5. Kramer, M.; Marquis, H.; Frigaard, P. Performance evaluation of the Wavestar prototype. In Proceedings of the 9th European Wave and Tidal Conference (EWTEC), Southampton, UK, 5–9 September 2011.
6. Hansen, R.H.; Kramer, M.; Vida, E. Discrete displacement hydraulic power take-off system for the Wavestar WEC. *Energies* **2013**, *6*, 4001–4044. [CrossRef]
7. Ambühl, S.; Kramer, M.; Sørensen, J.D. Reliability-based structural optimization of WECs. *Energies* **2014**, *7*, 8178–8200. [CrossRef]
8. Yang, S.H.; He, H.Z.; Zhang, J.; Chen, H. Design and experimental study of raft array type wave energy generating system. *J. Mech. Eng.* **2016**, *11*, 57–62. [CrossRef]
9. Zhang, M.Y.; Yang, S.H.; He, H.Z.; Zhang, J.; Li., H. Modeling and simulation analysis of array type wavy energy generator. *Ocean Eng.* **2017**, *2*, 83–88.
10. Sheng, S.Y.; Zhang, Y.Q.; Wang, K.L.; Wang, Z.P.; Lin, H.J.; Ye, Y. Research on the wave energy generating device of Eagle1. *Ship Eng.* **2015**, *9*, 104–108.
11. Sheng, S.Y.; Zhang, Y.Q.; Wang, K.L.; Lin , H.J.; Wang, W.S.; Chen, A. Overview of the overall design of the Eagle device Wanshan. *Ship Eng.* **2015**, *37*, 10–14.
12. Wang, W.S.; Sheng, S.Y.; Ye, Y.; Zhang, Y. Frequency domain motion analysis and optimization design of eagle1 wave energy device. *Ocean Technol.* **2018**, *2*, 101–106.
13. Sheng, S.Y.; Zhang, Y.Q.; Wang, K.L.; Ye, Y.; You, Y. Research on generation system of eagle wave energy generator. *Renew. Energy Resour.* **2015**, *9*, 1422–1426.
14. Zhang, Y.Q.; Sheng, S.W.; You, Y.G.; Huang, Z.X.; Wang, W.S. Study of hydrodynamic characteristics of a sharp eagle WEC. *China Ocean Eng.* **2017**, *3*, 364–369. [CrossRef]
15. Sheng, S.Y.; Wang, K.L.; Lin, H.J.; Zhang, Y.; You, Y.; Wang, Z.; Chen, A.; Jiang, J.; Wang, W.; Ye, Y. Model research and open sea tests of 100 kW wave energy convertor sharp eagle Wanshan. *Renew. Energy* **2017**, *113*, 587–595. [CrossRef]
16. Hu, Y.; Yang, S.H.; He, H.Z.; Chen, H.; Zheng, S.; Zhang, D. Hydrodynamic performance analysis of semi-submersible multi-body wave power plant. *J. Ocean Eng.* **2019**, *38*, 1–13.
17. Zheng, S.M.; Zhang, Y.H.; Zhang, Y.L.; Sheng, W.A. Numerical study on the dynamics of a two-raft wave energy conversion device. *J. Fluids Struct.* **2015**, *58*, 271–290. [CrossRef]
18. Babarit, A.; Hals, J.; Muliawan, M.J.; Kurniawan, A.; Moan, T.; Krokstad, J. Numerical benchmarking study of a selection of wave energy converteors. *Renew. Energy* **2012**, *41*, 44–63. [CrossRef]
19. Babarit. A database of capture width ratio of WECs. *Renew. Energy* **2015**, *80*, 610–628. [CrossRef]
20. Jefferys, E.R. Device characterization. In *Power from Sea Waves*; Count, B., Ed.; Academic Press: London, UK, 1980; pp. 413–438.

Journal of
Marine Science and Engineering

MDPI

Article

Analysis of Piled Concrete Foundation for a 3-MW Class Offshore Wind Turbine along the Southwest Coast in Korea

Hyun-Gi Kim [1], Bum-Joon Kim [2] and Kwang-Ho Lee [3,*]

[1] Department of Energy Plant Engineering, Catholic Kwandong University, Gangnung 25601, Korea; khk2696@nate.com

[2] Department of Civil and Environmental System Engineering, Konkuk University, Seoul 05029, Korea; undertakerii@naver.com

[3] Department of Civil Engineering, Catholic Kwandong University, Gangnung 25601, Korea

* Correspondence: klee@cku.ac.kr; Tel.: +82-33-649-7637

Received: 2 March 2020; Accepted: 14 March 2020; Published: 20 March 2020

Abstract: Concrete foundations have received attention as offshore wind turbine support structures because of their various advantages. However, because of the lack of information on structural analysis and the design method of complex marine environmental loads, concrete foundations cannot be applied on actual sites. Therefore, the structure behavior mechanism and concrete reinforcement design need to be evaluated based on soil-structure interactions. Herein, an efficient method for analysis of piled concrete foundations (PCFs) is presented, and the stability of PCF structures is evaluated under environmental conditions of the coast in Korea for a 3-MW wind turbine. Three analytical parameters for PCF models were defined to consider soil-structure interaction. The results of each model were compared with the displacement, stresses, and natural frequencies. Using the analysis results, a prestressing reinforcement design for concrete foundations was proposed. Quasi-static analysis showed that maximum displacement was sufficiently small and the maximum stresses did not exceed the allowable stresses. PCF showed excellent dynamic performance and structural stability. In addition, stiffness of the soil spring model influenced the natural frequency rather than the stiffness of the pile type. Detailed analysis of the connections between piles and concrete need to be studied in the future.

Keywords: piled concrete foundation (PCF); 3-MW turbine model; quasi-static analysis; natural frequency analysis

---

1. Introduction

The amount of offshore wind energy has recently increased across the globe as one of the renewable energy sources. The Korean Government has also started to invest in the offshore wind turbine industry and will move forward with projects to develop offshore wind farms in Korea. In 2010, the Korea Government Ministry of Knowledge Economy presented the Offshore Wind Turbine Roadmap that included a 2.5-GW class offshore wind farm on the southwest coast of Korea with an investment of 10.2 trillion KRW [1]. A private sector-led project was started to build a large-scale 5-GW wind farm off the coast of southwestern Jeolla Province in Korea. A 1-GW class onshore wind farm and a 4-GW class offshore wind farm are planned to be constructed by 2023 with the aim of reducing carbon dioxide emissions by 5.4 million tons.

Meanwhile, many studies focusing on offshore wind turbines in Korea have been completed. Oh et al. [2,3] studied the feasibility of the offshore wind resources around the Korean Peninsula for a 100-MW class offshore wind farm and performed an assessment of the wind energy potential at a demonstration site in Korea to estimate the economic benefits and establish reliable design criteria. Kim et al. [4] carried out research on the selection of the optimal site for an offshore wind farm around the Korean Peninsula by comparing economic benefits of the development project to the entire cost. From this research, it was concluded that the construction of an offshore wind farm along the coast near Buan-gun and Yeonggwang-gun in Jeolla Province was preferred from an economic and site applicability standpoint. Oh et al. [5] preliminarily evaluated the offshore wind turbine monopile foundation dimensions based on the site conditions near the island Wi-do in the West Sea of Korea by considering aerodynamic loads and hydrodynamic loads. They presented foundation design methods and optimal dimensions of the monopile foundation for the selected site. For the feasibility study of the 5 MW offshore wind turbine substructures in Korea's southwest coast, Shi et al. [6] compared the dynamic responses of different substructures, including the monopile, jacket, and multipile, by performing an eigenvalue analysis and a coupled aero-hydro-servo-elastic simulation.

There are many types of support structures, such as the steel monopile, steel jacket, and the concrete gravity-based structure (GBS). The most used type is the steel monopile, which has a transition piece to aid installation. Recently, however, the concrete GBS has been preferred because of its cost efficiency for large-capacity wind turbines in the deep sea [7–9]. However, a new type of structure that consists of concrete and steel materials is being developed due to the limitation of GBS. In December 2012, BAM Energie [10] presented a cutting edge concrete gravity base foundation suitable for larger offshore wind turbines in deeper waters. It consists of a concrete caisson and steel shaft; the base is cast onshore and then installed offshore using standard vessels, thus eliminating the need for heavy lifting equipment. Kim [11] proposed a new hybrid type of cost-efficient support structure composed of a concrete tripod and steel shaft. To validate the new structure, a preliminary design concept was applied. Despite these efforts, that kind of structure was difficult to apply to the southwest coast in Korea where weak soil is too deep to sustain the GBS structure. To overcome the disadvantages of the existing type of offshore wind turbine support structure, Kim et al. [12] suggested a new hybrid concrete support structure. The suggested type was supported not only by a gravity-type foundation, but also by driven piles, and it was possible to reduce the weight of the structure. Stream function wave theory and the environmental conditions in the Southwest Coast in Korea were applied, and the hybrid support structure showed enough structural stability to be applied to real site conditions in Korea. Unlike steel fabricated of isotropic material, reinforced concrete is a composite material in which the relatively low tensile strength and ductility of the concrete are counteracted by the inclusion of reinforcement with a higher tensile strength or ductility. For design of the concrete structure, the ultimate member force is calculated through structural analysis.

In this study, an efficient method for the analysis of a piled concrete foundation (PCF) is presented, and the stability of the PCF structure is evaluated under the environmental conditions of the coast near the Jaeun island in Jeonnam-do of Korea. Three analytical models for PCF were defined to consider the interaction between soil and structure. The results of each model were compared with the displacement, stresses, and natural frequencies. Using the results of analysis, the prestressing reinforcement design for the concrete foundation was carried out. The displacements and rotation angles, end bearing capacity, and pulling capacity were examined for evaluation of the pile. A pile foundation installed in soft ground must be secured to support various loads of the upper structure, so that a desired power can be stably obtained through a turbine. Moreover, natural frequency analysis was performed to investigate the resonance occurrence in the PCF structure. The study aims to provide an efficient analysis method and prestressing reinforcement design method through analysis results of piled concrete foundations for offshore wind turbines and present design references for the future.

2. Theoretical Background

2.1. Wave Force

Airy wave theory is a relatively simple wave theory with a small amplitude, and the wave curve is a sine curve. The wave height is assumed to be smaller than the wave length in deep water and the water depth in shallow, and thus, it is called the small amplitude wave theory. In the small amplitude wave theory, the waves used in the calculation of velocity and acceleration are periodically repeated with the same wave height and wave length; thus, the equation of motion is linearized. If the x-direction is defined as the horizontal direction and z-direction as the vertical direction, surface wave elevation from the still water level and the water particle velocity can be expressed by Airy wave theory as follows:

$$\eta(x,t) = \frac{H}{2}\cos(kx - \omega t) \tag{1}$$

$$u(x,z,t) = \frac{\pi H}{T}\frac{\cosh[k(h+z)]}{\sinh kh}\cos(kx - \omega t) \tag{2}$$

$$w(x,z,t) = \frac{\pi H}{T}\frac{\sinh[k(h+z)]}{\sinh kh}\sin(kx - \omega t) \tag{3}$$

where x and z are Cartesian coordinates with $z = 0$ at the still water level (positive upwards), H is wave height, T is wave period, h is water depth, η is the free water surface, u and w are velocity components in the x, z directions, respectively, k is the wave number, ω is wave frequency defined by $2\pi/T$, and t is the time. The horizontal and vertical accelerations of the fluid particle, a_x and a_y, can be calculated by differentiating u and w over time t, and Equations (2) and (3) can be applied to calculate the acceleration as follows:

$$a_x(x,z,t) = \frac{2\pi^2 H}{T}\frac{\cosh[k(h+z)]}{\sinh kh}\sin(kx - \omega t) \tag{4}$$

$$a_y(x,z,t) = \frac{2\pi^2 H}{T}\frac{\sinh[k(h+z)]}{\sinh kh}\cos(kx - \omega t) \tag{5}$$

Wave force can be calculated by Morison equation as follows:

$$F_T = \frac{1}{2}\rho C_D D|u|u + \frac{\pi D^2}{4}\rho C_M a_x \tag{6}$$

where F_T is the total force, ρ is the density of water, C_D and C_M are the drag and inertia coefficients, respectively, and D is the diameter of the vertical cylinder. The first term in Equation (6) is the drag component and the second term is the inertia component.

2.2. Soil Spring Stiffness

To calculate the soil spring stiffness, a soil model called the Winker model can be used, and the spring stiffness in the horizontal and vertical directions can be calculated [13]. First, the horizontal soil spring stiffness at any depth is estimated as follows:

$$K_h = k_s \Delta_Z Z \tag{7}$$

where Z denotes any depth in soil and ΔZ denotes the spacing between the springs at a depth Z. k_s is the modulus of subgrade reaction at a depth Z, and this can be calculated from some methods, such as constant distribution, linear distribution, equation by Matlock and Reese, equation by CSN 73 1004, and an equation from Vesic [14]. The vertical soil spring stiffness can be classified into end bearing stiffness K_{vb} and skin friction resistance stiffness K_{vf}, as shown in Figure 1. The vertical end bearing stiffness K_{vb} can be calculated using Equation (8).

$$K_{vb} = k_s \frac{D}{2} L \left(\frac{D}{D^*} \right) \tag{8}$$

where L is the pile length, D is the diameter of pile, and D^* is the nominal diameter of the pile. The vertical skin friction resistance stiffness K_{vf} can be calculated using the equation given by Pender [15].

$$K_{vf} = 1.8 E_{s-tip} \zeta \lambda^{0.5-\frac{\lambda}{\zeta}} \alpha, \tag{9}$$

where ζ is the pile ratio (L/D) and λ is the pile-soil stiffness ratio. To get the soil spring stiffness at the bottom of the pile, the end bearing stiffness and skin friction resistance stiffness need to be combined, as presented in Figure 1.

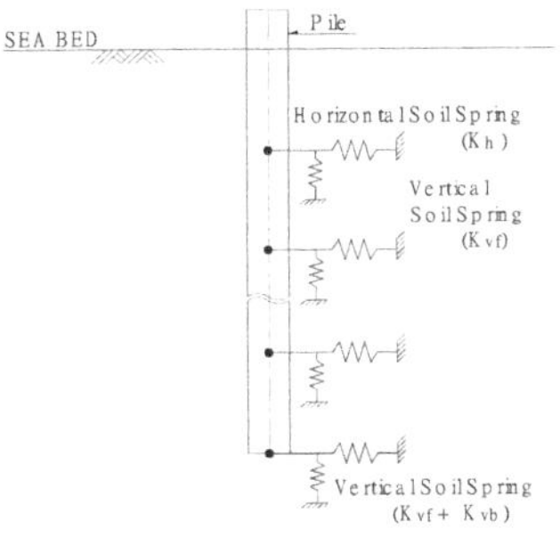

Figure 1. Model for soil springs.

3. 3-MW Class Piled Concrete Foundation Model

The PCF structure to support a 3-MW wind turbine was determined for the offshore conditions of the coast near Jaeun island in Jeollanam-do of Korea. The dimensions and material properties of the PCF structure are presented in Figure 2 and Table 1, respectively. The thickness of the steel shaft is 40 mm, including corrosion thickness, and the outer diameter of the shaft to be connected to the tower of the 3-MW wind turbine is 4.5 m. The diameter of the concrete foundation is 6.0 m, the outer diameter of the shaft is 4.5 m, and the thickness is 0.75 m. The diameter of the bottom of the concrete foundation is 11 m, excluding the concrete sleeve, which is 1.0-m thick. For the pile, which has a 2.0-m diameter, the concrete-sleeve has a 3.0-m diameter and 0.5-m thickness. To determine the heights of the shaft and concrete structure, the datum level (DL), mean higher high water level (MHHW), and maximum wave height are considered. MHHW is DL (+) 4.86 m and maximum wave height is DL (+) 12.64 m. Thus, the heights of the shaft and concrete structure are DL (+) 16.82 m and DL (+) 9.0 m, respectively, to install the steel platform above the maximum wave height of DL (+) 12.46 m. Construction cost per 1 MW turbine capacity of PCF was calculated based on the dimensions of the structure and construction site and was compared to the cost of the jacket structure calculated using the same site conditions. The cost of manufacturing and construction of 10 PCF structures are calculated as 4.47 and 23.89 billion KRW respectively, and construction cost per 1 PCF was 2.84 billion KRW. Therefore, construction costs per 1 MW of PCF structure is 0.945 billion KRW and this is indicated a cost difference of approximately 7% when comparing with the cost of jacket 1.013 billion KRW.

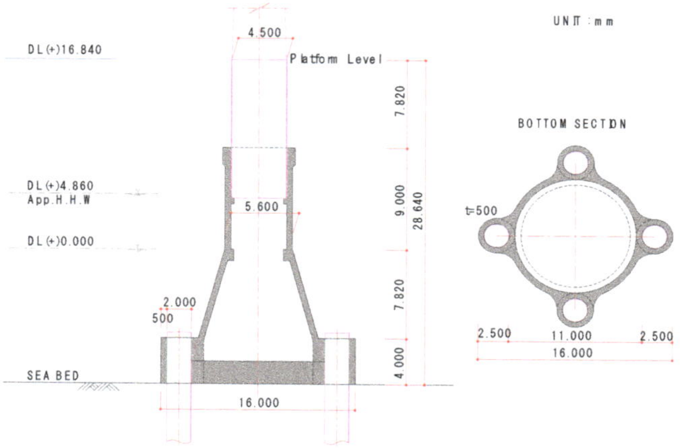

Figure 2. Dimensions of piled concrete foundation (PCF).

Table 1. Material properties.

Material	Properties
Steel (for shaft)	Specification and Grade: ASTM 572 Elastic Modulus: 200,000 MPa Unit Weight: 77.0 kN/m^3 Steel Strength: 345 MPa
Concrete	Design Strength: 45 MPa Design Strength for Grout part: 40 MPa ((More than 30 MPa)
PS Tendon	Ultimate Strength: fpu = 1900 MPa Yield Strength: fpy = 1600 MPa Nominal Cross-Sectional Area of Strand: Ap = 138:700 mm^2

4. Load Conditions and Load Combinations

4.1. Local Site Description

The coast near Jaeun island that was selected for analysis of the PCF structure is located in Jeolla Province of Korea, as shown in Figure 3. This area has sufficient wind conditions, and the range of wind velocity is from 7.0 to 7.5 m/s. Water depths are from 20 to 40 m, and the potentials for power generation by offshore wind turbines in this area are 4 GW and 6 GW within 20 m and 40 m water depths, respectively. The soil conditions are as shown in Table 2. The soil is sandy or clay soil. On account of the above conditions, Jeonnam-do announced a 5-GW wind farm project in 2009, and this project is currently progressing. For this project, the economic feasibility study of the selected site for the offshore wind farm was conducted from 2012 to 2014. From January 2013 to February 2014, wind condition data were measured by an offshore meteorological tower, and the mean wind velocity was found to be 7 m/s.

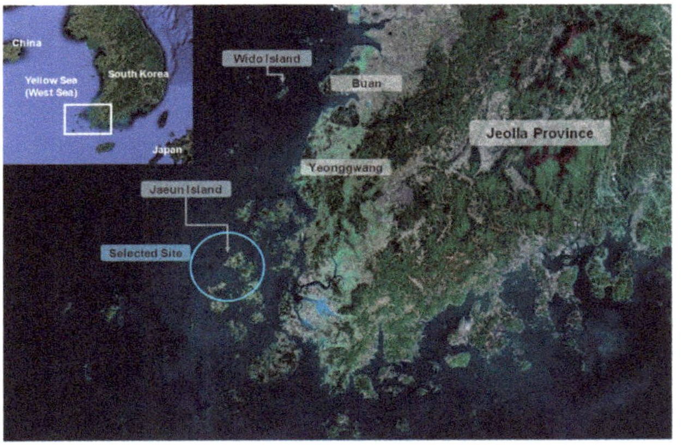

Figure 3. Selected site for the 5-GW wind farm project.

Table 2. Soil conditions.

Stratum	Depth (m)	γ_t (kN/m^3)	γ_{sub} (kN/m^3)	C (kN/m^3)	ϕ (deg)
Sand	0.0~14.0	17.7	10	-	20
Clay	14.0~16.0	16.7	9	10	-
Sand	16.0~25.0	17.7	10	-	30
Sand	25.0~34.0	17.7	10	-	35
Sand	34.0~37.0	17.7	10	-	25
Sand	37.0~44.0	17.7	10	-	30
Sand	44.0~58.0	17.7	10	-	30
Weathered soil	58.0~60.0	17.7	10	-	35
Weathered rock	60.0~64.5	18.7	11	-	35

4.2. Dead Load and Turbine Load

The self-weight, including the steel shaft, platform, and concrete support structure, affects the whole structure as a dead load. This dead load can be determined from the properties of the construction material. The weight of the steel shaft can be calculated by adding 5% of the self-weight to consider the connecting reinforcement in the shaft. We applied 32.634 kN of the self-weight to take account for the platform. For the turbine load, the 3-MW wind turbine of Doosan heavy industry (WinDS3000) was considered; the specifications are shown in Table 3 [16]. Tower base loads were calculated based on the data from the Jeju island offshore wind farm which has a tower height of 62.4 m and a wind speed (50 year extreme, 10 min) of 46.9 m/s.

Table 3. Specifications of WinDS3000 wind turbine.

Rating	3 MW	Rated Wind Speed	13 m/s
Rotor orientation	Upwind	Cut-out wind speed	25 m/s
Blade	3 blades	Nacelle mass	130,000 kg
Rotor diameter	91.3 m	Hub mass	30,000 kg
Hub height	80 m	Blade mass	100,000 kg
Cut-in wind speed	4 m/s	Tower mass	157,000 kg

4.3. Environmental Load

As mentioned above, the targeted site for analysis of the PCF structure is along the coast near Jaeundo in Jeollanam-do of Korea. The environmental load was calculated based on the marine conditions at a water depth of 11.80 m, maximum wave height of 12.64 m, wave period of 13.67 s, current velocity of 2.52 m/s by tide, current velocity of 6.0 m/s by wind, and maximum wind speed of 70.0 m/s. To analyze the PCF under environmental loads, structural analysis software MIDAS Civil [17] was used. MIDAS Civil is the general-purpose structural analysis software that has been used in worldwide and integrated solution system for civil engineering including static analysis, dynamic analysis, etc. However, the modules to calculate wave force and current force are not included in the MIDAS software. Thus, wave force and current force were calculated from the equations suggested by Det Norske Veritas [18], and these forces were applied to the structure as distributed loads in the MIDAS software.

Depending on the location such as deep water or shallow water, an adequate wave theory was selected for the computation of wave kinematics such as velocity and acceleration. Using the graph for finding validity of the wave theories proposed by Le Méhauté [19], a suitable wave theory can be selected the based on the relationship between dimensionless relative depth $h/(gT^2)$ and dimensionless wave steepness $H/(gT^2)$. From this graph, the wave condition in the targeted site was in the area above the shallow water breaking limit H/h=0.78 with $h/(gT^2)$ =0.00644 and $H/(gT^2)$ = 0.0069. Thus, the slamming force should be considered, and calculation of slamming force is classified to three cases according to sea level. In these three cases, the case where the wave force was the largest was determined. In the case of the highest astronomical tide (HAT), Morison's equation with Airy theory in Equation (6) is required. In the case of mean sea level (MSL) and lowest astronomical tide (LAT), the equation of slamming force with Airy theory in Equation (10) is recommended.

$$\begin{cases} F_s = \frac{1}{2}\rho C_S D u^2 \\ C_s = 5.15\left(\frac{D}{D+19s} + \frac{0.107s}{D}\right) \end{cases} \tag{10}$$

From the selected site conditions, HAT was 17.66 m (11.8 m + 4.86 m + 1.0 m), MSL was 14.73 m (11.8 m + 2.43 m + 0.5 m), and LAT was 11.80 m. The largest wave force was calculated in the case of LAT; 1.0 was applied for the drag coefficient, and 2.0 was applied for the inertia coefficient to calculate wave forces based on the recommendation from DNV-RP-C205 [18]. Similarly, Morison equation was used to calculate the current force, and current velocity was applied by using Equation (11).

$$v_z(z) = v_{tide}\left(\frac{h+z}{h}\right)^{\frac{1}{7}} + v_{wind}\left(\frac{h_0+z}{h_0}\right) \tag{11}$$

where v_{tide} and v_{wind} are current velocity by tide and wind, respectively. In this study, v_{tide} =2.52 m/s, v_{wind} =6.0 m/s and 50 m were used.

Total wave force F_{total} along the height calculated by adding all of the forces (drag F_d, inertia F_i, slamming $F_{slamming}$, and current forces $F_{current}$) are plotted in Figure 4. Maximum wave force occurred when the wave period was 7T/8 period (11.916 s). Here, 0 m is the highest position where wave and current is applied in the vertical axis.

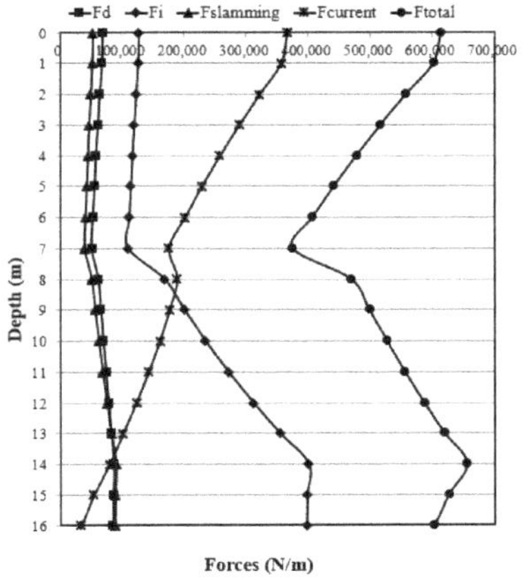

Figure 4. Distributions of wave and current forces.

4.4. Load Combinations

On the offshore wind turbine structure, environmental loads such as the wave load, current load, wind load, and turbine load due to the turbine located at the top of the tower were applied, and these loads are appropriately combined by the service limit state to perform the structural analysis. In the case of prestressing tendon, it is necessary to design using serviceability limit state(SLS), and the material factor γ_m for the SLS is taken as 1.0. In this analysis, a total of 12 load cases were applied and defined by the design conditions and the weather conditions according to IEC 61400-1 International Design Standards [20], as presented in Table 4. DLC6.1ab, DLC6.2da, DLC6.2ka, DLC6.2fb, DLC6.2kb, and DLC6.2db are parked and idling situations; DLC2.2cc and DLC2.3bc are power generation and fault situations; DLC1.4af and DLC1.3a are power production situations; DLC3a is parked and fault conditions; and DLC8.1al is transport, assembly, maintenance, and repair.

Table 4. Design load cases.

Design Situation	DLC	Safety Factor
Power production	1.3a	1.35
	1.4af	1.35
Power production plus loss of electrical grid connection	2.2cc	1.35
	2.3bc	1.10
	1.4af	1.10
Parked (standing still or idling)	6.1ab	1.35
	6.2da	1.10
	6.2ka	1.10
	6.2db	1.10
	6.2fb	1.10
	6.2kb	1.10
	3a	1.00
Transport, assembly, maintenance, and repair	8.1al	1.505

5. Structural Analysis and Design

5.1. Structural Modeling

For the structural analysis of the PCF structure, modeling was implemented by the structural analysis software MIDAS Civil, as shown in Figure 5. The steel shaft and concrete substructure were modeled using frame elements, and the pile and concrete sleeve were also modeled as frame elements to examine the behavior of the structure due to the effects of the pile. Modeling is performed using the frame element, therefore it is necessary to connect the concrete foundation element and concrete sleeve element, and an effective connection method needs to be considered.

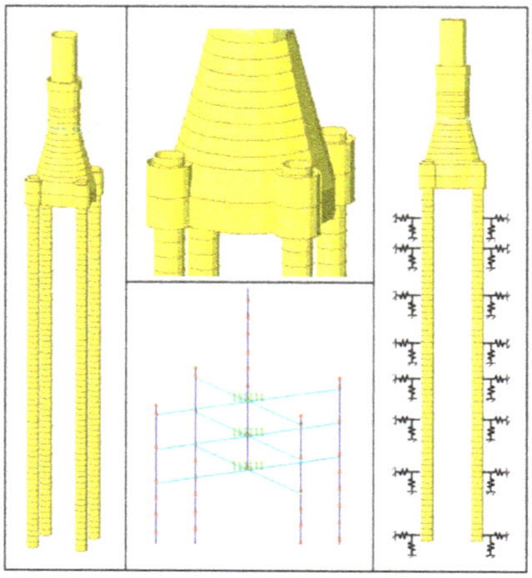

Figure 5. Modeling of PCF.

Assuming that the concrete sleeves and piles are sufficiently composited with a length of 4 m, the rigidity of the concrete foundation itself is high. Therefore, as shown in Figure 5, the concrete foundation element and the concrete sleeve element are connected using a rigid link. To apply the behavior of the pile according to the soil conditions, the soil is modeled as an elastic foundation, and the horizontal and vertical elastic springs are applied to the pile. For the pile, a reverse circulation drill (RCD) pile with a diameter of 2 m, which is used for marine construction, is applied. The soil where the PCF structure is located is composed of sandy soil and weathered soil up to a depth of 60 m, as shown in Table 2, therefore piles were penetrated to the weathered soil layer to achieve the end bearing capacity of the pile. Thus, the total length of the pile was 60 m, and the diameter and thickness were 1.8 m and 0.05 m, respectively. A 1-m deep pile was modeled as a frame element, and a total of 60 frame elements were used to model the pile, with horizontal and vertical elastic springs applied to the center of each pile. The spring stiffnesses were calculated according to the soil conditions in Section 2.2 and Equations (7)–(9), and the vertical and spring stiffness for each layer are shown in Table 5.

Table 5. Spring stiffness.

Layer	Horizontal Spring Stiffness (N/m)	Vertical Spring Stiffness (N/m)
1st	2.251×10^{9}	1.398×10^{7}
2nd	1.847×10^{9}	6.465×10^{6}
3rd	5.756×10^{9}	2.492×10^{7}
4th	9.557×10^{9}	3.389×10^{7}
5th	1.406×10^{10}	3.688×10^{7}
6th	1.783×10^{10}	4.385×10^{7}
7th	3.098×10^{10}	5.781×10^{7}
8th	2.976×10^{10}	1.296×10^{8}

In this study, three analytical parameters (i.e., boundary conditions, type of pile, and location of stress concentration) are considered for structural models, and each model is defined as shown in Table 6.

Table 6. Definition of structural models.

Analytical Parameters	Model Name	Model
Boundary conditions	BC-FB	Fixed foundation
	BC-PL	Piled foundation
Type of pile	PL-RCD	RCD piled foundation
	PL-ST	Steel piled foundation
Positions of stress concentration	SS-CF	Concrete foundation
	SS-PS	Concrete sleeve

5.2. *Quasi-Static Analysis*

The support structure is subjected to horizontal loads due to turbine and environmental loads, therefore the maximum displacement occurs in the horizontal direction. To evaluate the influence of the use of piles on the PCF structure, the horizontal displacement of the piled foundation model (BC-PL model) was compared to that of the model using the fixed boundary condition instead of the pile (BC-FB model). The maximum horizontal displacements of the two models according to the load case are shown in Figure 6, and the deformed shapes are shown in Figure 7.

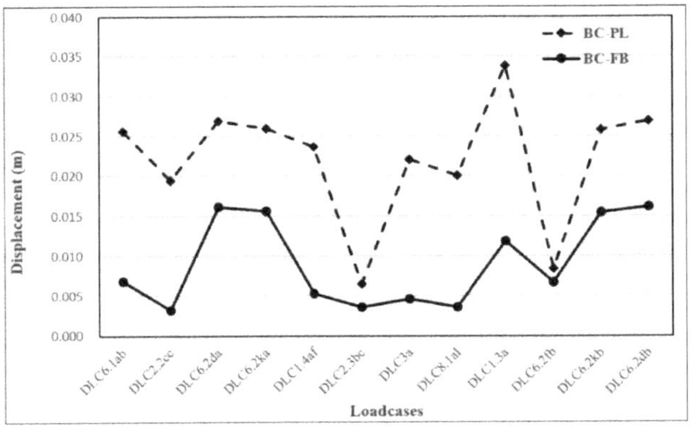

Figure 6. Comparison of displacements of BC-PL model and BC-PB model.

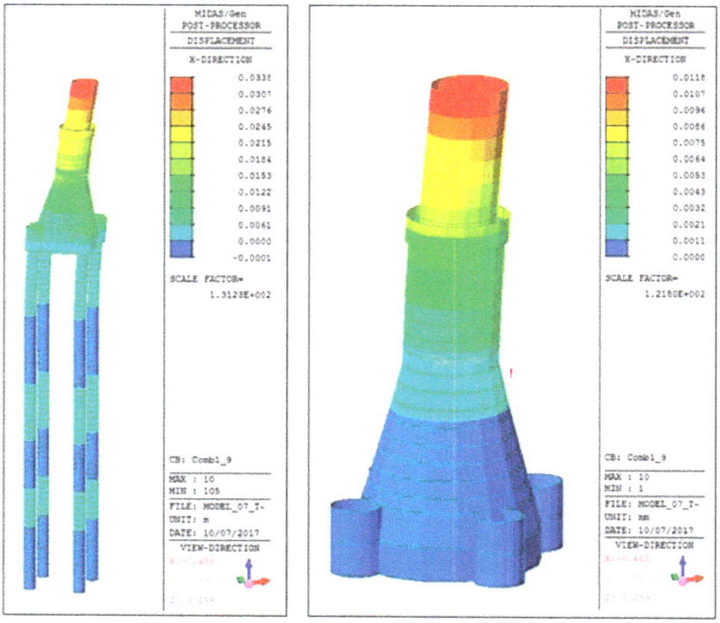

(**a**) Piled model (**b**) Fixed model.

Figure 7. Maximum displacement of structural models.

The maximum displacement occurred at the top of the support structure in both models. The maximum displacements were 0.0167 m in the case of the BC-FB model at DLC1.3a, and 0.0347 m in the case of the BC-PL model at DLC1.3a. Comparing the displacements of the two models, it can be seen that the displacement of the BC-FB model is about 46.40% of the displacement of the BC-PL model. As shown in Figure 7, the fixed model does not cause the displacement of the bottom of structure due to the fixed boundary condition.

However, because the BC-PL model has the displacement at the bottom of the pile and structure due to the use of the soil spring, two maximum displacements can be different by about 46.40%. Therefore, it is an accurate method to implement the analysis of the support structure by modeling the pile because the displacement difference occurs depending on the use of the pile when evaluating the behavior of the pile support structure. Notwithstanding the fact that structural analysis is performed using fixed boundary conditions due to the uncertainty of the boundary condition between soil and foundation, this result shows that the piled foundation model is required for the analysis of the global behavior of the structure. The maximum displacement of 0.0625 m in the support structure is 0.22% of the total length of the support structure of 28.640 m, and the maximum displacements of the 40-m long monopile and tripod structure in the study of Chen et al. [21] are 0.4864 m and 0.3687 m, respectively. Thus, it can be concluded that the maximum displacement occurring in the PCF structure is sufficiently small to be safe for displacement.

In order to evaluate whether the PCF structure is safe under the turbine load and environmental load, the generated stress and the displacement were evaluated. The stresses that occurred in the shaft, which is steel, are as shown in Figure 8, and the stresses of the BC-FB model and the BC-PL model were compared. The stresses in the two models were almost the same, and the maximum compressive stress and the maximum tensile stress were −47.900 MPa and 40.900 MPa in DLC6.2da, respectively. Stress distribution on the shaft is shown in Figure 9, and maximum stress was found at the center of the shaft. The allowable stress of the steel can be used as the allowable stress of SM400 steel of 185 MPa as proposed by the Harbor and Fishery design criteria [22], and the material factor γ m for tubular structures is 1.1 according to DNV-OS-J101. The steel shaft is safe because the maximum stress of −47.900 MPa does not exceed the allowable stress.

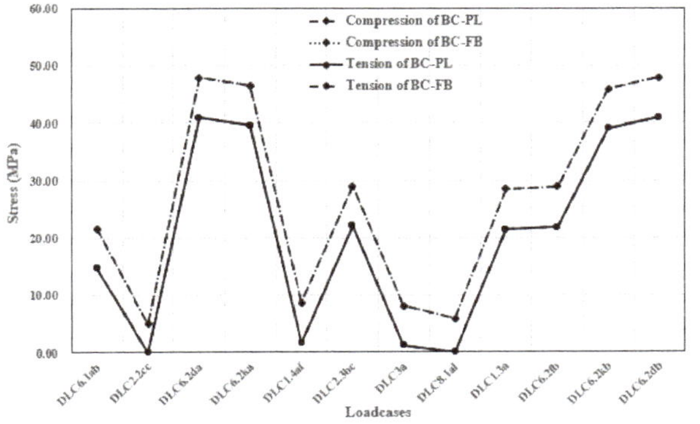

Figure 8. Comparison of stresses in shaft of piled model (BC-PL) and fixed model (BC-FB).

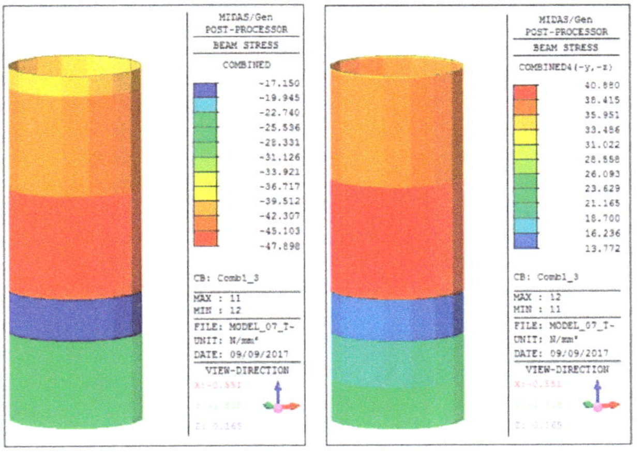

(a) Compression (b) Tension.

Figure 9. Stress distributions of steel shaft on DLC6.2da3.

To evaluate the stresses occurring in the concrete support structure, they were compared to the allowable compressive and tensile stresses presented in the structural concrete design code [23]. The allowable compressive and tensile stresses of concrete are $0.60f_{ck}$ and $0.60\sqrt{f_{ck}}$, respectively, and those stresses were calculated to be 27 MPa and 4.226 MPa, respectively, when the applied concrete design strength of 45 MPa was used.

As in the case of displacement, the generated stresses for the pile model are compared to those of the model using fixed boundary conditions, and the maximum stresses according to each load case are shown in Figure 10. The stresses in the two models show that the same stresses occur in each load case, except for the tensile stress of DLC2.2cc and DLC6.2fb. The maximum tensile stress of DLC2.2cc occurred in the concrete sleeve part, and it was found that the maximum tensile stress occurred differently depending on whether the pile was used. Therefore, it is necessary to model and analyze the structure with pile modeling for the stress check. Since the maximum tensile stress occurs in some load cases, further study about the concrete-sleeve is needed through a detailed analysis. The maximum compressive and tensile stresses in the PCF structure occurred in the case of DLC6.2da, and the values were −7.910 MPa and 6.340 MPa, respectively. The stress distribution for this load case is shown in Figure 11. It can be seen that the maximum stress occurs at the part where the concrete cone and the cylinder shape are connected (tapered section). Except for DLC2.2cc, the maximum stress occurred at the same part, and for DLC2.2cc, the maximum tensile stress occurred at the concrete sleeve.

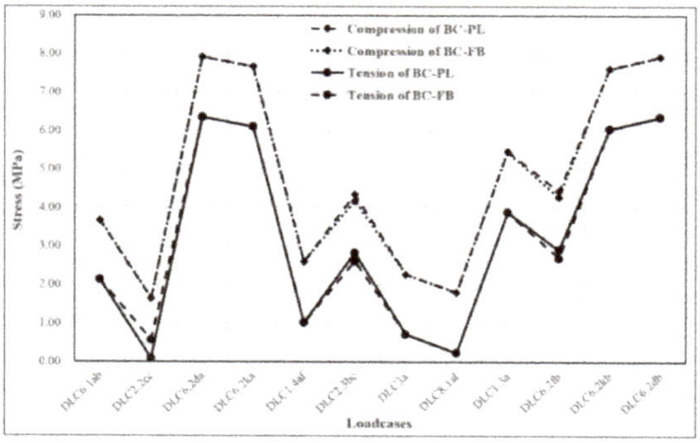

Figure 10. Comparison of stresses in concrete of BC-PL model and BC-FB model.

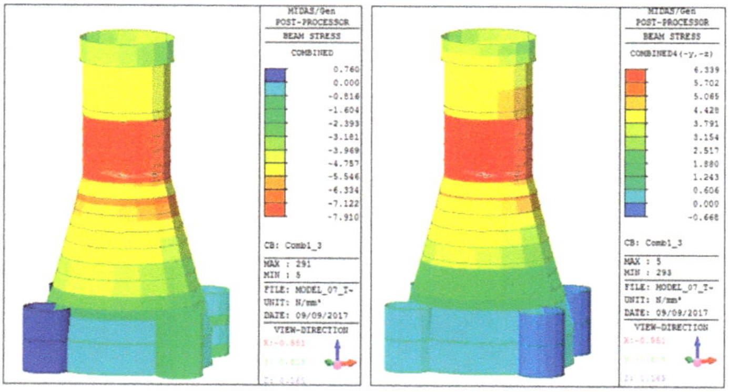

(a) Compression (b) Tension.

Figure 11. Stress distributions on DLC6.2da.

Figure 12 shows the stresses of the concrete sleeves for all load cases. When the stress is compared to the stress at the tapered section, it can be seen that the compressive stress is a maximum of 61.59%, and the tensile stress is about 90.22% when DLC2.2cc is excluded. This means that the bending moment generated by the load is supported by the shape and strength of the concrete support structure itself, but the pile penetrated into the soil has a huge influence on supporting the load. Thus, it is necessary to consider the stresses occurring in the concrete sleeve, as well as the concrete support structure for the stress evaluation, and additional reinforcement is necessary when the occurred stress exceeds the allowable stress.

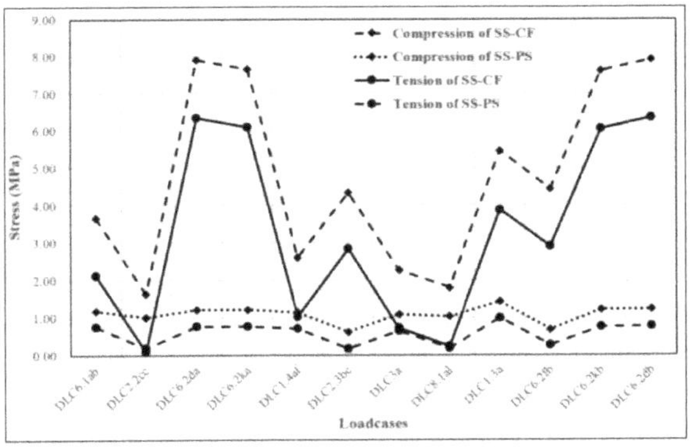

Figure 12. Comparison of stresses in concrete of SS-CF model and SS-PS model.

First, the maximum compressive and tensile stresses in the concrete sleeves were −1.43 MPa and 0.991 MPa, respectively. Compared to the allowable compressive stress and the tensile stress, it can be seen that the stresses did not exceed the allowable values. For the stresses of the entire concrete structure, the maximum compressive stress does not exceed the allowable compression, but in the case of tensile stress, it is 6.340 MPa, exceeding 4.226 MPa. This means that the safety of the support structure is not ensured due to damage, such as fracture in the case of tension, while it is safe in compression due to the applied loads. Therefore, it is necessary to reinforce the concrete foundation. The support structure is examined in the following section by applying prestressing steel.

5.3. Post-Tensioning Design

From the results of stress, it was found that prestressing steel is required to reduce tensile stress to less than the allowable stress. Prestressing steel was chosen from the Korean Industrial Standards for prestressing wire, and pre-stressing strand, KS D 7002 [24], and SWPC7B of 19 prestressing strand, which has a 0.01524 m diameter and low relaxation were used. The ultimate strength (f_{pu}) and yield strength of steel are 1900 MPa and 1600 MPa, respectively, and the nominal cross-sectional area (A_p) is 138.700 mm^2. The number and arrangement of tendons have to be determined by considering the size of the outer and inner diameter of the top of the concrete structure. The outer and inner diameters are 5600 mm and 4600 mm each, therefore the area of that part is 8.01×10^6 mm^2 and the number of tendons is 30. The tendon arrangement is presented in Figure 13.

To apply the effect of prestressing steel to concrete, prestressing force and effective stress have to be calculated based on the prestressing strand and the number of tendons. For prestressing force, 72% of the ultimate strength, 1368 MPa, was used. From this strength, the prestressing force of one strand can be calculated by multiplying 1368 MPa by the nominal cross-sectional area 138.700 mm^2, and the prestressing force of one tendon is the sum of the prestressing force of 19 strands. For the effective stress, 20% loss of prestressing was applied. Calculation procedure and values of the prestressing force are presented in Table 7. The calculated effective stress by prestressing steel was 9.36 MPa, as presented in Table 7. With this value, total stresses on the structure after applying the prestressing steel could be calculated, and these stresses were evaluated by comparing it with the allowable stresses. From the quasi-static analysis, maximum tensile and compressive stress were 9.026 MPa and −7.55 MPa, respectively. Thus, the total stresses could be calculated, as presented in Figure 14, and it is noted that the maximum tensile stress was changed to compressive stress because an effective stress of −9.36 MPa was bigger than the occurred stress of 9.02 MPa. Consequently, it is determined that the total stresses on the structure are safe enough by applying prestressing steel.

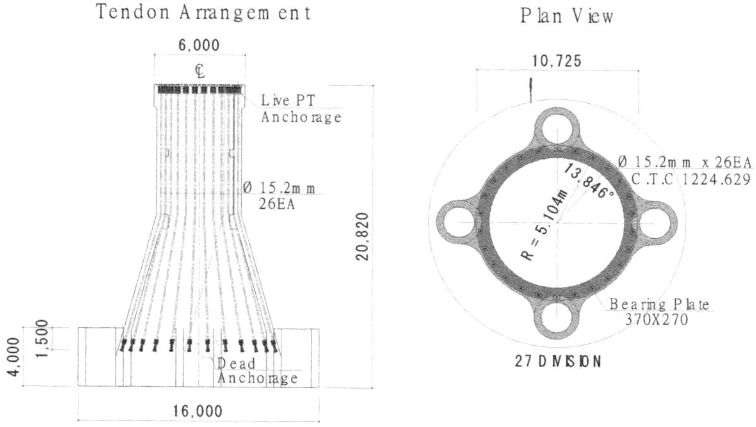

Figure 13. Prestressed tendon arrangement.

Table 7. Prestressing force.

Forces and Stresses	Equations and Values
1 strand force	1368 MPa × 138.700 mm^2 = 189,742 N
1 tendon force	189,742 N × 19 strands = 3,605,090 N
30 tendon force	3,605,090 N × 30 tendons = 108,152,940 N
Stress	108,152,940 N/8.01e+06 mm^2 = 11.702 MPa
Effective stress	11.702 MPa × 0.8 = 9.36 MPa

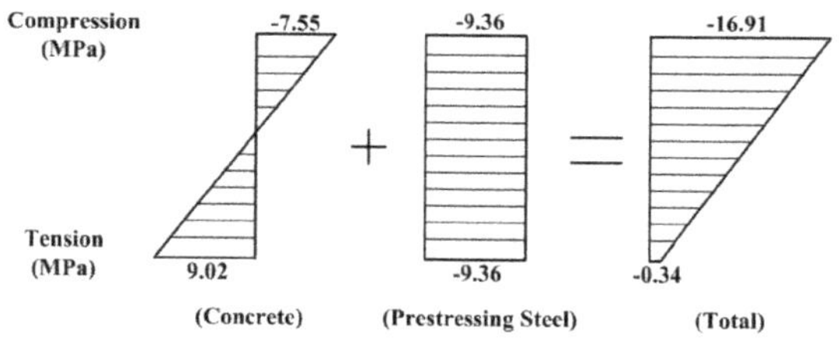

Figure 14. Total stresses on structure.

5.4. Pile Evaluation

Due to the turbine and the environmental loads acting on the support structure, the deformation, compressive stress, and tensile stress occurring in the piles are applied, so it is important to evaluate the stability and bearing capacity of the pile accordingly. Thus, the displacements and rotation angles at the top of the piles and the end bearing and pulling capacities of the piles were examined for the stability evaluation of the piles. According to Upwind final report WP 4.2 [25] and DNV-OS-J101 [26], the pile displacement at the seabed should be less than 0.1 m and the rotation angle should be less than 0.5°. The maximum displacements and rotation angles according to the load cases are shown in Table 8. The maximum displacement and rotation angle occurred at 0.0075 m and 0.0371°, respectively, at DLC1.3a. These results are less than 0.1 m and 0.5°; thus, the pile is safe for displacement and rotation.

Table 8. Maximum deflections, rotations, allowable capacities, and loads of the piles.

Load Cases	Max. Deflection (m)	Max. Rotation (deg)	Allowable Bearing Capacity (kN)	Allowable Pullout Capacity (kN)
			830,516	818,653
			Compressive Force	Tensile Force
DLC6.1ab	0.0066	0.0313	5337	11,641
DLC2.2cc	0.0058	0.0266	3254	9235
DLC6.2da	0.0056	0.0261	4537	10,545
DLC6.2ka	0.0061	0.0287	4249	10,261
DLC1.4af	0.0064	0.0305	4767	11,200
DLC2.3bc	0.0038	0.0152	-	4831
DLCe3a	0.0061	0.0288	4043	9849
DLC8.1al	0.0058	0.0270	3229	9897
DLC1.3a	0.0075	0.0371	8052	14,490
DLC6.2fb	0.0037	0.0145	-	5340
DLC6.2kb	0.0061	0.0286	4235	10,219
DLC6.2db	0.0057	0.0266	4541	10,551

To evaluate the end bearing and pulling capacities of the piles, they are compared to the allowable bearing capacities according to the soil conditions. The allowable bearing and pullout capacities can be calculated by dividing the ultimate bearing capacity by the safety factor. The safety factor is 1.5 according to the API design standard [27]. The compressive and tensile forces at the top of the pile according to the load cases are shown in Table 8, and the maximum values were 8052 kN and 14,490 kN for DLC1.3a, respectively. The end bearing and pulling capacities were calculated to be 1,229,715 kN and 1,227,980 kN, respectively; the allowable end bearing capacity is 819,810 kN, and the allowable pulling capacity is 818,653 kN. When comparing these allowable values to compressive and tensile forces at the top of the pile, it can be seen that the allowable values are larger than the applied loads. Therefore, it is concluded that the piles have sufficient end bearing and pulling capacities for the applied load and soil conditions.

5.5. Natural Frequency Analysis

To evaluate the dynamic response and stability of the PCF structure, natural frequency analysis was carried out. To account for the entire model of the structure, the tower was modeled as shown in Figure 15, and the turbine above the tower was applied as an additional mass. The properties of tower and turbine are presented in Table 3. The pile for the PCF structure can be considered as a steel pile or RCD pile, and the results of the three models (i.e., PL-ST, PL-RCD and BC-FB models) are compared to evaluate the effects of pile type and boundary conditions. The evaluation of the natural frequency was carried out using the natural frequency design range of the support structure, which has the 3-MW class turbine in Figure 16. The second and third modes rarely occur in a real structure, therefore the evaluation of first natural frequency is most important.

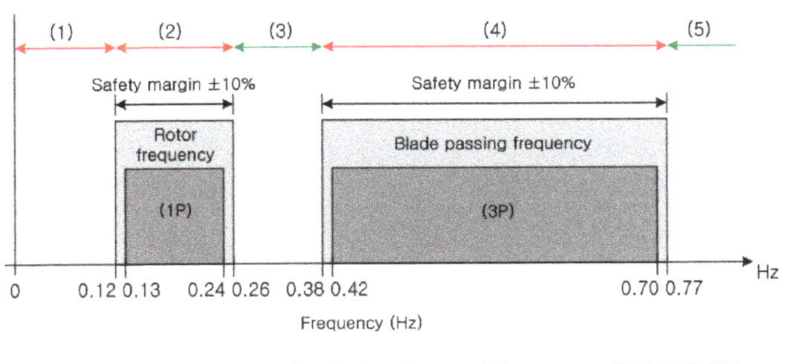

Figure 15. Design ranges for the fundamental frequency of WinDS3000.

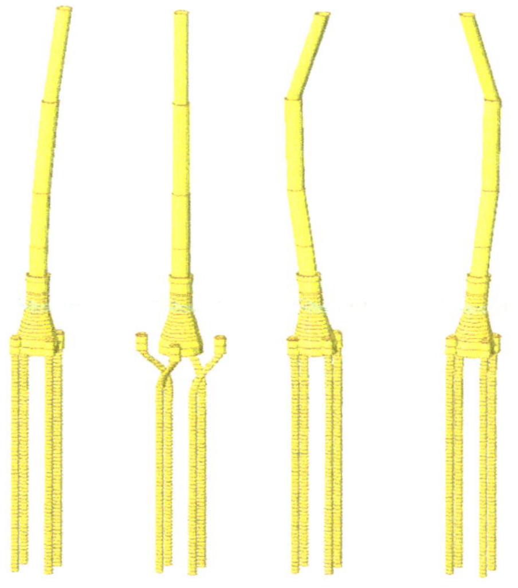

<1st mode> <2nd mode> <3rd mode> <4th mode>

Figure 16. Mode shapes of PCF.

A first natural frequency of the support structure must not be in rotor frequency range (1P) or in blade passing frequency ranges (3P) to avoid the resonance because the resonance induces large displacements and large stresses in the structure. Generally, in the case of the jacket and tripod support structures, the first mode natural frequency is designed to be located in the region between 1P and 3P. The mode shape according to frequency analysis is shown in Figure 15, and the frequencies and periods according to the models are shown in Table 9. The mode shapes of the PL-ST, PL-RCD, and BC-FB models were the same, and the first mode frequencies were 0.3519 Hz, 0.3561 Hz, and 0.3814 Hz, respectively. Two frequency values of the PL-ST model and PL-RCD model show a difference of about 1.1%, and they are all located between 1P and 3P, which is the region between the frequencies of the rotor and the blades. However, a frequency value of 0.3814 for the BC-FB model is in 3P region and resonance can occur. Therefore, when steel or RCD piles are used, it can be concluded that the PCF structure has similar and safe vibration characteristics to the existing support structure of the jacket or tripod. The RCD pile filled with concrete is more massive than the steel pile, therefore the frequencies between the two models are expected to show a large difference. However, the difference between frequencies was not significant. It is confirmed that the factor influencing the natural frequency is the pile model or the fixed model is affected by the pile type considering the pile stiffness.

Table 9. Results of natural frequency analysis.

Mode	PL-ST		PL-RCD		BC-FB	
	Frequency (Hz)	Period (sec)	Frequency (Hz)	Period (s)	Frequency (Hz)	Period (s)
1st	0.3519	2.8415	0.3561	2.8080	0.3814	2.6221
2nd	0.3519	2.8415	0.3561	2.8080	0.3814	2.6221
3rd	1.1818	0.8462	1.1838	0.8447	1.2079	0.8279
4th	1.5917	0.6282	1.6329	0.6124	1.9814	0.5047

6. Conclusions

In this study, structural analysis using general-purpose structural analysis software was performed to evaluate the adequacy of behavior mechanism for the 3-MW class piled concrete foundation. Three analytical parameters were defined to consider the interaction between the ground and the structure. Analysis was carried out on a fixed boundary application model that eliminated the uncertainty of the pile and concrete connections, and two models with different pile stiffness were proposed. By analyzing the behavior of the entire supporting structure and evaluating the stability of displacement, stress, and natural frequency, dominant mechanisms of the behavior were identified, and the impact of the influence factors was summarized. This study aims to provide well-documented results to help expand our understanding of piled concrete foundations and facilitate the application of this innovative foundation system. The following conclusions are drawn according to the structural analysis.

(1) From the quasi-static analysis, maximum stresses and displacements of a fixed foundation and piled foundation were evaluated. Stress results showed that the differences in the maximum stresses between the fixed foundation and piled foundation were as not large, but the maximum stress locations were different.

(2) As a result of structural analysis using three parameters, the fixed end boundary condition for the design of concrete member was found to be unreasonable because the overall stress and displacement was about 46.40% larger than the model using pile and soil spring.

(3) In the analytical model considering the soil-structure interaction, the stiffness of the soil spring model according to the type of soil rather than the pile type influenced the natural frequency. The boundary conditions of the pile and concrete connections must be considered when evaluating the natural frequencies, which is the first mode of frequency considered in the design.

(4) Tensile stress of the concrete structure was designed to be sufficiently reinforced by prestressing tendon.

(5) Further details on the construction method of PCF should be obtained by performing detailed studies on the connection between pile and concrete.

Author Contributions: Conceptualization, H.-G.K. and K.-H.L. Methodology, H.-G.K. and B.-J.K. Data curation, H.-G.K. and B.-J.K. Writing—original draft preparation, H.-G.K. and K.-H.L. Writing—review and editing, K.-H.L. Supervision, H.-G.K. and K.-H.L. All authors have read and agreed to the published version of the manuscript.

Funding: This research was funded by the research support project of Catholic Kwandong University.

Conflicts of Interest: The authors declare no conflict of interest.

References

1. Korea Government Ministry of Knowledge Economy (MKE). Offshore Wind Turbine Roadmap 2010. Available online: http://www.motie.go.kr (accessed on 10 January 2020).
2. Oh, K.Y.; Kim, J.Y.; Lee, J.S.; Ryu, K.W. Wind Resource Assessment around Korean Peninsula for Feasibility Study on 100 MW class Offshore Wind Farm. *Renew. Energy* **2012**, *42*, 217–226. [CrossRef]
3. Oh, K.Y.; Kim, J.Y.; Lee, J.K.; Ryu, M.S.; Lee, J.S. An Assessment of Wind Energy Potential at the Demonstration Offshore Wind Farm in Korea. *Energy* **2012**, *46*, 555–563. [CrossRef]
4. Kim, J.Y.; Oh, K.Y.; Kang, K.S.; Lee, J.S. Site Selection of Offshore Wind Farms around the Korean Peninsula through Economic Evaluation. *Renew. Energy* **2013**, *54*, 189–195. [CrossRef]
5. Oh, K.Y.; Kim, J.Y.; Lee, J.S. Preliminary Evaluation of Monopile Foundation Dimensions for an Offshore Wind Turbine by Analyzing Hydrodynamic Load in the Frequency Domain. *Renew. Energy* **2013**, *54*, 211–218. [CrossRef]
6. Shi, W.; Han, J.; Kim, C.; Lee, D.; Shin, H.; Park, H. Feasibility Study of Offshore Wind Turbine Substructures for Southwest Offshore Wind Farm Project in Korea. *Renew. Energy* **2015**, *74*, 406–413. [CrossRef]
7. Peire, K.; Nonneman, H.; Bosschem, E. Gravity Base Foundations for the Thornton Bank Offshore Wind Farm. *Terra Aqua* **2009**, *115*, 19–29.
8. Esteban, M.D.; López-Gutiérrez, J.S.; Negro, V. Gravity-Based Foundations in the Offshore Wind Sector. *J. Mar. Sci. Eng.* **2019**, *7*, 64. [CrossRef]
9. Thomsen, J.H.; Forsberg, T.; Bittner, R. Offshore Wind Turbine Foundations-the COWI Experience. In Proceedings of the 26th International Conference on Offshore Mechanics and Arctic Engineering, San Diego, CA, USA, 10–15 June 2007; p. 29567.

10. Energy, Oil & Gas Magazine. Integrated Solutions: BAM Energie. 2013. Issue 100. Available online: http://www.energy-oil-gas.com (accessed on 10 January 2020).
11. Kim, H.G. A Study on the Static and Dynamic Behaviors for Hybrid Support Structure of Offshore Wind Turbines. Ph.D. Thesis, Konkuk University, Seoul, Korea, 2013.
12. Kim, H.G.; Kim, B.J.; Kim, K.D. New Development of Hybrid Concrete Support Structure with Driven Piles for Offshore Wind Turbines. *J. Korean Soc. Steel Const.* **2013**, *25*, 307–320. [CrossRef]
13. Fleming, K.; Weltman, A.; Randolph, M.; Elson, K. *Piling Engineering*, 3rd ed.; CRC Press: Boca Raton, FL, USA, 2008; ISBN 978-0-415-26646-8.
14. Anjos, G.; Cunha, R.R.; Kuklik, P.; Brouček, M. Numerical Evaluation of Bored Piles in Tropical Soils by Means of the Geo-Technical Engineering GEO4 Fine Software. In Proceedings of the III European Conference on Computational Mechanics Solids, Structures and Coupled Problems in Engineering, Lisbon, Portugal, 5–8 June 2006.
15. Pender, M. A Seismic Pile Foundation Design Analysis. *Bull. N. Z. Natl. Soc. Earthqu. Eng.* **1993**, *26*, 49–160.
16. Park, J.; Kim, J.; Shin, Y.; Lee, J.; Park, J. 3 MW Class Offshore Wind Turbine Development. *Curr. Appl. Phys.* **2010**, *10*, S307–S310. [CrossRef]
17. MIDAS IT. *MIDAS/CIVIL Computer Program*; MIDAS IT: Seoul, Korea, 2019.
18. DNVGL. *Recommended Practice DNVGL-RP-C205: Environmental Conditions and Environmental Loads*; Det Norske Veritas: Oslo, Norway, 2019.
19. Le Méhauté, B. *An Introduction to Hydrodynamics and Water Waves*; Springer-Verlag: New York, NY, USA, 1976; ISBN 978-3-642-85567-2.
20. International Electrotechnical Commission (IEC). *Wind Turbines—Part 1: Design Requirements*, 4rd ed.; IEC 61400-1:2019-02; IEC: Geneva, Switzerland, 2019.
21. Chen, D.; Huang, K.; Bretel, V.; Hou, L. Comparison of structural properties between monopile and tripod offshore wind-turbine support structures. *AIME* **2013**, *3*, 943–950. [CrossRef]
22. Ministry of Maritime Affairs and Fisheries. *Korea Harbor and Fishery Design Criteria*; Korea Port and Harbours Association: Seoul, Korea, 2014.
23. Korean Concrete Institute. *Korean Structural Concrete Design Code*; Korean Concrete Institute: Seoul, Korea, 2012.
24. Korean Standards Association. *Uncoated Stress-Relieved Steel Wires and Strands for Prestressed Concrete (KS D 7002)*; Korean Standards Association: Seoul, Korea, 2019.
25. De Vries, W.; Vemula, N.K.; Passon, P.; Fischer, T.; Kaufer, D.; Matha, D.; Schmidt, B.; Vorpahl, F. *Final Report WP4.2: Support Structure Concepts for Deep Water Sites*; Project UpWind D4.2.8; TU-Delft: Delft, The Netherlands, 2011; p. 019945.
26. Det Norske Veritas. *Offshore Standard: Design of Offshore Wind Turbine Structures Draft (DNV-OS-J101)*; Det Norske Veritas: Oslo, Norway, 2014.
27. American Petroleum Institute. *American Petroleum Institute Recommended Practice for Planning, Designing and Constructing Fixed Offshore Platforms—Working Stress Design (API RP2A-WSD)*; American Petroleum Institute: Washington, DC, USA, 2014.

Journal of
Marine Science and Engineering

MDPI

Article

Investigation of Blade Tip Shape for Improving VAWT Performance

Yichen Jiang [1], Chenlu He [2], Peidong Zhao [1] and Tiezhi Sun [1,*]

1 School of Naval Architecture Engineering, Dalian University of Technology, Dalian 116024, Liaoning, China; ycjiang@dlut.edu.cn (Y.J.); zhaopeidong@mail.dlut.edu.cn (P.Z.)
2 School of Naval Architecture, Ocean and Civil Engineering, Shanghai Jiao Tong University, Shanghai 200240, China; chenlu.h@sjtu.edu.cn
* Correspondence: suntiezhi@dlut.edu.cn

Received: 25 February 2020; Accepted: 18 March 2020; Published: 22 March 2020

Abstract: Vertical axis wind turbine (VAWT) is a competitive power generation device due to structural simplicity, wind direction independence, no yaw mechanism required, easier maintenance, and lower noise emission. However, blade tip vortex will be generated at both ends of the blade during the rotation, resulting in torque loss and efficiency reduction. In this paper, computational fluid dynamics is used to study blade tip vortex and its reduction technique of a single-blade VAWT rotor in real scale. By monitoring the force and flow field at different heights of the blade, the influence ranges of tip vortex are obtained. The reduction effect of the bulkhead obtained from the blade profile curve is studied, and the size of the bulkhead is optimized. On the basis of adding the optimal bulkhead, the influence of the supporting strut is also explored. The joint action is obtained by changing the location of the supporting strut. The results show that the top supporting strut-bulkhead structure is the optimal position. The power-extraction efficiency of the rotor with this integrated structure is significantly improved at optimal tip speed ratios (TSRs) and higher TSRs.

Keywords: vertical axis wind turbine; tip vortex of blade; supporting strut; performance optimization; computational fluid dynamics

1. Introduction

Wind power is one of the solutions to energy problem as a clean energy source with large reserves in nature. According to the Renewables 2019 Global Status Report [1], in the whole year of 2018, the installed capacity of wind power in the world exceeded 51 gigawatt (GW), with a growth rate of 9%, and the total capacity reached to 591 GW. At present, the traditional horizontal axis wind turbine (HAWT) is more popular in the wind energy market due to its advanced development and high efficiency [2]. However, HAWT has difficultly achieving larger sizes because of the limitation of blade length and strength. On the other hand, vertical axis wind turbine (VAWT) has attracted more and more researchers' and manufacturers' attention due to its advantages of adaptiveness to wind direction, convenient installation and maintenance, and the ability to be expanded to larger sizes [3,4].

Improving the efficiency of VAWT is an important target in wind energy generation. Tip vortex is part of the three-dimensional unsteady flow, which constantly sheds in each blade revolution, decreasing the torque and causing structural vibrations. Kudela and Malecha revealed that the detachment of the vortex from the boundary layer leads to the sudden drop of the lift force, based on a vortex-in-cell model [5]. Hofemann and Simao et al. [6,7] experimentally detailed and qualified the evolution of the tip vortex in the near wake with 3D-Stereo Particle Image Velocimetry. Hamada et al. [8], Howell et al. [9], and Lam and Peng [10] compared the torque generated by blade between 2D and 3D computational fluid dynamics(CFD) models, and obtained the conclusion that the 3D effects significantly reduced the extraction rate of wind power. Yanzhao Y et al. [11] studied the effects of tip vortex on the wake of a two-blade VAWT with different tip speed ratios (TSRs) by CFD simulations. Balduzzi et al. [12] conducted a detailed 3D CFD analysis of H-type Darrieus single-blade rotor with an aspect ratio (AR) of 17.5. Studies have shown that the torque loss caused by the three-dimensional effect due to the finite blade length is equivalent to the blade length loss of 1.5c. Many researchers have made efforts to reduce the tip loss. Zanforlin and Deluca [13] proposed that AR > 0.8 should be considered in design to avoid the effect of tip vortex. Some others have proposed ways to change the blade shape for tip vortex reduction, such as a helical blade [14,15] and elliptical blade tip [16,17].The installation of a tip device is also a method to improve the turbine performance. Winglets and bulkheads are mostly used in experiments and simulations [16–19]. Considering the 3D effect caused by supporting strut, some research was undertaken in terms of the cross section and location of the strut [8,16,20–22]. The results show that the more streamlined hydrodynamic strut section provides lower strut drag.

To date, most of the 3D effect studies of VAWT focus on the medium or small turbines. In this paper, the 3D effects of blade tip vortex and supporting strut are studied by CFD numerical simulation with a large turbine model in real scale. The objectives of this study are to explore the methods of reducing tip vortex and optimizing the structure of the turbine, which could improve the performance of the VAWT.

The paper is organized as follows. Section 2 introduces the numerical methodology and some parameter settings. The independence verification of grid and time-step is also presented. Section 3 shows the numerical results and analysis. Section 3.1 analyzes the effect ranges of tip vortex in detail. In Section 3.2, a bulkhead is added to reduce tip vortex, and the size of the bulkhead is optimized. Section 3.3 considers the influence of supporting strut. Additionally, the optimal position and structural form of supporting strut are found based on the addition of the optimal bulkhead. Section 3.4 compares the gains on the presented structure with different TSRs.

2. Numerical Methodology

2.1. Computational Model

The subject of our study is a three-blade VAWT rotor whose rated power is 2.5MW. The rotor is part of an offshore floating twin VAWT system, as shown in Figure 1. The ratio of the height and the diameter for each rotor is larger than the conventional single-rotor VAWT. The main parameters and the model diagram of the rotor are shown in Table 1 and Figure 2, respectively. The large chord length of 7.2 m was chosen by directly scaling, based on our model-scale experimental results [23]. It is worth noting that this chord length may not be the optimal one for this rated power output due to the scale effects. In addition, a shorter blade chord will lighten the weight of the rotor, reduce the solidity of the rotor, and tend to enhance the power coefficient. Finding the optimal chord length, one may need to conduct a detailed investigation. Focusing on the objective of this work, the blade with the chord length of 7.2m was chosen to be the study subject. Considering the computational resources in real scale simulation, this paper simplified the model and only studied the single rotor with a single blade.

Table 1. Main parameters of the rotor.

Main Parameters	
Blade profile	NACA0018
Blade chord, c(m)	7.2
Rotor height, H(m)	107.8
Rotor diameter, D(m)	57.8

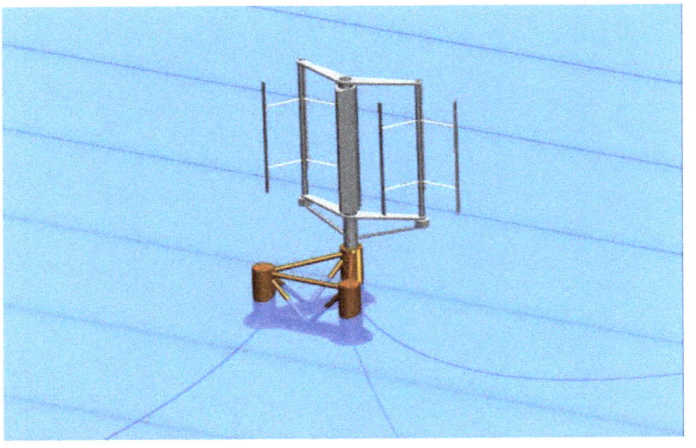

Figure 1. Concept design of an offshore floating twin vertical-axis wind turbine system.

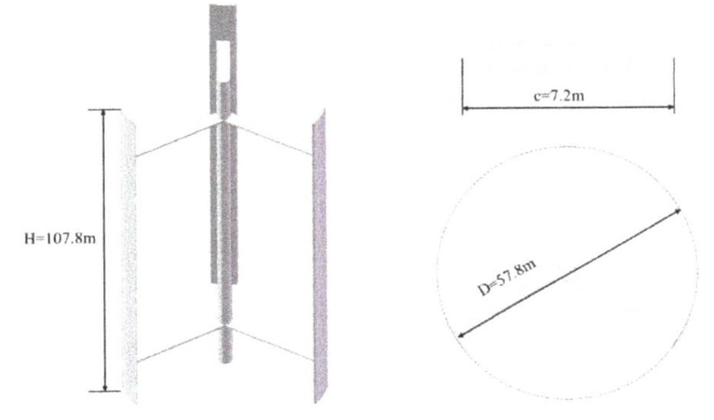

Figure 2. Model diagram of the rotor.

In the numerical calculation, we used the finite-volume Navier–Stokes solver in the STAR-CCM+ for simulation. In order to improve the accuracy and minimize the calculation cost, this study only carried out a semi-basin calculation, and divided the computational domain into external flow field and internal flow field, as shown in Figure 3. The internal flow field was a cylindrical region, including the rotor part of the VAWT, which rotated. The external flow field was a rectangular region, including the part outside the rotor. The center of the column in midspan was selected as the coordinate origin of the computational domain. The velocity inlet was located at the 3D position in the upstream area from the rotor center, and the pressure outlet was located at the 10D position in the downstream area. The domain of the entire flow field was 13D long, 8D wide, and 1.5H high. The cylinder diameter and the height of the rotor domain were 1.18D and 0.8H. These two computational domains were connected by interfaces.

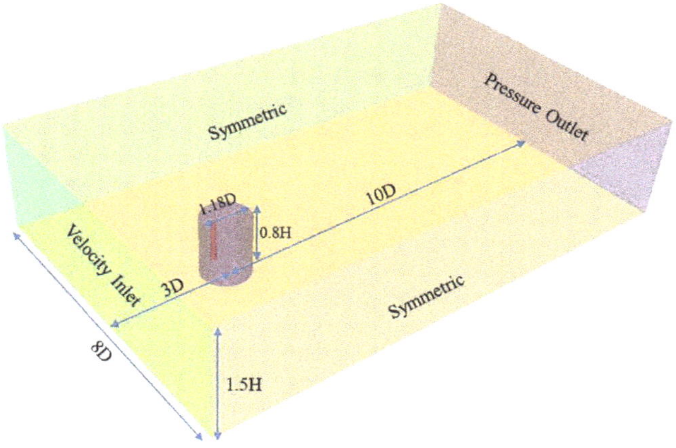

Figure 3. Computational domain with boundary conditions and sizes.

The surface of blade was set as wall, and the remaining boundaries of the domain were set as symmetrical plane. The turbulence model was SST $k-\omega$, commonly used in the numerical simulation of VAWT [11,13,21,24–26]. The wind speed at the velocity inlet was set at 10m/s, and the pressure at the outlet was set at 0 Pa. The blade tip speed ratio (TSR) was set at 3, so the rotation angular velocity of the rotor was 1.038rad/s.

The grid form adopted in this study was the cut volume grid. Two layers of wake refinement area were set in the outflow field whose basic size of the grid was 0.8 chord length. The grid size was 50% of the basic size in the first layer and 25% in the second layer. In order to get a better transition between the external flow field and the internal one, the basic size of the rotating domain was the same as the size in the second layer of refinement area. The condition of the blade was defined as no-slip, and 20 layers of prismatic boundary layer cells were generated with growth ratio of 1.2. Local refinement was carried out on the blade surface. The total number of grids was 5,694,103. The computational mesh of the whole domain and near the blade are shown in Figure 4. Using implicit unsteady method, a time step was set at 0.0168s, namely the time for the rotor turning 1 °.

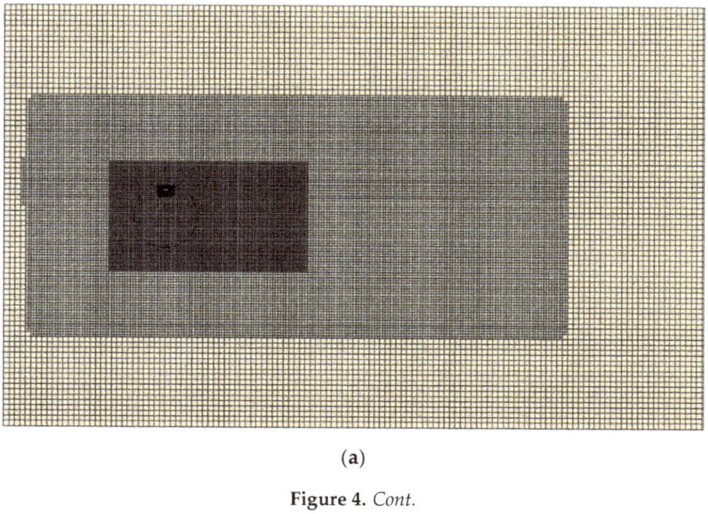

(a)

Figure 4. *Cont.*

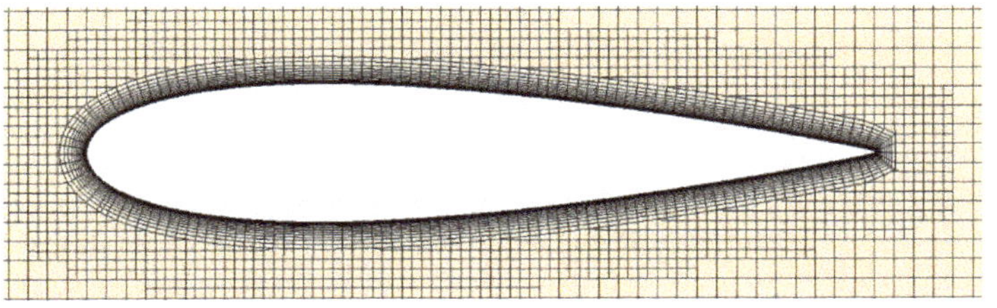

(b)

Figure 4. (a) Computational mesh of the domain; (b) computational mesh near the blade surface.

2.2. Mesh Convergence

Another two sets of mesh with coarser division and finer division were obtained by changing the basic size. The three mesh sets are shown in Table 2. The instantaneous torque coefficient C_t is defined by Equation (1), where $h = \frac{H}{2}$ due to the semi-basin simulation. The blade instantaneous torque curves of the three mesh sets are shown in Figure 5. According to Equation (2), we could obtain the work done by torque using the area under the curve. The deviation of the work done obtained by medium mesh and fine mesh was 2%, much less than the deviation of medium and coarse mesh. Moreover, the torque curves of medium and fine mesh showed more similar variation. Thus, the medium mesh was considered to be converged and such set was used for subsequent simulations.

$$C_t = \frac{T}{\frac{1}{2}\rho V^2 DhR}, \tag{1}$$

$$W = \frac{\pi}{180}\int_{\theta_1}^{\theta_2} Td\theta, \tag{2}$$

Table 2. Mesh size and quantity.

Mesh	Basic Size (m)	Total Number
Coarse	6.79	2,659,364
Medium	4.80	5,694,103
Fine	3.39	12,621,404

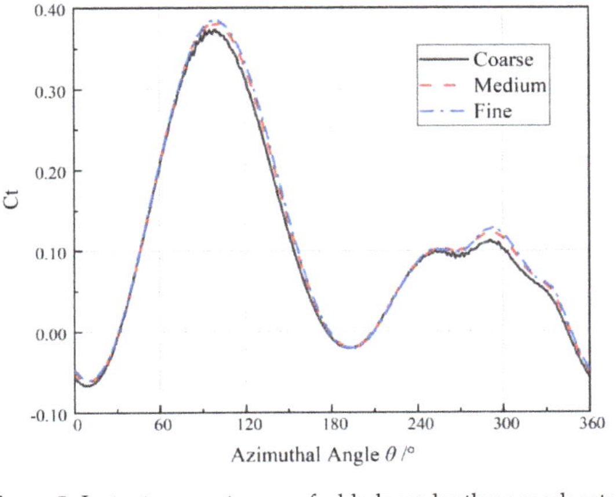

Figure 5. Instantaneous torque of a blade under three mesh sets.

2.3. Time-Step Convergence

Balduzzi et al. [27] noted that in most studies of the VAWT simulation, the time-step is set to the time in which the rotor makes a rotation between 0.5° and 2°. Three groups of time-step were used to calculate, respectively, the rotor turning 1°, 2°, and 3°. For the 1°and 2° time settings, the difference of the torque area under the curves (Figure 6) were less than 1‰, so the 2° turning for a time-step converged.

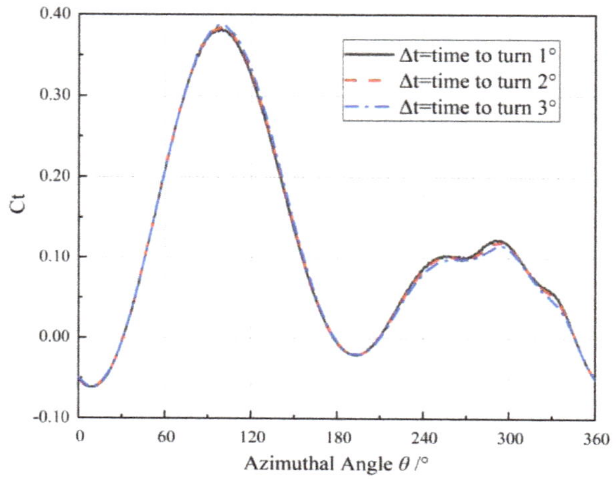

Figure 6. Instantaneous torque of a blade under three time-step sets.

2.4. Numerical Model Validation against Model-Scale Experimental Data

Before further investigating the numerical simulation results, it is necessary to check the reliability of abovementioned modelling techniques against experimental data. However, there is no experimental data available for full-scale 2.5MW VAWT. The model-scale experimental measurements were adopted to validate our numerical model. More information on the wind tunnel experiments can be found in the previous publication [23]. Y+ distribution and mesh generation along the blade of the numerical model is shown in Figure 7. The average of the measured torque acting on the wind turbine rotor under the wind speed of 7.7 m/s are plotted in Figure 8, together with the predictions of CFD simulations based on the abovementioned modelling techniques. It can be seen that the torque predicted by numerical simulation tends to be slightly larger than the experimental value at low rotational speeds and slightly smaller at higher rotational speeds. Overall, the simulated torque curve agrees well with the one obtained from the experiment, and is therefore considered valid.

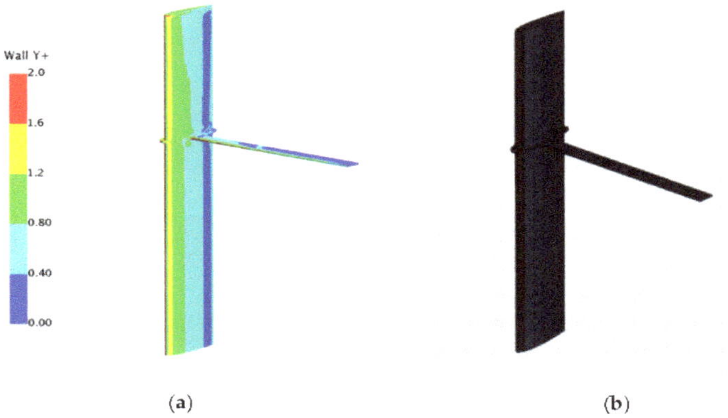

Figure 7. (**a**) Y+ distribution along the blade; (**b**) numerical mesh on the blade.

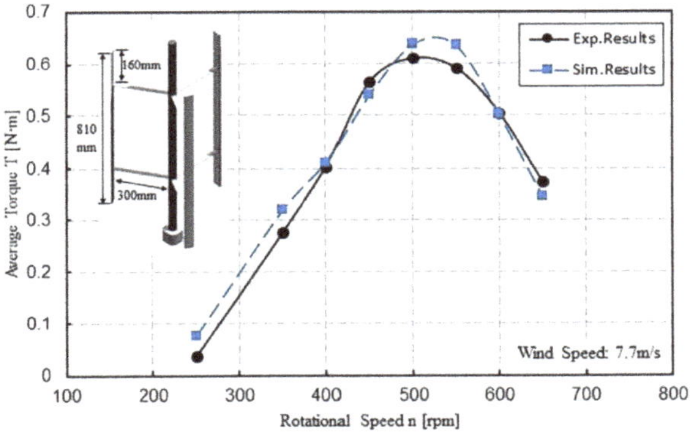

Figure 8. Experimentally obtained and numerically simulated torque as a function of rotational speed of an isolated vertical axis wind turbine (VAWT).

3. Results and Numerical Analysis

3.1. Influence of Tip Vortex

In order to better calculate and monitor the influence range of tip vortex on the blade surface, the blade length was divided based on the chord length. Airfoil sections with a height of 0.05m were intercepted at positions 0.05c, 0.25c, 0.5c, 1c, 2c, 3c, 4c, 5c, 6c, and 7c from the tip of the blade. As shown in Figure 9, the initial position angle of blade is 0°. From 0° to 180° is upwind region in a rotating cycle (upstream), and 180° to 360° is downwind region (downstream).

The instantaneous torque curves of the fourth cycle were selected for analysis, as shown in Figure 10. The ordinate is the instantaneous torque coefficient per 0.05m height section, which is defined by Equation (3). Here, T_z denotes the instantaneous torque per unit section. It can be seen that the torque at section 0.05c is negative for most of the cycle. In the upwind region, the closer the section is to the blade tip, the smaller the torque contribution will be. The sections below 3c from the tip of the blade are slightly affected by tip vortex. In the downwind area between 300° and 340°, the torque curve shows a steep fall. It can be observed from the three-dimensional vortex diagram that this steep drop is caused by the collision between the blade and the vortex shedding in the previous cycle, as shown in Figure 11.

$$C_{tz} = \frac{T_z}{\frac{1}{2}\rho V^2 D h_z R}, \quad (3)$$

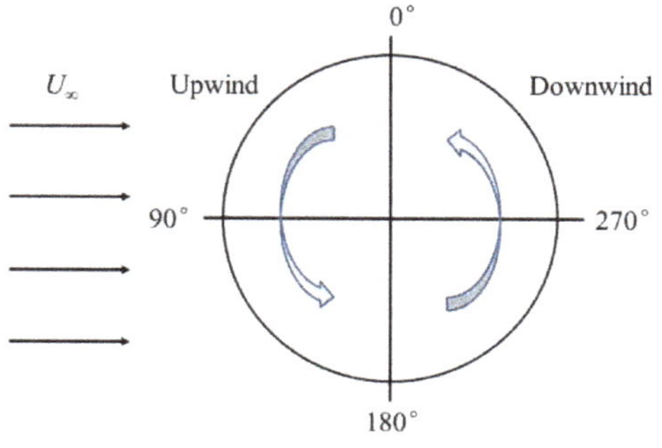

Figure 9. Upwind and downwind paths during a rotation.

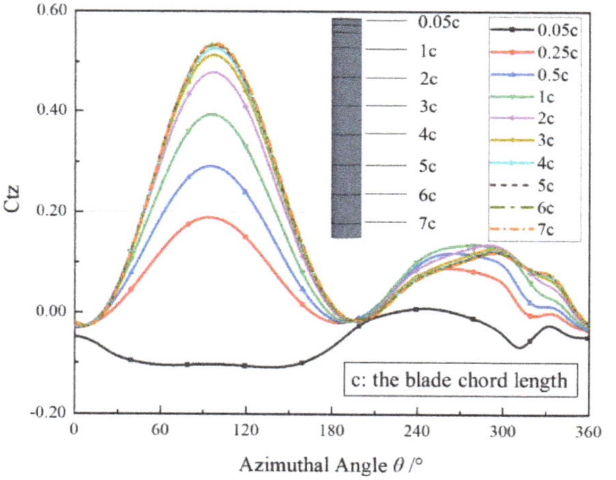

Figure 10. Instantaneous torque of sections at different heights during a rotation.

These instantaneous torque curves can be divided into two modes according to the downwind variations at different heights. The distances of 0.05c, 0.25c, and 0.5c from the blade tip are in the first mode, and distances of 1c to 7c from the blade tip are in the second mode. In the first mode, the torque is greatly reduced due to severe influence of tip vortex and severe damage of flow field. Additionally, the closer the section is to the blade tip, the greater the torque reduction will be. It can be said that for height above the 0.5c-section, tip vortices are more likely to disturb the flow field, as shown in Figure 12. However, in the second mode, there is a larger torque of section closer to the blade tip in a specific azimuthal angle range, which could be explained as follows. The geometrical relationship between angle of attack and blade azimuthal angle is shown in Figure 13, and the lift coefficient (angle of attack of NACA0018) is shown in Figure 14. The data is obtained by applying the Xfoil software, and published on the website, http://airfoiltools.com/polar/details?polar=xf-naca0018-il-1000000. To make sure the predicted install angle is adequately accurate, we also conducted our 2D numerical simulations to predict the lift coefficient corresponding to several angles of attack of NACA0018 airfoil.

The predicted results are compared with the published data in Figure 14. The predicted lift coefficient is slightly lower, but trends of two results are close, and both results show that the stall angle of NACA0018 at high Reynolds number is around 17°. From Figure 13, we can obtain the geometrical relationship between angle of attack and azimuthal angle, as shown in Equation (4). For simplicity, the induction factor a was ignored [28]. It can be derived that (224.3°, 281.7°) in the downwind is a zone for the blade stall by Equation (4). At a height close to the tip of the blade, due to the existence of tip vortex, a downwash angle is generated [12], which reduces the attack angle of the blade and fails to reach the critical angle of stall. Thus, in the range of 220° to 280° shown in Figure 10, the higher the height from 1 c to 7 c, the greater the torque value will be.

$$tan\alpha = \frac{U_\infty(1-a)sin\theta}{\omega R + U_\infty(1-a)cos\theta}, \quad (4)$$

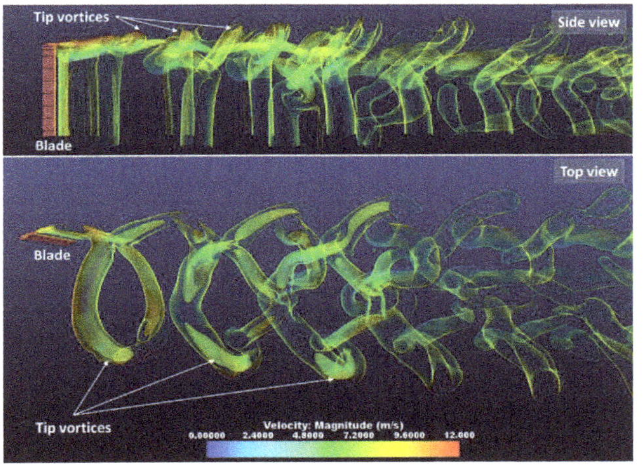

Figure 11. Collision between the blade and the vortex in the downwind.

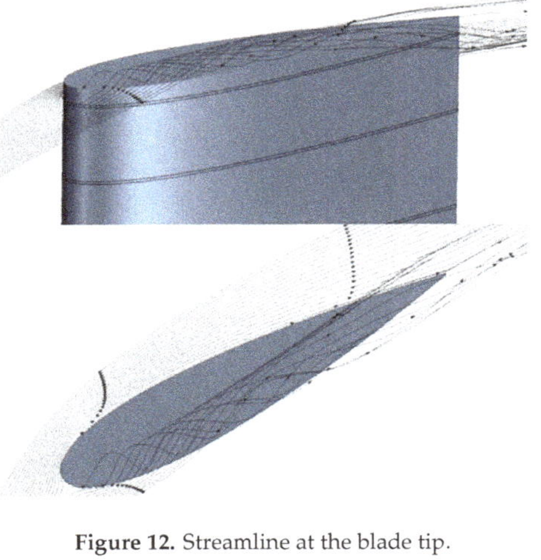

Figure 12. Streamline at the blade tip.

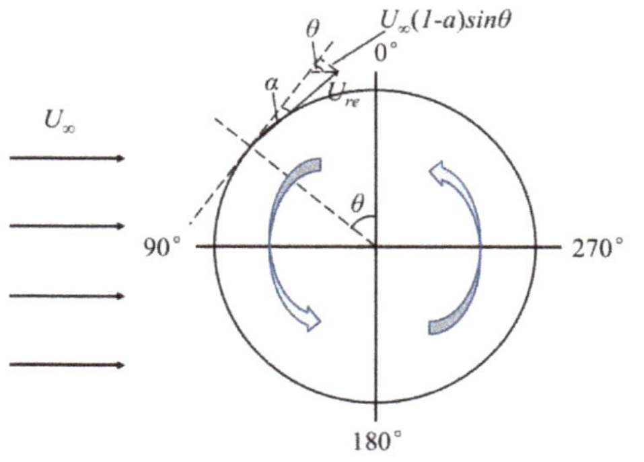

Figure 13. Geometrical relationship between angle of attack and blade azimuth. θ is the azimuthal angle, α is the angle of attack, a is the axial induction factor.

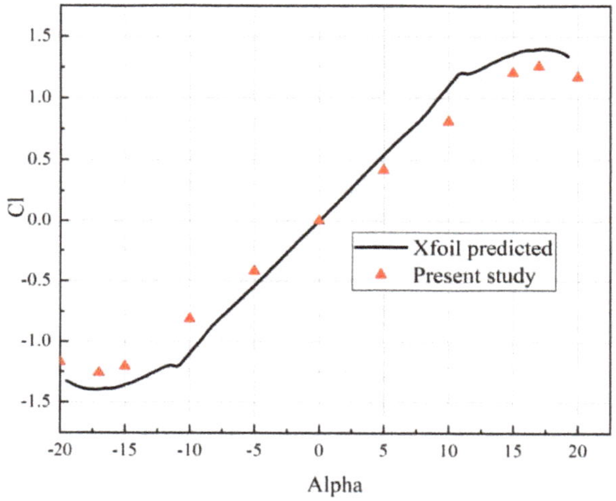

Figure 14. NACA0018 airfoil lift coefficient (angle of attack).

3.2. Bulkhead Inhibition of Tip Vortex

In order to restrain the influence of tip vortex on blade surface, a tip device was added. The bulkhead was obtained by expanding the blade profile curve. The edge of the bulkhead was 2m away from the blade surface, and the thickness was 0.2m. The edge was set as a round angle with a radius of 0.1m. Figure 15 shows the streamline at different heights of the blade sections when the bulkhead is added or not. The azimuthal angle is 88°.It can be seen from Figure 15a that the closer the height is to the top of the blade, the more disordered the streamline is. Figure 15b shows the streamline is smoother under the action of the bulkhead. The comparison of instantaneous torque at different heights is shown in Figure 16. It can be seen that the instantaneous torque of the blade with a bulkhead is much larger than that of the blade without a bulkhead above the height of the 1c-section. That is, the inhibition of the bulkhead on tip vortex is obvious above the height of 1c-section, and there is basically no effect below the height of 3c-section.

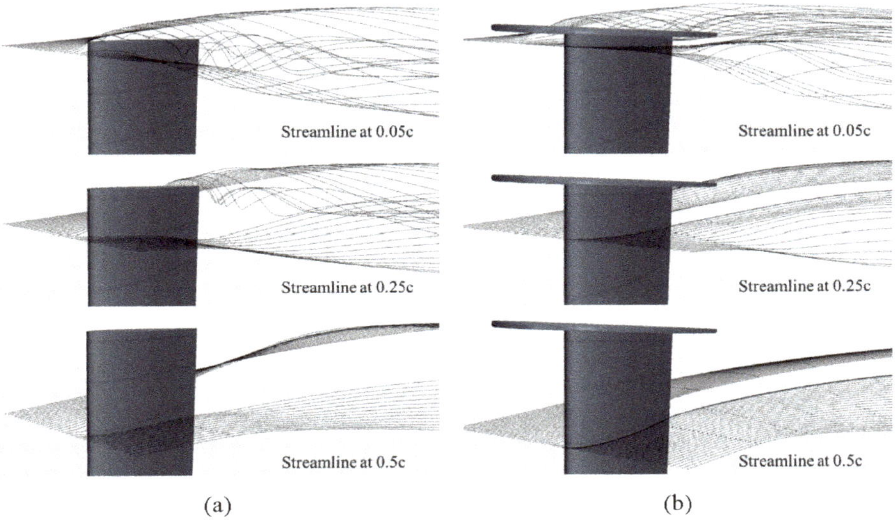

Figure 15. The streamline at different heights of the blade sections when the azimuthal angle is 88°: (**a**) without a bulkhead; (**b**) with a bulkhead.

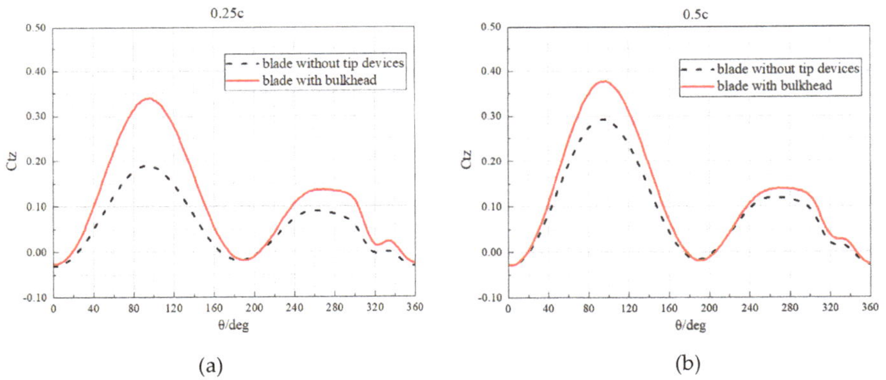

Figure 16. *Cont.*

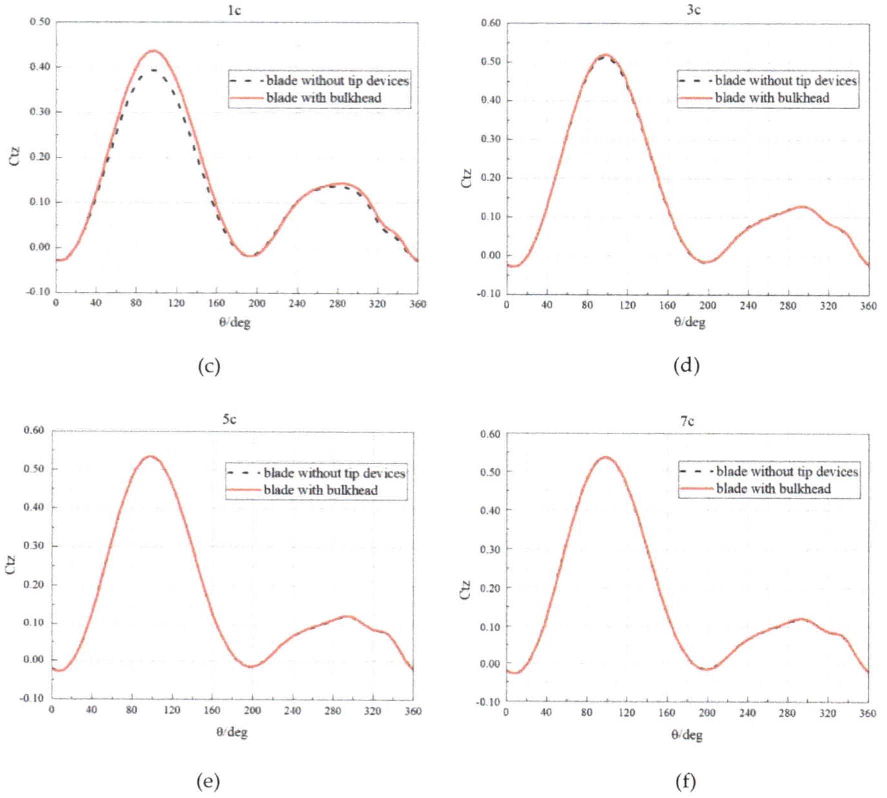

Figure 16. Instantaneous torque with or without a bulkhead at height of (**a**) 0.25c; (**b**) 0.5c; (**c**) 1c; (**d**) 3c; (**e**) 5c; (**f**) 7c.

In order to study the effect of bulkhead size on turbine performance, six cases (Δ = 0, 0.18c, 0.35c, 0.56c, 0.76c, and 1c) were selected for calculation, where Δ represents the distance between the edge of the bulkhead and the blade surface. The geometric models are shown in Figure 17.

In the simulation, the total torque, torque of the blade, and torque of the bulkhead were monitored, respectively. Work done by the blade with six sizes of bulkhead were calculated from the torque data and are shown in Figure 18. We used a non-dimensional quantity W' to represent the work W. The total work concludes both the bulkhead and the blade work. It can be seen from Figure 18 that, as the size of the bulkhead increases, the blade work shows an increasing trend. However, the larger the bulkhead size is, the greater the resistance will be. The growth trend of the total work starts to slow down after the corresponding point of Δ= 0.35c, and Δ= 0.35c is considered as the preferred size. At this time, the total work increases by 4.25% compared with the blade without a bulkhead.

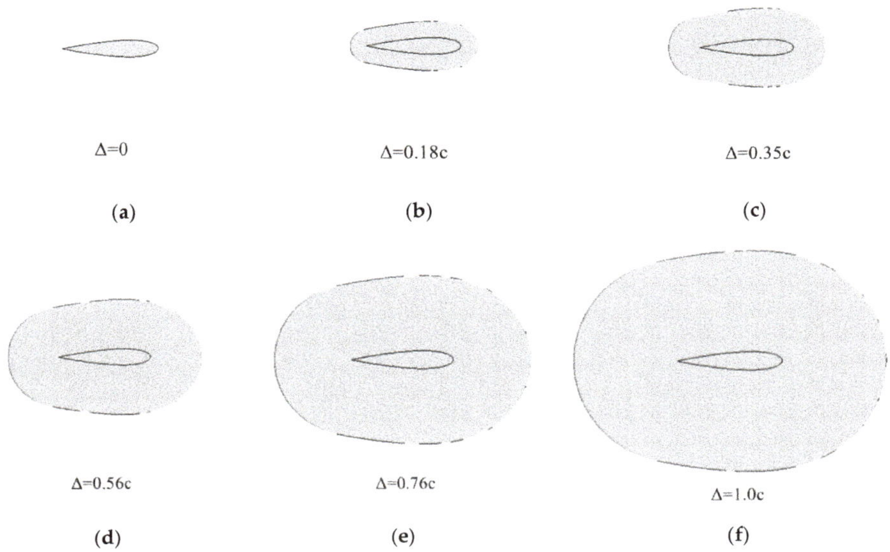

Figure 17. Perspective view of six bulkhead sizes: (**a**) Δ = 0; (**b**) Δ = 0.18c; (**c**) Δ = 0.35c; (**d**) Δ = 0.56c; (**e**) Δ = 0.76c; (**f**) Δ = 1.0c.

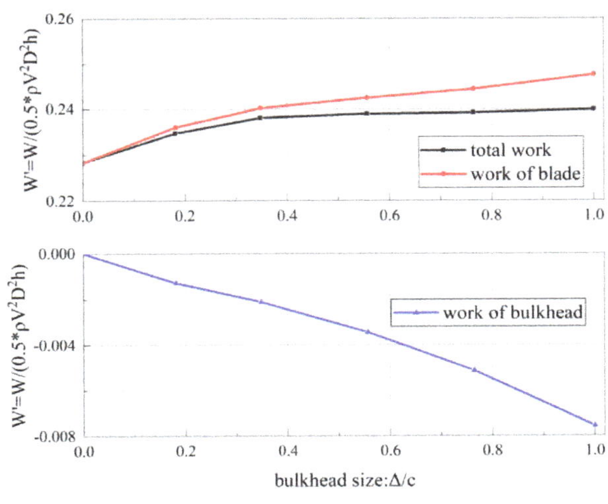

Figure 18. Total work, blade work, and bulkhead work under six sizes of bulkhead.

3.3. *Influence of Supporting Strut*

The supporting strut connecting the blade to the central column also produces a three-dimensional effect that affects the performance of the rotor. The cross section of supporting strut usually has several forms, such as rectangle, circle, and airfoil. In this part, a streamlined airfoil section was selected for the modeling and calculation.

The NACA0018 airfoil, the same as the blade, was selected for the supporting strut, and the chord length was 0.5c, as shown in Figure 19.

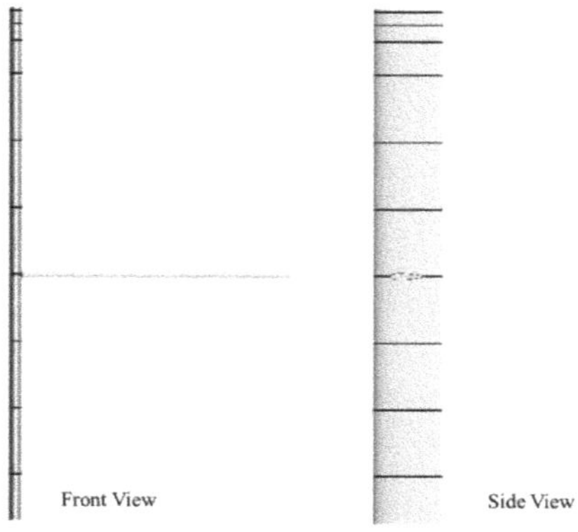

Figure 19. Blade-supporting strut rotor model.

On the basis of installing a bulkhead with size $\Delta = 0.35c$, four cases where the supporting strut was located at 4c, 2.5c, and 1c from the blade tip and at the top were selected for calculation, as shown in Figure 20.

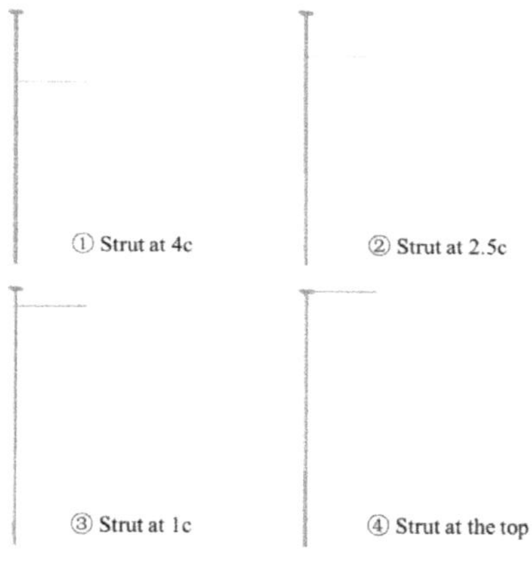

Figure 20. Supporting strut at different heights.

Table 3 lists the total work and the work of blade in each case, also presented by W'. When the supporting strut is 4c, 2.5c, and 1c away from the blade tip, it performs negative work. Therefore, the total work is less than the work generated by the blade. However, in the process of the supporting strut moving up, the difference between the total work and the work of blade becomes smaller and smaller, which indicates that the resistance of the supporting strut together with the bulkhead gradually decreases. When the strut is at the top, the total work is even greater than that of the blade part. As can be seen from Figure 21, a forward torque is generated when the strut is at the top. Such results can be attributed to the windward angle of attack generated by the incoming flow when the strut is at the top of the turbine, thus generating lift.

Table 3. Total work and work of the blade with a strut at different heights.

	Total Work (W′)	Increased Percentage	Work of Blade (W′)	Increased Percentage
Blade with strut at 4c	0.2215	-	0.2222	-
Blade with bulkhead and strut at 4c	0.2308	4.23%	0.2338	5.20%
Blade with bulkhead and strut at 2.5c	0.2306	4.11%	0.2334	5.02%
Blade with bulkhead and strut at 1c	0.2336	5.46%	0.2356	6.04%
Blade with bulkhead and strut at top	0.2415	9.05%	0.2404	8.20%

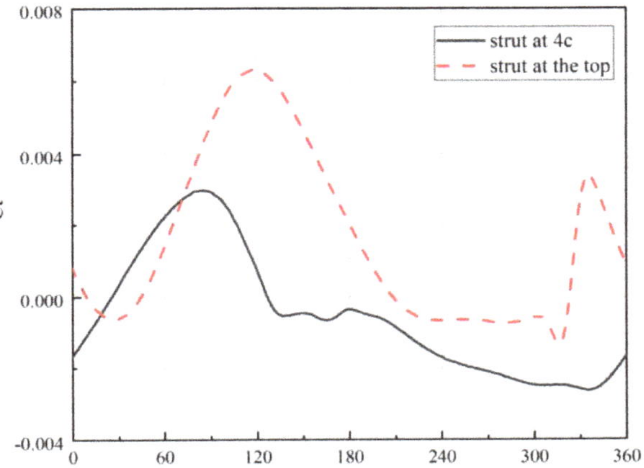

Figure 21. Instantaneous torque of the supporting strut at the top and at the height of 4c.

As can be seen from Figure 22, when the supporting strut is at the top, the torque value of the 0.05c-section to 0.5c-section increases compared with that when only the optimal bulkhead is installed. That is, the supporting strut enhances the effect of tip vortex reduction. From Figure 23, it can be seen that the streamline is smoother when the supporting strut is at the top. We call this structure the top supporting strut-bulkhead structure. Therefore, the supporting strut at the top has two advantages: firstly, it generates forward torque rather than resistance; secondly, it enhances the effect of reducing tip vortex. As can be seen from Table 3, the top supporting strut-bulkhead structure has a work performance gain of 9.05%.

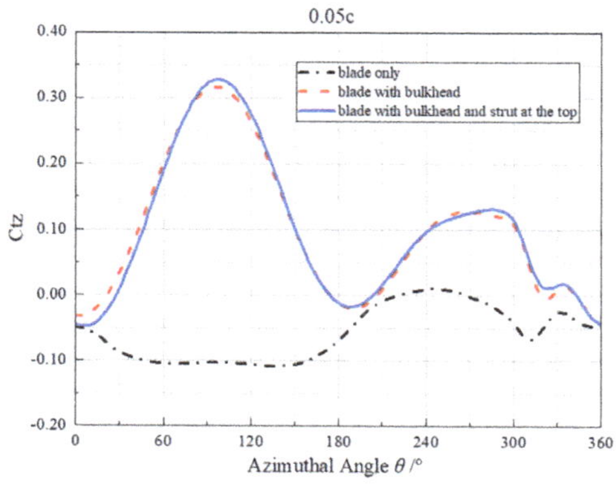

(a)

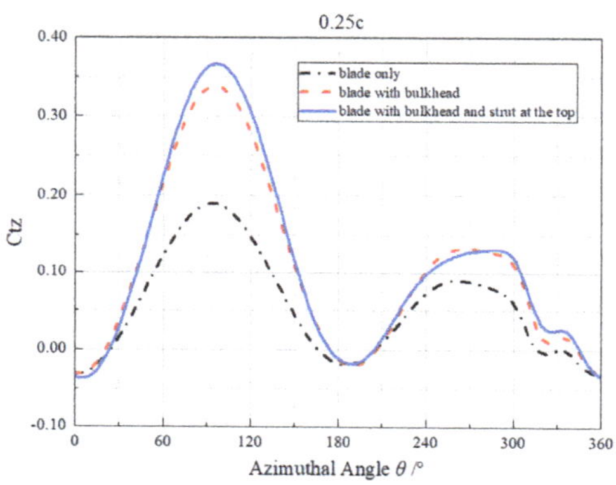

(b)

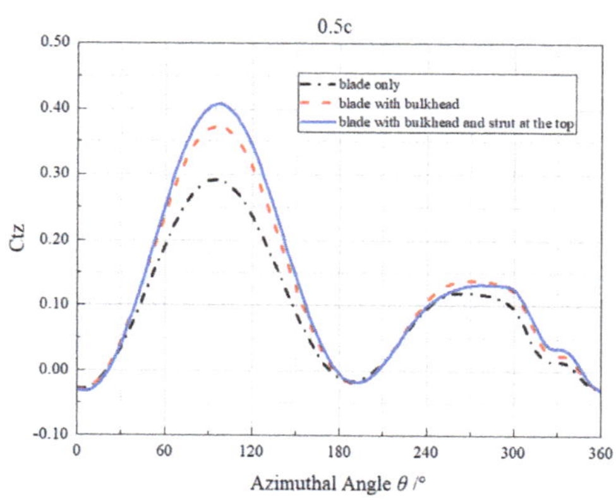

(c)

Figure 22. Comparison of instantaneous torque among three different blade tip shapes at height of (**a**) 0.05c; (**b**) 0.25c; (**c**) 0.5c.

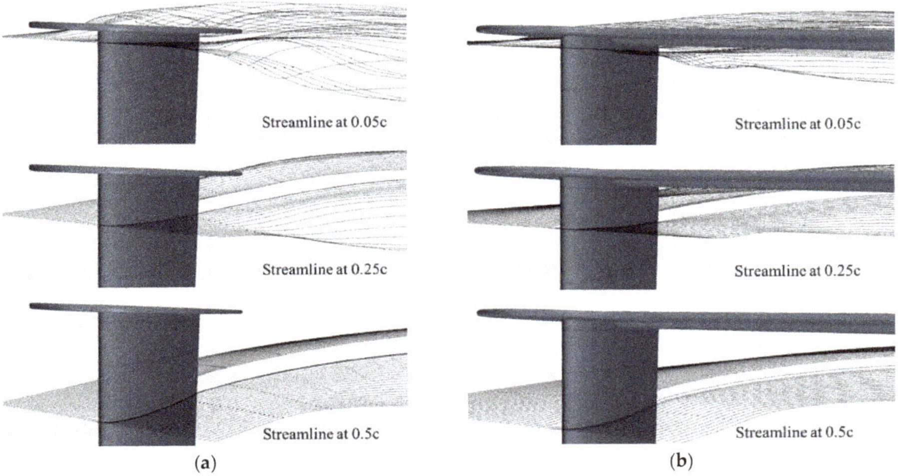

Figure 23. The streamline at different heights of the blade sections when the azimuthal angle is 88°: (**a**) with a bulkhead; (**b**) with a top supporting strut-bulkhead structure.

3.4. Performance of the Top Supporting Strut-Bulkhead Structure under Different TSRs

In the previous study, only the calculation and analysis at TSR = 3 were carried out. Since the aerodynamic performance varies under different TSRs [29], other cases were calculated under TSR = 1, 2, and 4. Figure 24 shows the work done by the whole structure under four TSRs. TSR = 1 and 2 are low tip speed ratios, while TSR = 3 and 4 are good and high ones. When TSR is low, blade stall is serious and the top supporting strut-bulkhead structure does not perform well. However, under higher TSRs, it can be seen that the work done by the top supporting strut-bulkhead structure is much larger. When TSR = 3 and TSR = 4, compared with conventional blade-and-strut structure, the work done by the proposed structure is increased by 9.05% and 10.48%, respectively.

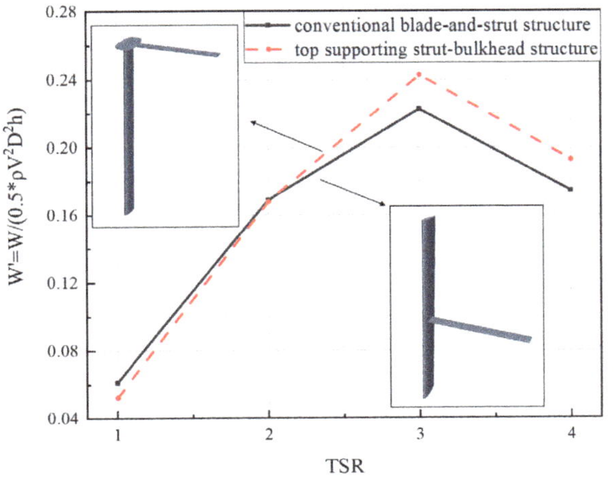

Figure 24. Work done by torque of the whole structure under four TSRs.

4. Discussion

In this paper, a numerical model for predicting the aerodynamic performance of a real-scale VAWT was established. The influence of blade tip vortex was analyzed, and the supporting strut at different heights was considered together with the action of the bulkhead. The structure of the rotor was optimized with the selected bulkhead size and strut position, which greatly improved the performance of the rotor.

First, the torque and flow field at different heights of the blade were monitored to analyze the influence ranges of tip vortex. In the upwind area of rotation, tip flow was mainly manifested as the destruction of flow field. The torque of the blade section less than 3c away from the tip of the blade decreased to different degrees, and the closer it was to the tip, the more obvious the reduction was. The overall torque of the blade section at a height less than 0.05c away from the blade tip was negative work.

Second, a bulkhead was added to reduce blade tip vortex. The effect was more obvious when the bulkhead was enlarged. However, a larger bulkhead could generate resistance, which limited the improvement of the total torque. When the bulkhead size increased to a certain extent, the total torque of the blade changed slowly. In the six cases of $\Delta =$ 0, 0.18c, 0.35c, 0.56c, and 0.76c selected in this study, it was calculated that $\Delta =$ 0.35c was the best extension size, and the torque power of the blade at this time increased by 4.25% compared with the blade without a tip device.

In addition, different heights of the supporting strut were considered in the calculation. When the position of the strut moved from the 4c-section to 1c-section, the total power of the rotor kept increasing. The supporting strut could reduce tip vortex when it was located at the blade tip together with the bulkhead. At this time, the total power increased by 9.05% compared with the conventional structure whose supporting strut was at the height of 4c-section and no bulkhead.

Moreover, the effect of the top supporting strut-bulkhead structure was compared under different tip speed ratios. At low TSRs, because of the serious stall, the supporting strut-bulkhead structure did not show performance gain. While at higher TSRs, the torque of the whole structure was significantly improved.

The design of this supporting strut-bulkhead structure provides an option for the design and optimization of large VAWT structures in the future. Further studies can consider the interaction between several blades and whether multi-blades will limit the installation of tip devices.

Author Contributions: Conceptualization, Y.J. and C.H.; Data curation, C.H.; Formal analysis, C.H.; Investigation, Y.J. and C.H.; Methodology, C.H. and P.Z.; Project administration, Y.J.; Resources, T.S.; Software, T.S.; Supervision, Y.J.; Validation, P.Z.; Visualization, C.H.; Writing—original draft, C.H.; Writing—review and editing, Y.J. and T.S. All authors have read and agreed to the published version of the manuscript.

Funding: Qingdao National Laboratory for Marine Science and Technology (QNLM2016ORP0402); National Natural Science Foundation of China (Grant No. 51639003, 51279030); and the Fundamental Research Funds for the Central Universities (DUT2017TB05).

Acknowledgments: The authors would like to thank Qingdao National Laboratory for Marine Science and Technology, National Natural Science Foundation, and the Fundamental Research Funds for the Central Universities.

Conflicts of Interest: The authors declare no conflict of interest.

References

1. REN21. Renewables 2019: Global Status Report. Available online: http://www.ren21.net/gsr-2019/ (accessed on 26 July 2019).
2. Möllerström, E.; Gipe, P.; Beurskens, J.; Ottermo, F. A historical review of vertical axis wind turbines rated 100 kW and above. *Renew. Sustain. Energy Rev.* **2019**, *105*, 1–13. [CrossRef]
3. Liu, J.; Lin, H.; Zhang, J. Review on the technical perspectives and commercial viability of vertical axis wind turbines. *Ocean. Eng.* **2019**, *182*, 608–626. [CrossRef]

4. Mohamad, A.; Mohd Amin, N.A.; Toh, H.T.; Abdul Majid, M.S.; Daud, R. Review of Analysis on Vertical and Horizontal Axis Wind Turbines. *Appl. Mech. Mater.* **2014**, *695*, 801–805. [CrossRef]
5. Kudela, H.; Malecha, Z.M. Eruption of a boundary layer induced by a 2D vortex patch. *Fluid Dyn. Res.* **2009**, *41*, 055502. [CrossRef]
6. Hofemann, C.; Ferreira, C.; van Bussel, G.; Kuik, G.; Scarano, F.; Dixon, K.R. 3D stereo PIV study of tip vortex evolution on a VAWT. In Proceedings of the European Wind Energy Conference and Exhibition EWEC, Brussels, Belgium, 31 March–3 April 2008; pp. 1–8.
7. Simao Ferreira, C.; Dixon, K.; Hofemann, C.; van Kuik, G.; van Bussel, G. VAWT in Skew: Stereo-PIV and Vortex Modeling. In Proceedings of the 47th AIAA Aerospace Sciences Meeting Including The New Horizons Forum and Aerospace Exposition, Orlando, FL, USA, 5–8 January 2009.
8. Hamada, K.; Smith, T.; Durrani, N.; Qin, N.; Howell, R. Unsteady Flow Simulation and Dynamic Stall Around Vertical Axis Wind Turbine Blades. In Proceedings of the 46th AIAA Aerospace Sciences Meeting and Exhibit, Reno, NV, USA, 7–10 January 2008.
9. Howell, R.; Qin, N.; Edwards, J.; Durrani, N. Wind Tunnel and Numerical Study of a Small Vertical Axis Wind Turbine. *Renew. Energy* **2010**, *35*, 412–422. [CrossRef]
10. Lam, H.F.; Peng, H.Y. Study of wake characteristics of a vertical axis wind turbine by two- and three-dimensional computational fluid dynamics simulations. *Renew. Energy* **2016**, *90*, 386–398. [CrossRef]
11. Yang, Y.; Guo, Z.; Zhang, Y.; Jinyama, H.; Li, Q. Numerical Investigation of the Tip Vortex of a Straight-Bladed Vertical Axis Wind Turbine with Double-Blades. *Energies* **2017**, *10*, 1721. [CrossRef]
12. Balduzzi, F.; Drofelnik, J.; Bianchini, A.; Ferrara, G.; Ferrari, L.; Campobasso, M.S. Darrieus wind turbine blade unsteady aerodynamics: A three-dimensional Navier-Stokes CFD assessment. *Energy* **2017**, *128*, 550–563. [CrossRef]
13. Zanforlin, S.; Deluca, S. Effects of the Reynolds number and the tip losses on the optimal aspect ratio of straight-bladed Vertical Axis Wind Turbines. *Energy* **2018**, *148*, 179–195. [CrossRef]
14. Alaimo, A.; Esposito, A.; Messineo, A.; Orlando, C.; Tumino, D. 3D CFD Analysis of a Vertical Axis Wind Turbine. *Energies* **2015**, *8*, 3013–3033. [CrossRef]
15. Cheng, Q.; Liu, X.; Ji, H.S.; Kim, K.C.; Yang, B. Aerodynamic Analysis of a Helical Vertical Axis Wind Turbine. *Energies* **2017**, *10*, 575. [CrossRef]
16. Islam, M.; Fartaj, A.; Carriveau, R. Analysis of the design parameters related to a fixed-pitch straight-bladed vertical axis wind turbine. *Wind Eng.* **2008**, *32*, 491–507. [CrossRef]
17. Amato, F.; Bedon, G.; Raciti Castelli, M.; Benini, E. Numerical Analysis of the Influence of Tip Devices on the Power Coefficient of a VAWT. *Proc. World Acad. Sci. Eng. Technol.* **2013**, *7*, 1035–1060.
18. Mishra, N.; Gupta, A.S.; Dawar, J.; Kumar, A.; Mitra, S. Numerical and Experimental Study on Performance Enhancement of Darrieus Vertical Axis Wind Turbine with Wingtip Devices. *J. Energy Resour. Technol.* **2018**, *140*, 121201. [CrossRef]
19. Laín, S.; Taborda, M.; López, O. Numerical Study of the Effect of Winglets on the Performance of a Straight Blade Darrieus Water Turbine. *Energies* **2018**, *11*, 297. [CrossRef]
20. Islam, M.; Amin, M.; Ting, D.; Fartaj, A. A New Airfoil for the Supporting Struts of Smaller-Capacity Straight-Bladed VAWT. In Proceedings of the 12th AIAA/ISSMO Multidisciplinary Analysis and Optimization Conference, Victoria, BC, Canada, 10–12 September 2008; p. 5874. [CrossRef]
21. Marsh, P.; Ranmuthugala, D.; Penesis, I.; Thomas, G. Three-dimensional numerical simulations of straight-bladed vertical axis tidal turbines investigating power output, torque ripple and mounting forces. *Renew. Energy* **2015**, *83*, 67–77. [CrossRef]
22. Marsh, P.; Ranmuthugala, D.; Penesis, I.; Thomas, G. Performance predictions of a straight-bladed vertical axis turbine using double-multiple streamtube and computational fluid dynamics models. *J. Ocean. Technol.* **2013**, *8*, 87–103.
23. Jiang, Y.; Zhao, P.; Stoesser, T.; Wang, K.; Zou, L. Experimental and numerical investigation of twin vertical axis wind turbines with a deflector. *Energy Convers. Manag.* **2020**, *209*, 112588. [CrossRef]
24. Castelli, M.; Pavesi, G.; Battisti, L.; Benini, E. Modeling Strategy and Numerical Validation for a Darrieus Vertical Axis Micro-Wind Turbine. In Proceedings of the International Mechanical Engineering Congress & Exposition 2010, Vancouver, BC, Canada, 12–18 November 2010.

25. Raciti Castelli, M.; Dal Monte, A.; Quaresimin, M.; Benini, E. Numerical evaluation of aerodynamic and inertial contributions to Darrieus wind turbine blade deformation. *Renew. Energy* **2013**, *51*, 101–112. [CrossRef]
26. Hasan, M.; Kabir, A.; Akib, Y.M. Dynamic stall investigation of two-dimensional vertical axis wind turbine blades using computational fluid dynamics. *Aip Conf. Proc.* **2019**, *2121*, 120003.
27. Balduzzi, F.; Bianchini, A.; Maleci, R.; Ferrara, G.; Ferrari, L. Critical issues in the CFD simulation of Darrieus wind turbines. *Renew. Energy* **2016**, *85*, 419–435. [CrossRef]
28. Rezaeiha, A.; Montazeri, H.; Blocken, B. Characterization of aerodynamic performance of vertical axis wind turbines: Impact of operational parameters. *Energy Convers. Manag.* **2018**, *169*, 45–77. [CrossRef]
29. Yi, M.; Jianjun, Q. Numerical Study of Flow Field and Aerodynamic Performance of Straight Bladed VAWT at Variable Tip Speed Ratios. *Open Mech. Eng. J.* **2015**, *9*, 1017–1024. [CrossRef]

Journal of
Marine Science and Engineering

MDPI

Article

Hydrodynamic Response of a Combined Wind–Wave Marine Energy Structure

Yapo Wang [1], Lixian Zhang [1], Constantine Michailides [2], Ling Wan [3] and Wei Shi [1,*]

1. State Key Laboratory of Coastal and Offshore Engineering, Deepwater Engineering Research Center, Dalian University of Technology, Dalian 116024, China; wangyapo@mail.dlut.edu.cn (Y.W.); 2248742388@mail.dlut.edu.cn (L.Z.)
2. Department of Civil Engineering and Geomatics, Cyprus University of Technology, Limassol 3036, Cyprus; c.michailides@cut.ac.cy
3. Newcastle Research & Innovation Institute, Newcastle University, Singapore 609607, Singapore; ling.wan@newcastle.ac.uk

* Correspondence: weishi@dlut.edu.cn; Tel.: +86-411-8470-8709

Received: 28 February 2020; Accepted: 2 April 2020; Published: 3 April 2020

Abstract: Due to the energy crisis and greenhouse effect, offshore renewable energy is attracting increasing attention worldwide. Various offshore renewable energy systems, such as floating offshore wind turbines (FOWTs), and wave energy converters (WECs), have been proposed and developed so far. To increase power output and reduce related costs, a combined marine energy structure using FOWT and WEC technologies has been designed, analyzed and presented in the present paper. The energy structure combines a 5-MW braceless semisubmersible FOWT and a heave-type WEC which is installed on the central column of the semisubmersible. Wave power is absorbed by a power take-off (PTO) system through the relative heave motion between the central column of the FOWT and the WEC. A numerical model has been developed and is used to determine rational size and draft of the combined structure. The effects of different PTO system parameters on the hydrodynamic performance and wave energy production of the WEC under typical wave conditions are investigated and a preliminary best value for the PTO's damping coefficient is obtained. Additionally, the effects of viscous modeling used during the analysis and the hydrodynamic coupling on the response of the combined structure are studied.

Keywords: wind–wave energy structures; floating offshore wind turbines; wave energy converters; viscous effects; hydrodynamic coupling

---

1. Introduction

Due to the exhaustion of fossil fuels and their pollution on the environment, people are trying to develop various clean and renewable energy technologies to replace conventional energy resources. Offshore wind energy has experienced rapid development [1] during the past ten years however it faces significant challenges to ensure commercial viability due to the policy of grid parity [2]. The ocean contains a variety of clean and renewable resources such as wind energy, wave energy and tidal energy, etc. Wave energy is still in the very early stages of development. The integration of offshore wind energy and wave energy power generation by sharing space, transmission infrastructure, operation and maintenance (O&M) activities and supporting structure will improve the utilization efficiency of marine renewable energy, but also reduce their levelized cost of energy.

Many scholars tried to develop different wind–wave combined systems. The EU funded Marine Renewable Integrated Application Platform (MARINA Platform) program [3] that proposed possible combined systems for use. For example, two combined systems have been proposed, which respectively consist of the same WindFloat semisubmersible turbine platform with a point-absorber-type wave energy converter (WEC) and an oscillating-water-column type WEC. Peiffer et al. [4] and Aubault et al. [5] investigated the coupled dynamic analysis by numerical and laboratory models. Muliawan et al. [6,7] proposed an integrated concept combining a spar-type floating wind turbine with a torus WEC (named as STC). The dynamic response characteristics of the combined structure in typical operational and extreme sea states were studied. Further, Wan et al. [8–10] studied the dynamic response characteristics of the STC under typical operational and extreme sea states based on numerical and experimental methods. Another combined structure combining a floating semisubmersible wind turbine and flap-type WECs (named as SFC) was proposed. Michailides et al. [11–13] systematically studied the integrated operation of the SFC by numerical simulation and physical model tests and it was found that the total produced power can be increased. Moreover, Ren et al. [14] studied a new combined structure consisting of a monopile type wind turbine and a heave-type wave energy converter (denoted as MWWC). However, the conceptual designs of combined structure based a floating semi-submersible wind turbine and WECs in deep water have been insufficient so far.

In the present paper, we introduce a combined marine energy structure consisting of a 5-MW braceless wind turbine developed in [15] with a heave-type WEC inspired by the STC system [16,17], which was proposed first by our team. At the first stage, it aims to investigate the dynamic response and its geometry and various parameters. Based on potential flow theory by means of ANSYS/AQWA [18], we developed a numerical model, and the best diameter and draft of the WEC are selected. Moreover, the effect of different power take-off (PTO) parameters on the performance of the WEC's produced power is investigated. A rational PTO damping and stiffness coefficient related with the power production is proposed. The effect of viscous load modeling and hydrodynamic coupling on the response of the combined structure is studied. Motions and dynamic responses of the combined structure under typical operational and extreme sea states are presented. The relative heave, semisubmersible pitch, horizontal contact force and PTO damping force of the combined structure, have an almost linear relationship with the wave height, while WEC surge and PTO produced power have an almost quadratic relationship under regular waves. Finally, a survival mode is examined in order to ensure combined concepts survivability in extreme wave conditions.

2. Theoretical Background Used for the Development of the Numerical Mode

2.1. Potential Flow Theory

For large-diameter elements, the existence of the structure affects the characteristics in the surrounding flow field. For the calculation of wave load, the three-dimensional potential flow theory is adopted [19,20], which generally assumes an ideal fluid and incompressible flow. The Laplace equation can be obtained with a velocity potential Φ based on the conservation of fluid mass and momentum, and is expressed as follows:

$$\nabla^2\phi(x,y,z,t) = 0 \tag{1}$$

The free wave surface, structure surface and subsea conditions of Φ can be expressed as below:

$$\frac{\partial^2\phi}{\partial t^2} + g\frac{\partial\phi}{\partial z} = 0,\ z = 0 \tag{2}$$

$$\frac{\partial\phi}{\partial n} = U_j\widetilde{n}_j \tag{3}$$

$$\frac{\partial\phi}{\partial z} = 0, z = -h \tag{4}$$

where U_j and $\widetilde{n}_j$ represent the generalized velocity and generalized normal vector of the corresponding motion mode of a point on the structure surface, respectively.

In the potential flow theory, the control equations and boundary conditions can be linearized, and the linear superposition principle can be used to decompose the velocity potential as below:

$$\phi(x, y, z, t) = \phi(x, y, z, t)^i + \phi(x, y, z, t)^d + \phi(x, y, z, t)^r \tag{5}$$

where $\phi(x, y, z,t)^i$ represents the incident wave velocity potential, $\phi(x, y, z,t)^d$ is the diffraction potential and $\phi(x, y, z,t)^r$ represents the radiation potential.

If the floating platform exhibits simple harmonic motion at the balanced position in static water, the velocity potential of the floating platform can be expressed as follows:

$$\phi(x, y, z, t) = \mathrm{Re}\{\varphi(x, y, z) \cdot \exp(-i\omega t)\} \tag{6}$$

$$\varphi(x, y, z, t) = \varphi^i(x, y, z, t) + \varphi^d(x, y, z, t) + \varphi^r(x, y, z, t) \tag{7}$$

The incident wave velocity potential can be expressed as follows:

$$\varphi(x, y, z, t)^i = -\frac{Ag}{\omega}\frac{\cosh[k(z+h)]}{\cosh kh} \cdot \exp(ik(x \cdot \cos\beta + y \cdot \sin\beta)) \tag{8}$$

where A, g, h, k and β represent the wave amplitude, gravitational acceleration, water depth, wavenumber and wave propagation direction, respectively.

The radiation potential that can be solved by the following governing equations and boundary conditions can be expressed as:

$$\varphi(x, y, z, t)^r = -i\omega \overline{x}_j \varphi_j^r(x, y, z) \tag{9}$$

The control equation, free wave surface condition, structure surface condition, subsea conditions and $r \to \infty$ can be expressed as follows:

$$\nabla^2 \varphi_i^r(x, y, z) = 0 \tag{10}$$

$$\frac{\partial \varphi_j}{\partial n} - v\varphi_j^r = 0,\ z = 0 \tag{11}$$

$$\frac{\partial \varphi_j^r}{\partial n} = \widetilde{n}_j \tag{12}$$

$$\frac{\partial \varphi_j^r}{\partial n} = 0,\ z = -h \tag{13}$$

$$\lim_{r \to \infty} \sqrt{r}\left(\frac{\partial \varphi_j^r}{\partial r} - ik\varphi_j^r\right) = 0 \tag{14}$$

where $\overline{x}_j$ is the motion amplitude in the motion mode of j, and φ_j^r represents the radiation potential in the motion mode of j.

$$r = \sqrt{x^2 + y^2} \tag{15}$$

Finally, the Green's function method is used to calculate the diffraction potential and is no elaborated further here.

2.2. Viscous Load Modeling

2.2.1. Morison Equation (Cylinders of the Semisubmersible Platform)

For small-diameter offshore structures, the Morrison equation is mainly used to calculate the hydrodynamic loads [19,21]. When the diameter of the structure is less than or equal to 0.15 times the wavelength, the existence of the structure has small effect on the wave field. The wave force applied on the structure is the sum of the drag force and the inertia force, which can be expressed as follows:

$$f_d = C_d \frac{\rho}{2} D\mu|\mu| \tag{16}$$

where C_d represents the drag force coefficient. In the paper, $C_d = 1.2$ is selected due to $H/d \leq 0.2$ and $d/L \geq 0.2$; H, d and L represent wave height, water depth and wave length, respectively; μ, D, and ρ are the incoming flow velocity, structure diameter and fluid density, respectively; and f_d represents drag force on a unit height of the structure.

2.2.2. Heave Mode Viscous Load Effects on the WEC

The viscosity can significantly affect the dynamic responses of the combined structure (especially the estimation of the PTO produced power), but the effect of viscous damping on the WEC cannot be calculated based on potential flow theory, which has a great difference with physical truth. Therefore, the viscous effect of the heave mode of the WEC has been additionally included, which is calculated as follows [22,23]:

$$D^{critical} = 2\sqrt{(M_{33} + m_{33}) \cdot K_{33}} \tag{17}$$

$$F^{viscous} = D^{critical} \times \eta \tag{18}$$

where M_{33}, m_{33} and K_{33} represent the mass, added mass at heave natural frequency and hydrostatic stiffness coefficient of the WEC in the heave direction and their values are 540.982 t, 422.3 t and 1515.4 kN/m, respectively; $F^{viscous}$, $D^{critical}$ and η are the viscous force, the critical damping and damping coefficient based on the viscous effects of the heave mode of the WEC. Based on the reference [24], a damping coefficient of 6–10% in heave direction is recommended using experimental study for different cylindrical semisubmersible platforms. Therefore, 8% of damping coefficient for WEC in heave direction is used in our simulation.

2.3. *Equation of Motion of the Combined Structure*

The simulation of the combined structure is carried out in the time domain. The hydrodynamic properties of the numerical modeling, such as the linear excitation forces, added mass and potential damping as well as the coupling terms of the two bodies, are obtained. In addition, the mechanical coupling (PTO of WEC) between the two bodies is modeled with both linear damper and spring. The mooring lines are also regarded as linear springs. Therefore, the equation of motion for the analysis of the combined structure can be expressed as follows [9]:

$$\begin{gathered}\begin{pmatrix} (M+m)_{11} & m_{12} \\ m_{21} & (M+m)_{22} \end{pmatrix}\begin{pmatrix} \ddot{x}_1(t) \\ \ddot{x}_2(t) \end{pmatrix} + \int_0^t \begin{pmatrix} h(t-\tau)_{11} & h(t-\tau)_{12} \\ h(t-\tau)_{21} & h(t-\tau)_{22} \end{pmatrix}\begin{pmatrix} \dot{x}_1(\tau) \\ \dot{x}_2(\tau) \end{pmatrix} d\tau \\ +\begin{pmatrix} (R)_{11} & (R)_{12} \\ (R)_{21} & (R)_{22} \end{pmatrix}\begin{pmatrix} x_1(t) \\ x_2(t) \end{pmatrix} = \begin{pmatrix} f_1^{wind}(t) \\ 0 \end{pmatrix} + \begin{pmatrix} f_1^{wave}(t) \\ f_2^{wave}(t) \end{pmatrix} + \begin{pmatrix} f_1^{viscous}(t) \\ f_2^{viscous}(t) \end{pmatrix} + \\ \begin{pmatrix} f_1^{interface}(t) \\ f_2^{interface}(t) \end{pmatrix} + \begin{pmatrix} F_1^{PTO}(t) \\ F_2^{PTO}(t) \end{pmatrix}\end{gathered} \tag{19}$$

$$f_1^{interface}(t) = -f_2^{interface}(t) \tag{20}$$

$$F_1^{PTO} = -F_2^{PTO} \tag{21}$$

where the subscripts 1 and 2 are denoting the semisubmersible platform and the WEC, respectively; M and m represent the structural mass matrix and the added mass, respectively; x, $\dot{x}$ and $\ddot{x}$ represent the displacement, velocity and acceleration of the bodies; $h(t\text{-}\tau)$ represents the retardation function; R_{ij} is the restoring coefficient matrix; $f^{wind}(t)$ and $f^{wave}(t)$ represent wind loads on the rotor and wave forces, respectively; $f_1^{viscous}(t)$ is the viscous force based on Morison equation applied on the semisubmersible and $f_2^{viscous}(t)$ is the viscous load applied on the WEC (Equation (21)); $f^{interface}(t)$ is the interface forces between the semisubmersible platform and the WEC including horizontal contact forces and vertical friction forces; and F^{PTO} represents the PTO force.

3. Characteristics of the Combined Structure

3.1. Design Parameters of the 5-MW Semisubmersible Wind Turbine and of the WEC

The combined model proposed in the present paper consists of a braceless 5-MW semisubmersible wind turbine and a heave-type WEC. The conceptual model and the main design parameters of the 5-MW semisubmersible wind turbine and the WEC are shown in Figure 1 and Table 1, respectively. In the model, the WEC can be moved up and down relative to the center column of the semisubmersible, and the wave energy is produced by the PTO at the same time. The relative movement is constrained by the mechanical connection between the center column and the WEC in a direction perpendicular to the center column [25]. As a result, the utilization efficiency of marine renewable energy is improved through shared power generation platforms and shared transmission systems, effectively reducing the unit power generation cost [6,13]. For the numerical simulation, viscous forces of the columns of the semisubmersible are considered as viscous drag loads using Morison equation in the horizontal direction.

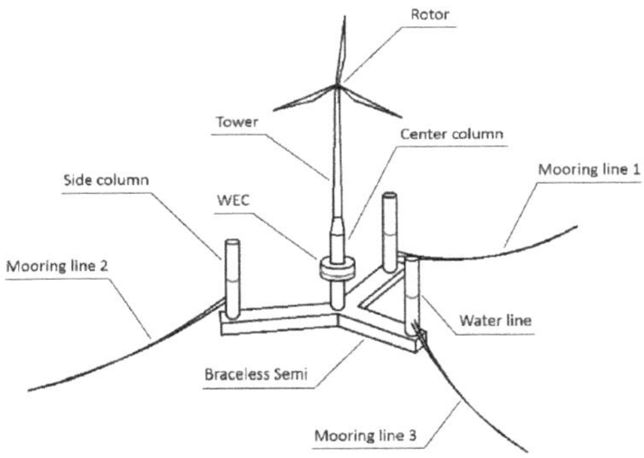

Figure 1. Conceptual sketch of the combined structure.

Table 1. Main design parameters of the combined structure.

	Parameters	Values
Wind turbine (NREL 5 MW)	Rotor-Nacelle-Assembly	350 t
	Hub height	90 m
	Tower mass	347.46 t
semisubmersible platform	Semisubmersible mass	9738 t
	Diameter of the central column	6.5 m
	Diameter of the three side columns	6.5 m
	Operating draft	30 m
	Water displacement	10,298 m^3
	Water depth	200 m
WEC device	Outer/Inner diameter	16 m/8 m
	Height/Draft	8 m/3.5 m
	Mass	463.5 t
	Water displacement	452.2 m^3
	Center of mass	(0, 0, −1 m)

3.2. Power-Take-Off (PTO) System

The combined structure has been modeled as two rigid bodies consisting of a semisubmersible platform and a WEC, which are connected by a PTO. The PTO has been simplified as a system of heave-direction linear dampers (B_{PTO}) and linear springs (K_{PTO}) and is modeled with the use of corresponding fender features in ANSYS/AQWA (Figure 2).

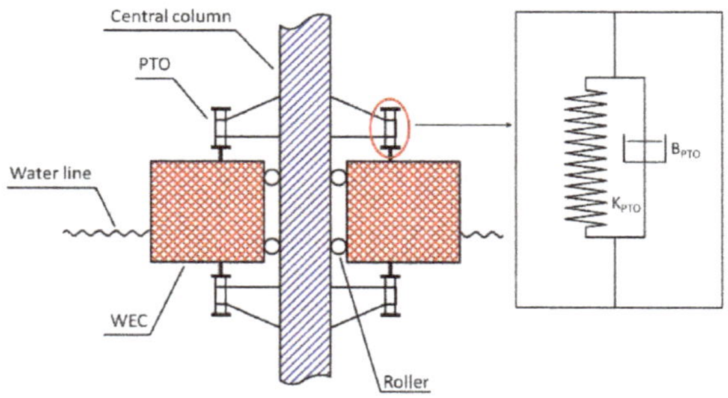

Figure 2. Connection detail between the column of semisubmersible and the wave energy converter (WEC).

For the PTO system [9], the PTO force includes the linear damping force and the linear spring force, which can be expressed as:

$$F_{PTO} = B_{PTO} \cdot (\dot{x}_2 - \dot{x}_1) + K_{PTO} \cdot (x_2 - x_1) \tag{22}$$

where B_{PTO} and K_{PTO} are the linear damping stiffness coefficient and the linear spring stiffness coefficient, respectively.

Moreover, the produced power by the relative heave motion between the column of the semisubmersible and the WEC is calculated as follows:

$$P = B_{PTO} \cdot (\dot{x}_2 - \dot{x}_1) \cdot (\dot{x}_2 - \dot{x}_1) \tag{23}$$

4. Results and Discussion

4.1. Model Parameters Best Selection

4.1.1. Preliminary Determination of Structural Design Parameters of the WEC

The modification of the structural parameters of the WEC will directly affect its hydrodynamic performance, resulting in changes in the produced wave energy efficiency. By evaluating the motion responses of the combined structure, a rational outer diameter and draft can be obtained.

Five different outer diameters of the WEC ranging from 12.0 m to 20.0 m are examined. To obtain the best WEC in terms of PTO produced power, WEC's cost and structure's safety, some factors such as gravity of the WEC and the horizontal contact forces between the column of the semisubmersible and the WEC have been considered. The values of those parameters are calculated under a regular wave with wave height 2 m and wave period 9 s. It is noted that wind loads are not considered here. To reduce the cost and increase safety, the best selection of outer diameter of the WEC can be obtained by Equation (24) and Figure 3. It is noted that we did not solve the mathematical optimization problem in this paper. Based on the curves in Figure 3, a rational value of 16 m is selected for the rest of the paper since all the curves are close to their peak for 16 m.

$$\theta = \frac{P}{\alpha F_h + \beta G} \tag{24}$$

where θ reflects a performance indicator of the WEC; P, F_h and G represent the PTO produced power, horizontal contact force and WEC weight, respectively, and α and β are related to the cost and safety of the WEC. With the comparison of different parameters with various groups (α = 0.4, β = 0.6; α = 0.3, β = 0.7; α = 0.2, β = 0.8), it is found that the best selection will be achieved when the outer diameter is 16 m. In this paper, we assume that cost reduction (related to β) is more important and the safety is less important (related to α) for the combined structure. Therefore, α = 0.3, β = 0.7 is assumed in this study.

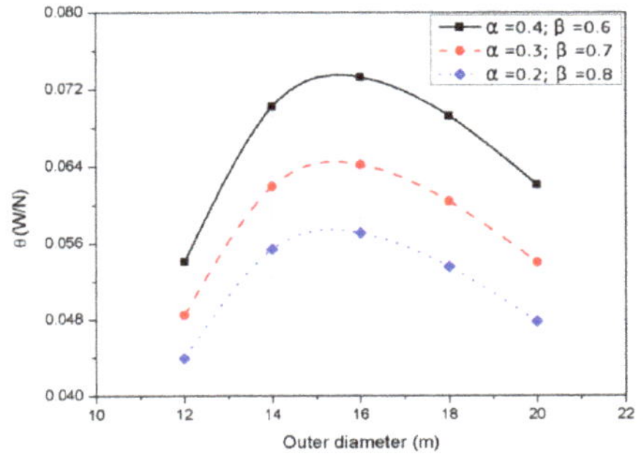

Figure 3. θ values for different outer diameters of the WEC.

Moreover, the effect of different drafts of the WEC on the responses of the semisubmersible are presented in Figure 4 and are slightly different when the draft is 3.0 m and 3.5 m. However, considering the possible slamming effects due to water entry and exit, which can destroy the WEC, choosing the value of the draft (3.5 m) is more reasonable. Therefore, the selection of diameter 16 m and draft 3.5 m is followed in the rest of the paper.

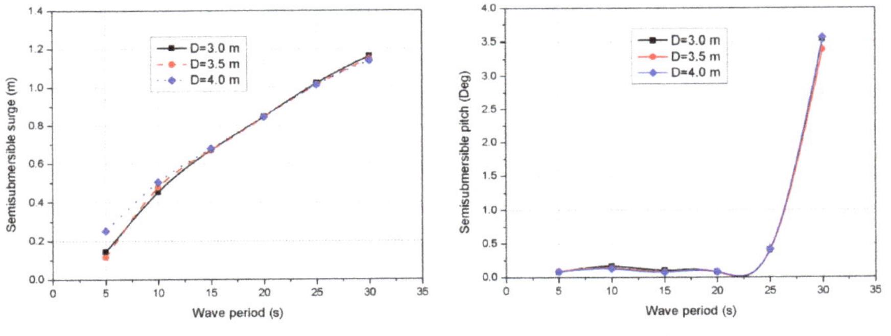

(a) (b)

Figure 4. Effects of the daft on the WEC: (**a**) semisubmersible surge; (**b**) semisubmersible pitch.

4.1.2. Determination of Rational B_{PTO} and K_{PTO} Parameters

Before determination of linear damping and stiffness coefficients, the effect of different wave periods (B_{PTO} = 2000 kNs/m, K_{PTO} = 100 N/m was arbitrary selected) on the PTO produced power has been investigated under regular waves with wave height 2 m. The values of produced power are shown in Figure 5. It indicates that there is a maximum produced power when the wave period is around 9 s because the WEC was designed to have a natural period in heave around 7–2 s.

The B_{PTO} and K_{PTO} parameters can significantly affect the dynamic responses of the combined structure. To illustrate mutual effects of linear damping and stiffness coefficients on the PTO produced power of the WEC, different B_{PTO} and K_{PTO} values are simulated with wave height 2 m and wave period 9 s. The results of produced power are shown in Figure 6; it can be seen that the value of produced power will be the largest when the B_{PTO} coefficient is 1500 kNs/m with an invariable K_{PTO}. The value of produced power increases when the B_{PTO} is smaller than 1500 kNs/m, while the values will decrease when the B_{PTO} is larger than 1500 kNs/m. Moreover, for same B_{PTO} values, produced power decreases as K_{PTO} increases, which is more significant when its value is over 1000 N/m. Therefore, a best B_{PTO} coefficient of 1500 kNs/m and a K_{PTO} coefficient of 1 N/s are selected for the rest of the paper, respectively.

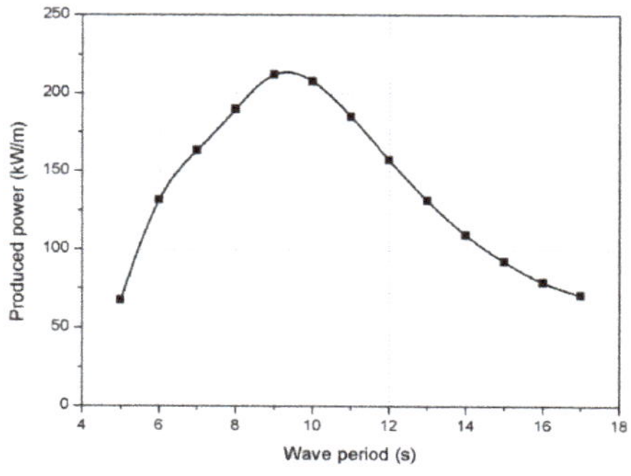

Figure 5. Performance of the WEC with different wave periods.

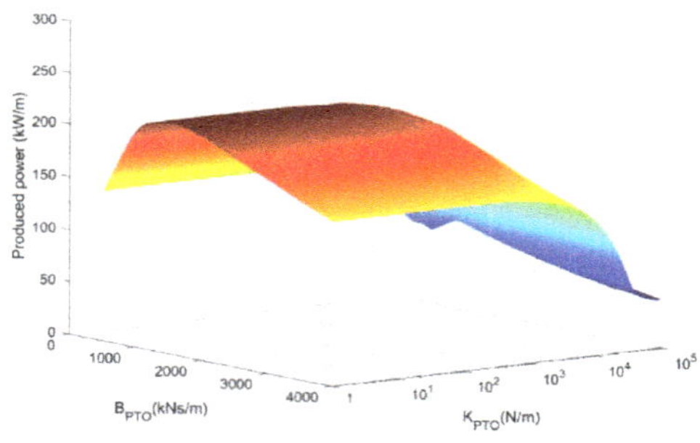

Figure 6. Mutual effect of different K_{PTO} and B_{PTO} values on the WEC performance.

4.2. Viscous Effects on the Heave Motion of the WEC

By ignoring viscous effects, the produced power may be overestimated when there is a large relative motion between the semisubmersible and the WEC. The dynamic responses of the combined structure with or without considering viscous effects have been presented for regular waves with a wave height 2 m and a wave period 9 s. In Figure 7, the dynamic responses considering viscous effects are smaller than those without considering them. The relative heave motion with viscous effect has a peak value when wave period is 11 s, which is different from that without viscous effect. However, the radiation force, PTO damping and produced power, a peak value exists when the period is 9 s. Moreover, the ratio which can quantitatively describe viscous effects on the combined structure between those responses (Equation (25)) always increases with increasing period.

$$\varepsilon = \frac{\lambda}{v} \times 100\% \tag{25}$$

where ε is a ratio, and λ and v represent the dynamic responses with and without viscous effects.

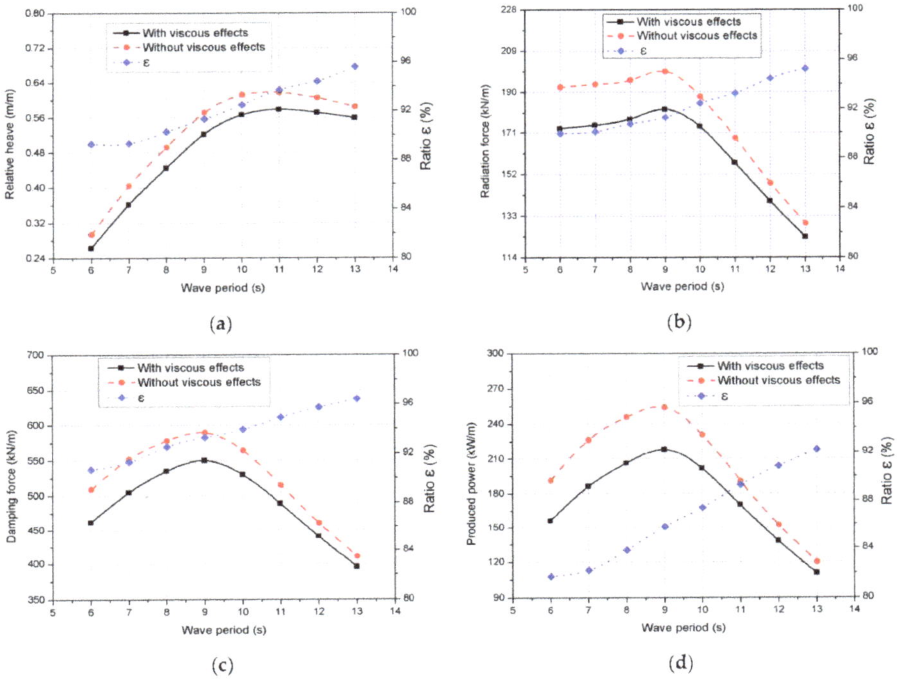

Figure 7. Comparison of responses of the WEC with and without viscous effects: (**a**) relative heave; (**b**) radiation force; (**c**) damping force; (**d**) produced power.

4.3. The Hydrodynamic Coupling Effect on the Combined Structure

In the flow field, in addition to the direct interaction between the semisubmersible and WEC, there is also an interaction between the two bodies due to the radiation and diffraction effects of the combined structure on the flow field. The radiation and diffraction potentials acting on the semisubmersible and WEC can interact when considering hydrodynamic coupling [26,27], which is proved by comparing diffraction forces and radiation forces of the semisubmersible and the WEC with considering hydrodynamic coupling with those without considering hydrodynamic coupling (Figure 8). It can be seen that the diffraction forces (Figure 8a) and radiation forces (Figure 8b) acting on the semisubmersible have same trends for different wave periods when considering hydrodynamic coupling. There is a little increase of the diffraction force and radiation force acted on the semisubmersible considering hydrodynamic coupling. The largest difference for diffraction force and radiation force is approximately 3.8% and 3.6% larger than those without considering hydrodynamic coupling at wave period of 11 s. However, for WEC, the effects of the hydrodynamic coupling are significant on the diffraction forces (Figure 8c) and radiation forces (Figure 8d). And the diffraction force acted on the WEC considering hydrodynamic coupling is approximately 52.2% larger than that without considering hydrodynamic coupling at the wave period of 8 s and the radiation force is 55.4% larger than that without considering hydrodynamic coupling at the period of 11 s. Therefore, motion and dynamic responses of the WEC should be put more emphasis compared with the semisubmersible for the combined structure.

As hydrodynamic loads with considering hydrodynamic coupling are different from those without considering hydrodynamic coupling the relevant responses of the combined structure with wave height 2 m are presented and studied. The WEC surge motion (Figure 9a) and the relative heave motion (Figure 9b) considering hydrodynamic coupling are not always larger compared to those without considering hydrodynamic coupling. Additionally, the amplitudes of the relative heave motion (Figure 9b) and WEC pitch motion (Figure 9c) have similar trends, but WEC pitch motion considering hydrodynamic coupling is always larger than the case without considering hydrodynamic coupling. Figure 9d shows the wave period considering hydrodynamic coupling (approximately 9 s) is larger than that without considering hydrodynamic coupling (approximately 8 s). The largest produced power considering hydrodynamic coupling is smaller than that without considering hydrodynamic coupling. Same conclusions with PTO damping force (Figure 9e) are presented.

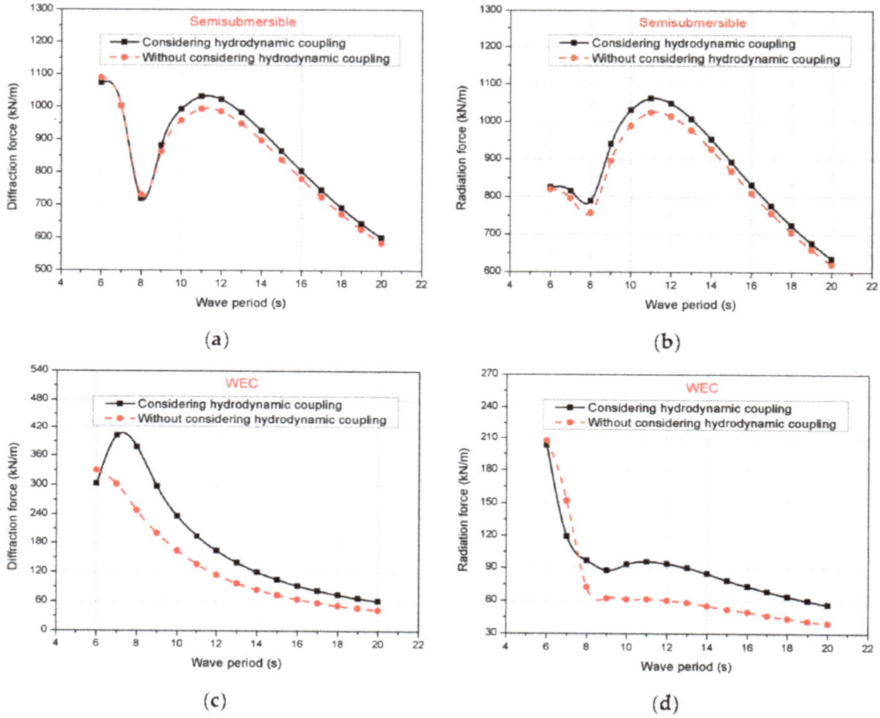

Figure 8. Comparison of diffraction forces and radiation forces of the combined structure between considering and without considering hydrodynamic coupling: (**a**) diffraction forces for semisubmersible; (**b**) radiation forces for semisubmersible; (**c**) diffraction forces for WEC; (**d**) radiation forces for WEC.

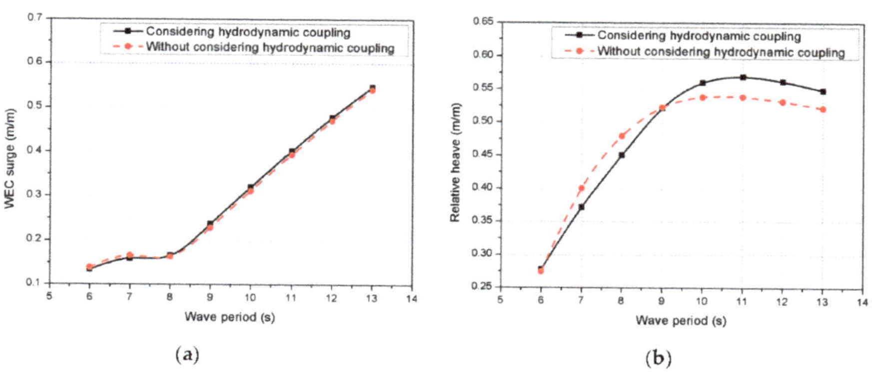

Figure 9. *Cont.*

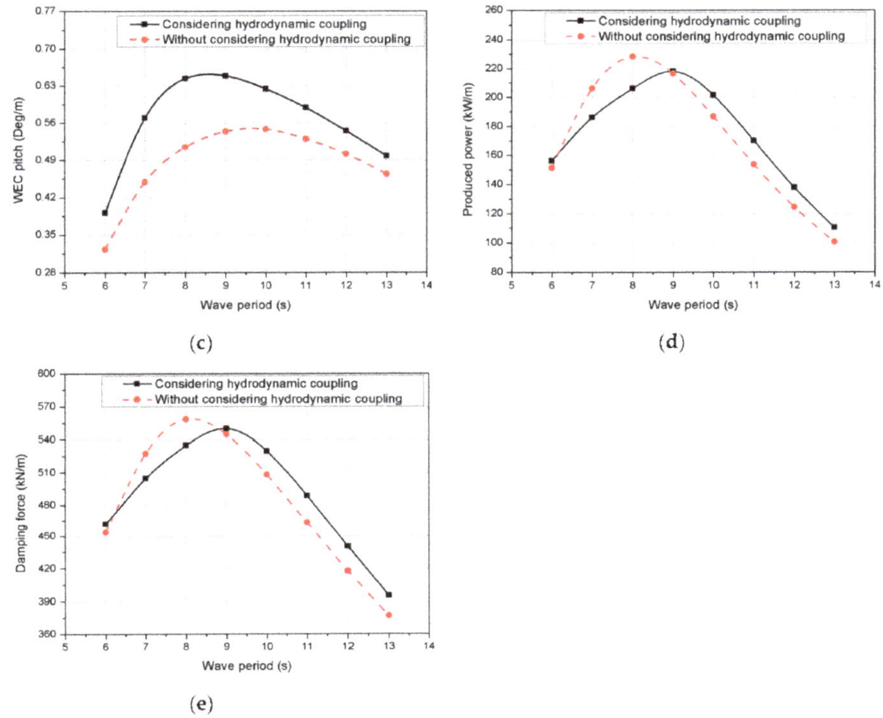

Figure 9. Comparison of responses of the combined structure between considering and without considering hydrodynamic coupling cases under different wave periods (H = 2 m): (**a**) WEC surge; (**b**) relative heave; (**c**) WEC pitch; (**d**) produced power; (**e**) damping force.

4.4. Dynamic Responses of the Combined Structure under Different Type Environmental Conditions

4.4.1. Regular Waves

Regular wave sea conditions with H = 2 m and T = 9 s are simulated in the time domain (Figure 10) without wind excitation loads [28]. In Figure 10a, the maximum relative heave velocity of the combined structure is smaller than 0.375 m/s while the minimum absolute value is larger than 0.375 m/s, which is an explanation why the peaks of produced power (Figure 10c) are not obtaining same values (Equation (23)). In Figure 10b, the PTO damping force induced by fenders in AQWA mode is presented. The phase of the forces is similar with relative heave, which means the damping forces are negative or positive when the WEC is up or down. Moreover, the capture width ratio, which is used to evaluate the WEC capability of absorbed wave power, is calculated by Equations (26) and (27) [29,30] as follows:

$$\overline{P}_{wave} = \frac{\rho g^2}{32\pi} H^2 T \tag{26}$$

$$\sigma = \frac{\overline{P}_{PTO}}{\overline{P}_{wave}} \times 100\% \tag{27}$$

where $\overline{P}_{wave}$ and $\overline{P}_{PTO}$ are the average wave power per unit width and average PTO produced power, respectively, and σ denotes the capture width ratio.

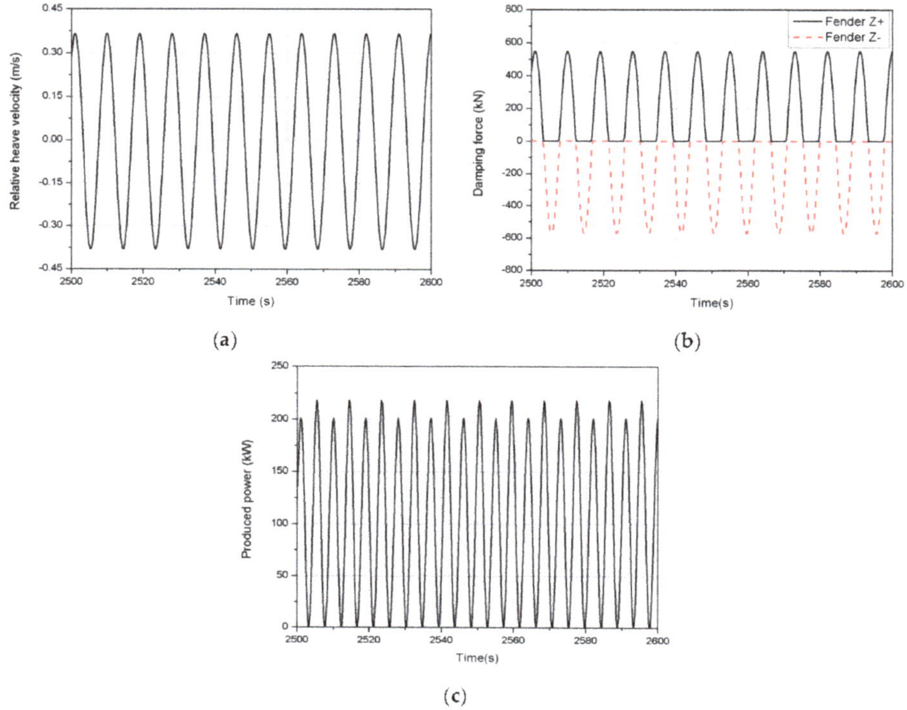

Figure 10. Time series of different responses of the combined structure under regular waves H = 2 m and T = 9 s: (**a**) relative heave velocity; (**b**) damping force; (**c**) produced power.

To obtain the characteristics of the combined structure, the responses of the proposed combined structure are compared with similar work done by Wan et al. [9], Gao el al. [31] and Ren et al. [32] (Figure 11). Attention should be paid that in our work a semisubmersible platform is used while a spar platform was used in Wan and Gao's work and a tension leg platform (TLP) was used Ren's work. In Figure 11a,b, platform surge and WEC heave have similar tendency compared with Wan et al.'s work. The wave energy capture width ratio is compared with STC [9,31] and [32] in Figure 11d. The difference of the capture width ratio is caused by two reasons. The first one is the difference of PTO damping and stiffness coefficient, which result in the difference for the best wave period and produced power. The second reason is that the relative heave motions between the semisubmersible platform and WEC are much less than those between the spar and the TLP platforms and their WECs. This is caused by the heave motion difference of the spar, TLP and semisubmersible platforms. According to Equation (23), produced power of the combined structure will be smaller compared with those power produced by STC [9,31] and [32]. Special attention should be made that Wan's results are based on the irregular wave case. It is stated that the comparisons are presented for having an idea about the well comparable response of the combined structure proposed in the present paper.

Figure 12 shows that the amplitudes of the responses, such as the relative heave, semisubmersible pitch, horizontal contact forces and PTO damping forces of the combined structure, almost linearly increase with the wave height, while the amplitudes of semisubmersible surge and PTO produced power have an approximately quadratic relationship with the wave height.

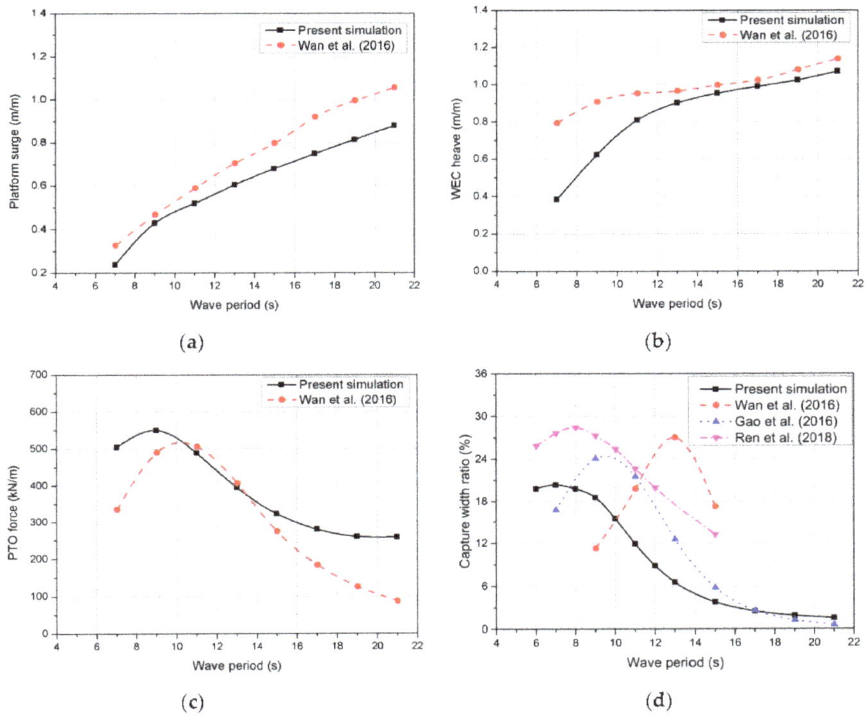

Figure 11. Responses of the combined structure under different wave periods with H = 2 m: (**a**) platform surge; (**b**) WEC heave; (**c**) power take-off (PTO) force; (**d**) capture with ratio.

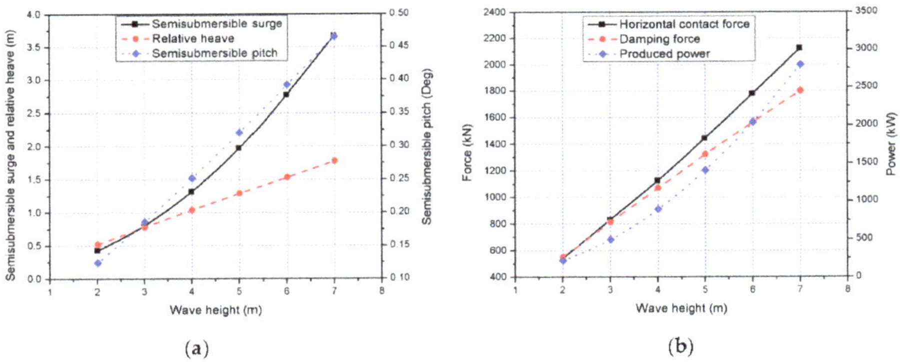

Figure 12. Responses of the combined structure under different wave heights T = 9 s: (**a**) semisubmersible surge, relative heave and semisubmersible pitch; (**b**) horizontal contact force, damping force and produced power.

4.4.2. Irregular Wave and Wind Conditions

The sea conditions of certain sea areas in Guangdong Province, China whose location of the offshore wind farm is presented in Figure 13 have been selected, as shown in Table 2. In the paper, the aerodynamic loads from wind flow are simplified as a mean constant thrust load based on NREL 5 MW wind turbine thrust force-wind speed curve (Figure 14). The JONSWAP spectrum is used to simulate the irregular wave in this section. Comparisons of the time series and power spectral density (PSD) of motion and dynamic responses for LC1 (H_s = 2 m, T_p = 9 s) and LC2 (H_s = 2 m, T_p = 9 s and U_{wind} = 17 m/s) are presented in Figures 15 and 16, respectively. Figure 15 shows that the semisubmersible surge and pitch motions have big differences due to wind loads. However, the values of the relative heave, PTO damping force and PTO produced power have slight differences.

Figure 13. Location of the reference offshore site.

Table 2. Sea states of the reference sea areas in Guangdong Province, China.

Sea States	Wave Type	Wave Height H_s (m)	Wave Period T_p (s)	Wind Speed U_{wind} (m/s)
LC 1	Irregular	2.0	9	0
LC 2	Irregular	2.0	9	17
LC 3	Irregular	3.0	10	24
LC 4	Irregular	8.6	15	31.2

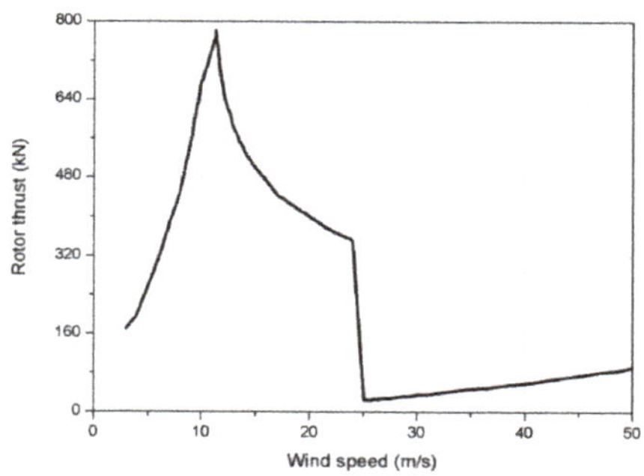

Figure 14. Rotor thrust curves at different wind speeds.

The natural frequencies (surge, heave and pitch) of the semisubmersible could be identified through free decay tests (Table 3). In Figure 16, PSD of motion and dynamic responses of the combined structure under LC1 have same trends compared with those under LC2. In Figure 16a, a resonance is around 0.073 rad/s, which is contributed by surge natural frequency. At this frequency, the surge motion in LC2 is enlarged due to additional wind load compared with LC1, which can also be confirmed from Figure 15a. The second peak (around 0.7 rad/s) in Figure 16a presents the contribution from the wave frequency to the surge response. Figure 16c shows that the pitch motion response is dominant by pitch resonance frequency (0.218 rad/s). Resonances from surge natural frequency (0.073 rad/s) and wave frequency (0.7 rad/s) could also be identified from pitch motion response. For heave motion in Figure 16b, the wave frequency response is dominated in the range between 0.500 rad/s to 0.800

rad/s, in which WEC is designed to capture the energy from incident wave. For PTO damping force (Figure 16d), the surge natural frequency and wave natural frequency could be clearly identified from load responses.

Table 3. Natural frequencies (rad/s) of the semisubmersible.

Surge	Heave	Pitch
0.07545	0.2543	0.2108

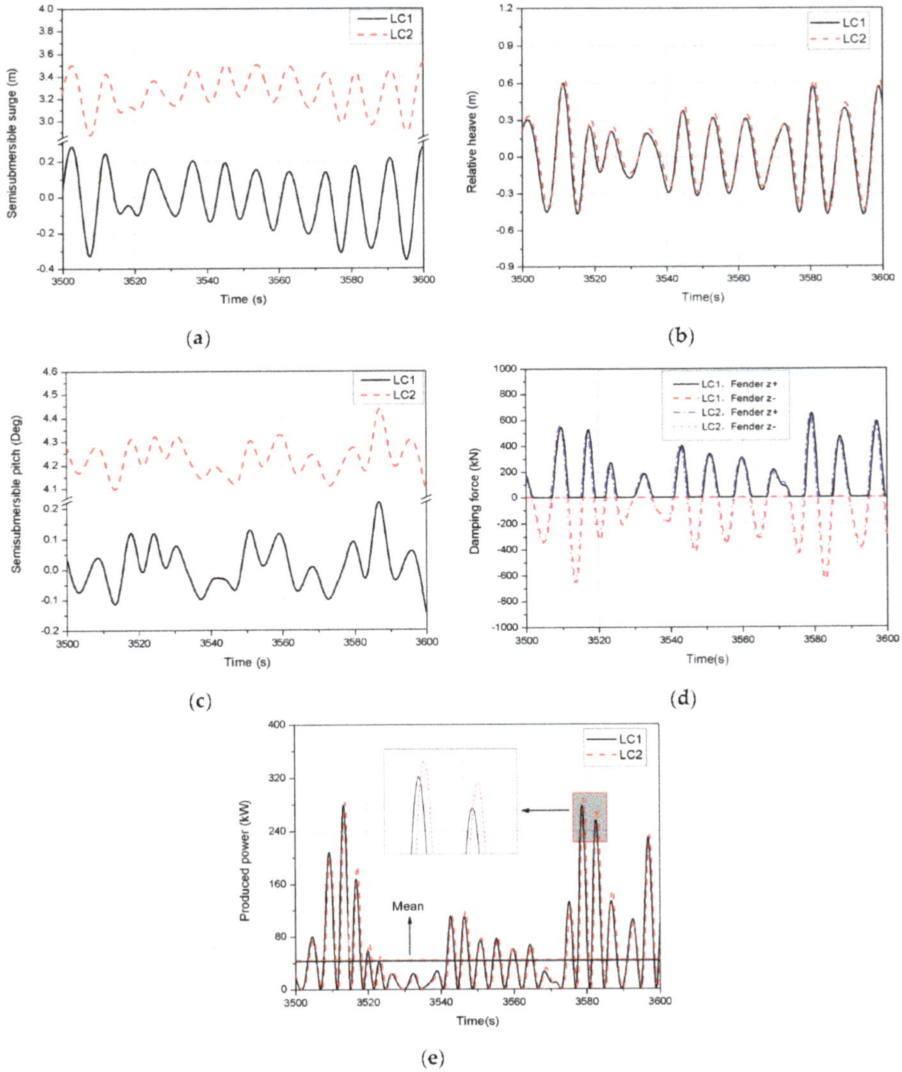

(a) (b) (c) (d) (e)

Figure 15. Comparisons of the time series of the responses for LC1 and LC2 of: (**a**) semisubmersible surge; (**b**) relative heave; (**c**) semisubmersible pitch; (**d**) damping force; (**e**) produced power.

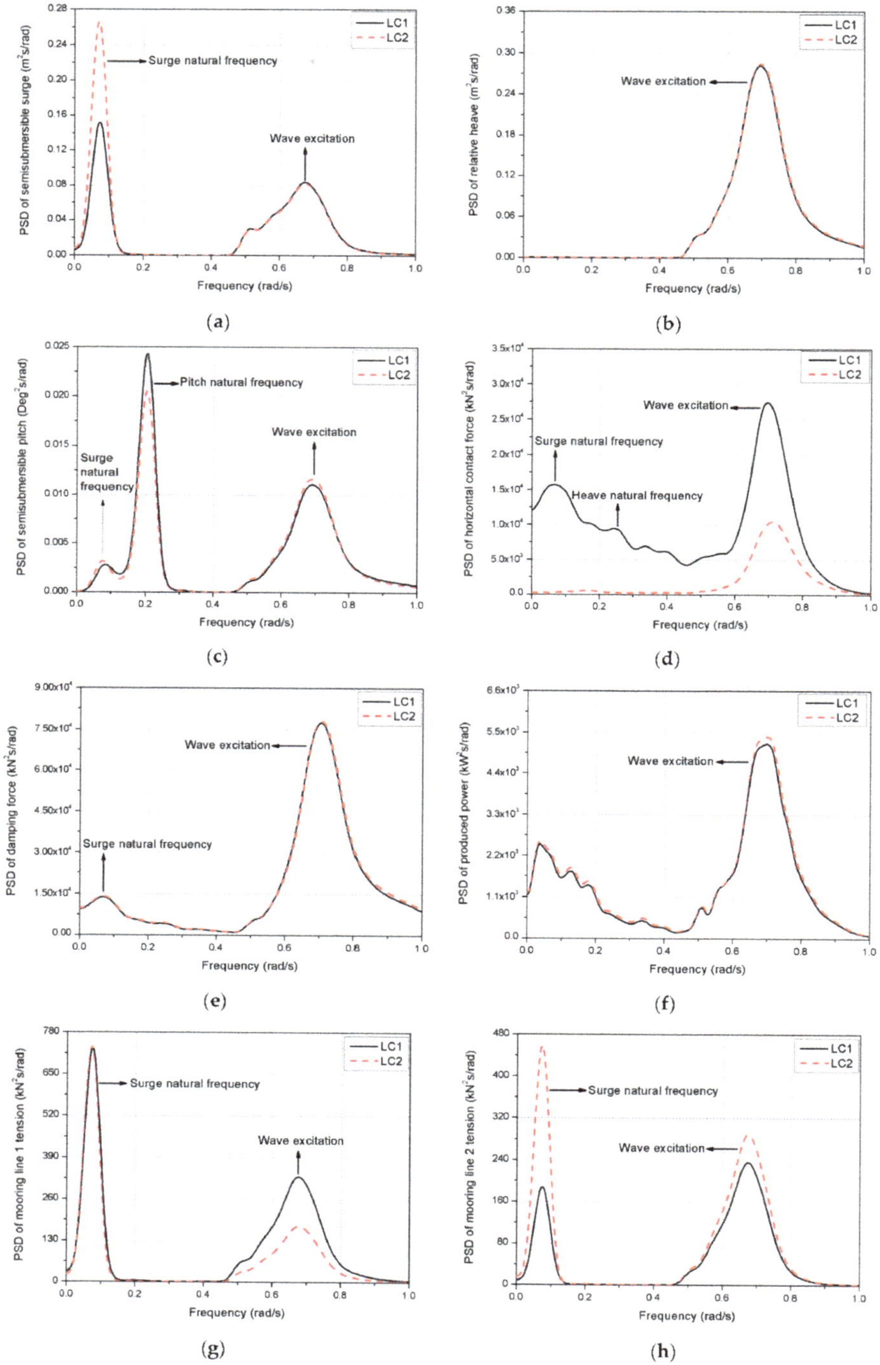

Figure 16. Comparisons of the PSD of the motion and dynamic responses for LC1 and LC2: (**a**) semisubmersible surge; (**b**) relative heave; (**c**) semisubmersible pitch; (**d**) horizontal contact force; (**e**) damping force; (**f**) produced power; (**g**) mooring line 1 tension; (**h**) mooring line 2 tension.

The statistical values of the motion and dynamic responses of the combined structure under LC3 are shown in Figure 17. In Figure 17a, we can see that the maximum surge is less than 3.5 m and maximum pitch is less than 3.5 degrees. The maximum relative heave is larger than 1.25 m, while the absolute value of the minimum relative heave is smaller than 1.25 m, which may be caused by the lift effect induced by strong wave action [22]. Compared with LC1, it finds that large mean value of surge and pitch motion are caused by wind loads and positive surge and pitch moment presents in LC3. Figure 17b shows that the horizontal contact force between the semisubmersible and the WEC is larger than the wave force acting on the WEC due to the relative acceleration of the two bodies. The mooring line 1 tension force is larger than that of mooring line 2 because the drift of the combined structure in downwind direction due to the wind load. The mean PTO produced power is around 50 kW with maximum power around 1245 kW under LC3.

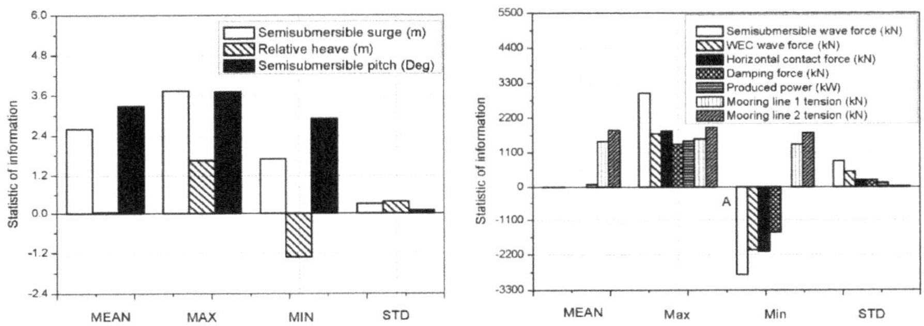

(a) (b)

Figure 17. Statistical values of the responses for LC3: (**a**) statistic of information for semisubmersible surge, relative heave and pitch; (**b**) statistic of information for semisubmersible wave force, WEC wave force, horizontal contact force, damping force, produced power, mooring line 1 tension and mooring line 2 tension.

Moreover, in Figure 18, PSD of motion responses (surge, relative heave and pitch) and dynamic responses (horizontal contact force, PTO damping force, mooring line 1 tension and mooring line 2 tension) of the combined structure with considering hydrodynamic coupling have similar trend compared with those without considering hydrodynamic coupling. For surge motion (Figure 18a), the surge resonant motion is dominant at 0.073 rad/s, which is the surge natural frequency, followed by the wave frequency response in the range 0.400 rad/s to 0.800 rad/s. A large decrease of the response at the surge frequency exhibits if the hydrodynamic coupling is considered. Similar as LC1 and LC2 (Figure 16b), the most significant contribution to the relative heave motion comes from the wave frequency (Figure 18b). Considering hydrodynamic coupling will give a small increase in heave motion response near wave frequency, which is similar as in surge motion response. This enhancement of the surge and pitch responses near wave frequency can also be identified with wave period of 10 s in Figure 9a,b. For pitch motion (Figure 18c), the wave frequency response and pitch resonant response are very comparable, meanwhile, surge resonance response also visible in this motion. The pitch response considering hydrodynamic coupling at the surge natural frequency and wave frequency is larger than that without considering hydrodynamic coupling, opposite influence can be identified at pitch natural frequency. Figure 18d,e shows that the wave frequency responses are also dominant in the same range for horizontal contact force and damping force and the hydrodynamic coupling will contribute the increase of the response in the wave frequency range. For mooring line tension force (Figure 18g,h), PSD of mooring line 1 tension with surge resonance frequency occupy a dominant position compared with the value of mooring line 2 tension. Due to the hydrodynamic coupling effect, both responses of mooring line 1 and line 2 reduce at the surge.

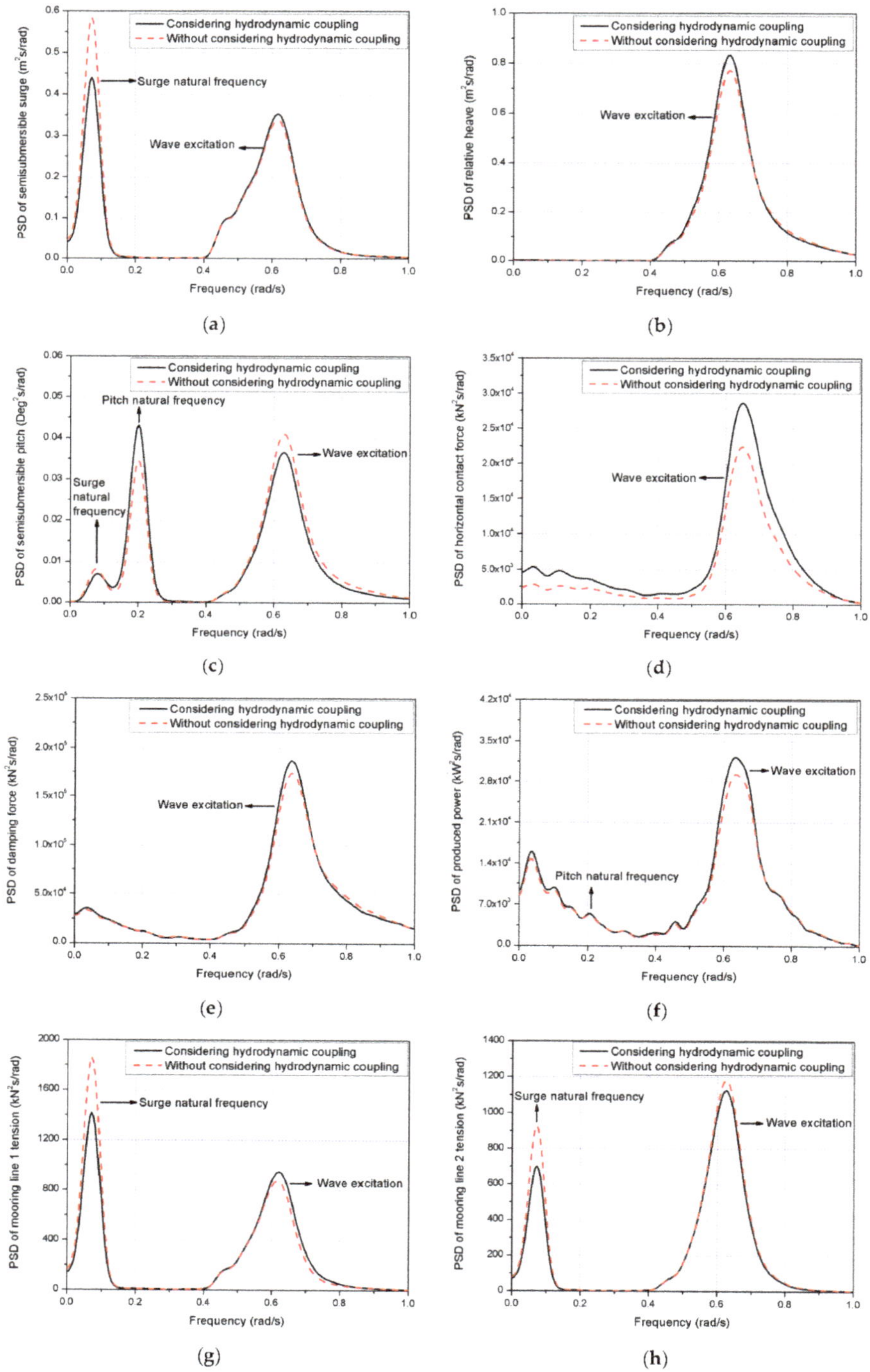

Figure 18. Power spectral density (PSD) of motion and dynamic responses for LC3: (**a**) semisubmersible surge; (**b**) relative heave; (**c**) semisubmersible pitch; (**d**) horizontal contact force; (**e**) damping force; (**f**) produced power; (**g**) mooring line 1 tension; (**h**) mooring line 2 tension.

4.5. Dynamic Responses of the Combined Structure in Extreme Sea Conditions

For the LC4 with extreme sea conditions (wave height 8.6 m, wave period 15 s and wind speed of 31.2 m/s), the PTO system is released, the wind turbine is parked, and the WEC is locked mechanically to the semisubmersible at the mean sea level as a survival mode. In this case, the two bodies are locked and can move together, which can be confirmed from the relative heave motion between WEC and semisubmersible in Figure 19. The statistical value from LC4 (mean, maxi mum, minimum and STD) are presented in Figure 20. In this load case, maximum heave motion of the combined structure is approximate 6 m (Figure 20a), in which WEC may bear huge loads due to slamming forces. The mean surge and pitch responses of the combined structure in LC4 is much less than those responses in LC3 (Figure 17a). Particularly, the maximum and minimum pitch response is much less than those values in LC3 for operating conditions because the thrust load in LC4 is much smaller than the thrust load in LC3 (Figure 14). Figure 20b shows that the maximum wave force of the combined structure is smaller than the vertical contact force between the semisubmersible column and the WEC, while the horizontal contact force is far smaller than the wave force and vertical contact force, which proves the locked mode of the combined structure results to a reasonable response.

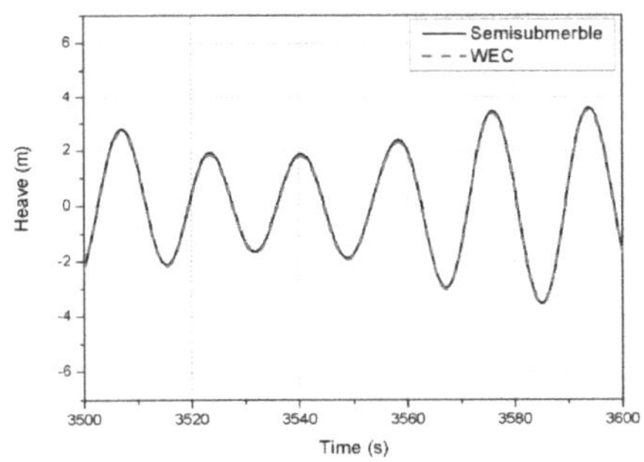

Figure 19. Relative heave of the combined structure under the locked mode.

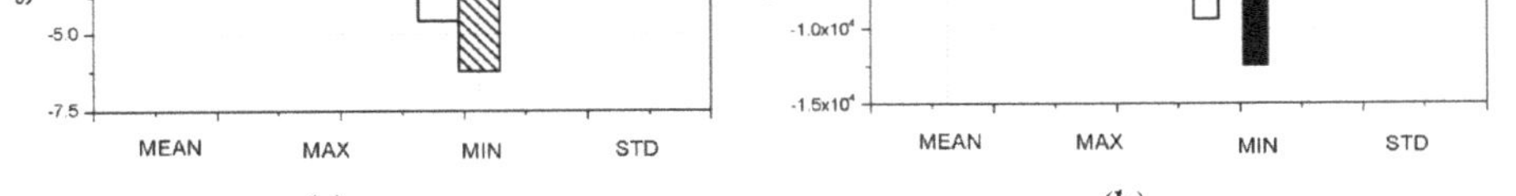

(a) (b)

Figure 20. Statistical values of the responses for the extreme sea state LC4: (**a**) statistic of information for semisubmersible surge, heave and pitch; (**b**) statistic of information for wave force of the combined structure, horizontal contact force, vertical contact force, mooring line 1 tension and mooring line 2 tension.

Figure 21 presents the PSD of the motion response and mooring line forces of the combined structure and pure semisubmersible platform under the extreme sea condition LC4. PSD of surge of the combined structure have same trends compared with pure semisubmersible platform and the response is dominant by the wave frequency (Figure 21a). For heave motion (Figure 21b), the response of pure semisubmersible is dominated by the wave frequency. However, for the combined structure, the response around the frequency of 0.35 rad/s is significantly larger because this frequency is the heave natural frequency of the combined structure and is near to the range of wave frequency. The shifting of the heave natural frequency of the combined structure from the pure semisubmersible (0.254 rad/s) is due to the attachment of the WEC to the semisubmersible. The surge resonant motion of the combined structure (Figure 21c) is dominant at 0.21 rad/s, which is the pitch natural frequency, followed by the wave frequency response. But the surge motion of the pure semisubmersible is dominated by the wave frequency response. Figure 21d,e shows that PSD of mooring line 1 tension and mooring line 2 tension of the combined structure are not always identical with only pure semisubmersible platform in high peak frequency region, which is caused by the coupling effect of heave resonance and incident wave for the combined structure. The most significant contribution to the mooring line 1 tension for both the combined structure and the semisubmersible platform comes from the wave frequency range. A clear peak response from the heave natural frequency for the combined structure could be identified for mooring line 1 tension. For mooring line 2 (Figure 21e), the most significant contribution to the mooring line tension for the pure semisubmersible comes from the wave frequency range from 0.300 rad/s to 0.600 rad/s, while the low-frequency response does not contribute significantly to the total response. However, for the combined structure, the heave resonance response at approximately 0.35 rad/s dominates the total response for mooring line 2.

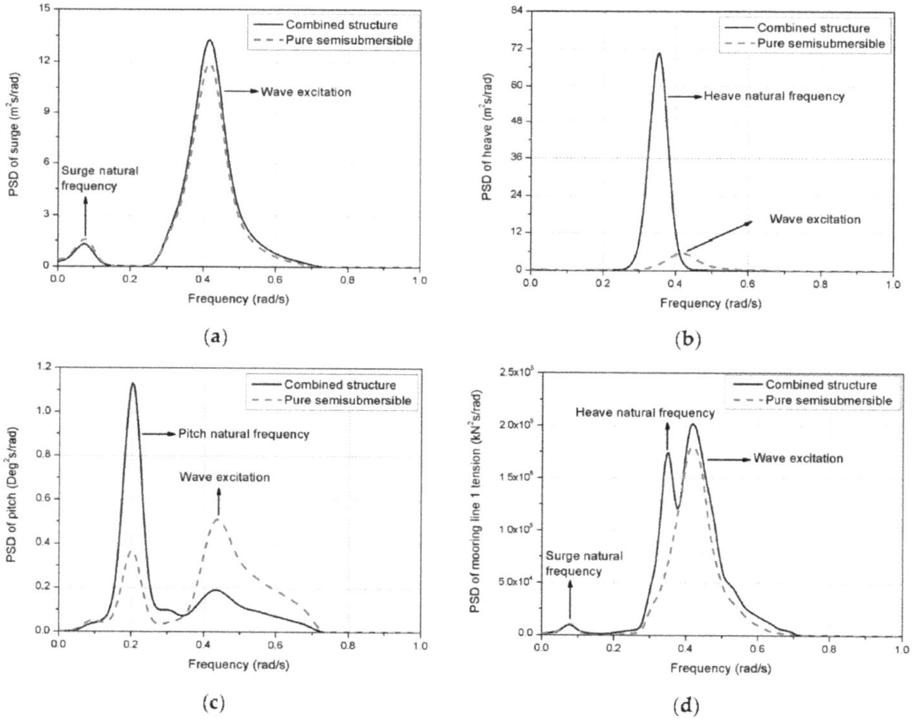

Figure 21. *Cont.*

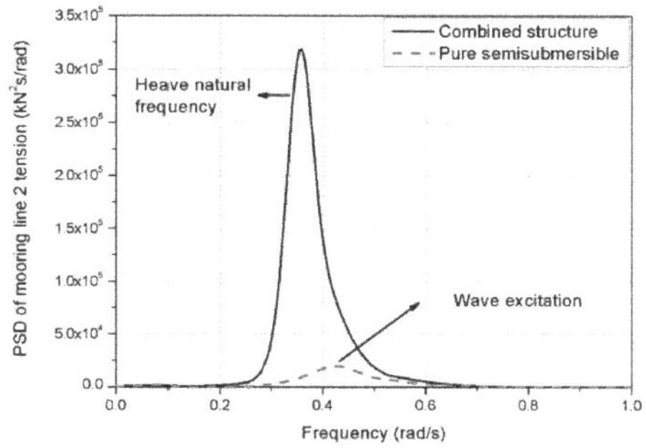

(e)

Figure 21. PSD of motion responses for the extreme sea state LC4: (**a**) surge; (**b**) heave; (**c**) pitch; (**d**) mooring line 1 tension; (**e**) mooring line 2 tension.

5. Conclusions

In this paper, a combined marine renewable energy structure consisting of a 5-MW braceless floating wind turbine and a heave-type wave energy converter (WEC) is presented. A numerical model of the combined structure that is capable of simulating its motion and dynamic responses under different sea states was developed and used. Different types of analysis have been performed for examining the hydrodynamic response of the combined structure. The main results are summarized as follows:

The outer diameter of the WEC is selected to be 16 m, in term of its economy, safety and PTO produced power. In order to avoid the slamming phenomena of the WEC as far as possible motion response, the draft of the WEC with 3.5 m is selected.

The K_{PTO} parameter has a significant impact on the performance of the WEC when its value is over 1000 N/m. A preliminary best value of the B_{PTO} parameter for the PTO of the combined structure is selected to be 1500 kNs/m with the perspective of PTO produced power.

The trends of hydrodynamic loads of the WEC with considering hydrodynamic coupling will change compared to those without considering hydrodynamic coupling, but responses of the combined structure have the same trends. The viscosity of the heave of the WEC will have an influence on the PTO produced power and dynamic responses of the combined structure.

Under regular waves, the amplitudes of the responses, such as the relative heave, semisubmersible pitch, horizontal contact force and PTO damping force of the combined structure, have an almost linear relationship with the wave height, while the amplitudes of WEC surge and PTO produced power have an almost quadratic relationship.

Under irregular waves, the surge and pitch of the semisubmersible with considering constant thrust wind have great effects compared to those without considering constant thrust wind, but there are slight effects for relative heave, damping force and produced power of the combined structure. Moreover, the maximum relative heave of the combined structure is larger than the absolute value of the minimum relative heave, and the wave forces acting on the WEC can be transferred to the semisubmersible by means of their contacted fenders. The PSD of motion responses and dynamic responses of the combined structure with considering hydrodynamic coupling have similar trends compared with those without considering hydrodynamic coupling.

Under the extreme sea conditions, the WEC was mechanically locked to the semisubmersible platform to withstand the large impact force on WEC as a survival mode. Heave natural frequency shifting could be identified and may coincide with wave frequency. The resonance in heave motion and mooring line 2 tension could be identified near the heave natural frequency. Other survival models should be proposed and investigated in future work. Experiments and validation for this integrated structure will also be carried out in the future.

Author Contributions: Conceptualization, W.S., L.Z., C.M. and L.W.; methodology, W.S., C.M. and L.W.; investigation, Y.W., L.Z., W.S. and C.M.; writing—first draft preparation, Y.W., L.Z., W.S. and C.M.; review and editing, W.S., C.M. and L.W.; supervision, W.S. and C.M.; project administration, W.S.; funding acquisition, W.S. All authors have read and agreed to the published version of the manuscript.

Funding: This research was funded by the National Natural Science Foundation of China (Grant No. 51709039). This work is also supported by the international collaboration and exchange program from the NSFC-RCUK/EPSRC with grant No. 51761135011. This work is also partially supported by LiaoNing Revitalization Talents Program (XLYC1807208), State Key Laboratory of Hydraulic Engineering Simulation and Safety, Tianjin University and the Fundamental Research Funds for the Central University (DUT19GJ209).

Acknowledgments: The authors would like to thank the Nianxin Ren from Hainan University for his valuable discussion of simulation in AQWA software.

Conflicts of Interest: The authors declare no conflict of interest.

References

1. European Wind Energy Association. *Offshore Wind in Europe Key Trends and Statistics 2019*; Wind Europe: Brussels, Belgium, 2020.
2. Cai, S.K. Grid parity speeds up the development of offshore wind power industry—the practitioner mission of offshore wind power in the next five years. *South Energy Constr.* **2019**, *2*, 7–15.
3. Marina Platform. Available online: https://www.msp-platform.eu/projects/marina-platform (accessed on 3 April 2020).
4. Peiffer, A.; Roddier, D.; Aubault, A. Design of a point absorber inside the WindFloat structure. In Proceedings of the 30th International Conference on Ocean, Offshore and Arctic Engineering (OMAE), Rotterdam, The Netherlands, 19–24 June 2011.
5. Aubault, A.; Alves, M.; Sarmento, A.; Roddier, D.; Peiffer, A. Modeling of an oscillating water column on the floating foundation WindFloat. In Proceedings of the 30th International Conference on Ocean, Offshore and Arctic Engineering (OMAE), Rotterdam, The Netherlands, 19–24 June 2011.
6. Muliawan, M.J.; Karimirad, M.; Moan, T. Dynamic response and power performance of a combined spar-type floating wind turbine and coaxial floating wave energy converter. *Renew. Energy* **2013**, *50*, 47–57. [CrossRef]
7. Muliawan, M.J.; Karimirad, M.; Gao, Z.; Moan, T. Extreme responses of a combined spar-type floating wind turbine and floating wave energy converter (STC) system with survival modes. *Ocean Eng.* **2013**, *65*, 71–82. [CrossRef]
8. Wan, L.; Gao, Z.; Moan, T. Experimental and numerical study of hydrodynamic responses on a combined wind and wave concept in survival modes. *Coast. Eng.* **2015**, *104*, 151–169. [CrossRef]
9. Wan, L.; Gao, Z.; Moan, T.; Lugni, C. Experimental and numerical comparisons of a combined wind and wave energy converter concept under operational conditions. *Renew. Energy* **2016**, *93*, 87–100. [CrossRef]
10. Wan, L.; Gao, Z.; Moan, T.; Lugni, C. Comparative experimental study of the survivability of a combined wind and wave energy converter in two testing facilities. *Ocean Eng.* **2016**, *111*, 82–94. [CrossRef]
11. Michailides, C.; Gao, Z.; Moan, T. Experimental study of the functionality of a semisubmersible wind turbine combined with flap-type Wave Energy Converters. *Renew. Energy* **2016**, *93*, 675–690. [CrossRef]
12. Michailides, C.; Gao, Z.; Moan, T. Experimental and numerical study of the response of the offshore combined wind/wave energy concept SFC in extreme environmental conditions. *Mar. Struct.* **2016**, *50*, 35–54. [CrossRef]
13. Michailides, C.; Luan, C.Y.; Gao, Z.; Moan, T. Effect of flap type wave energy converters on the response of a semi-submersible wind turbine in operational conditions. In Proceedings of the ASME 2014 33rd International Conference on Ocean, Offshore and Arctic Engineering (OMAE), San Francisco, CA, USA, 8–13 June 2014.
14. Ren, N.X.; Ma, Z.; Fan, T.H.; Zhai, G.J.; Ou, J.P. Experimental and numerical study of hydrodynamic responses of a new combined monopile wind turbine and a heave-type wave energy converter under typical operational conditions. *Ocean Eng.* **2018**, *159*, 1–8. [CrossRef]
15. Luan, C.; Gao, Z.; Moan, T. Design and analysis of a braceless steel 5-MW semi-submersible wind turbine. In Proceedings of the ASME 2016 35st International Conference on Ocean, Offshore and Arctic Engineering (OMAE 2016), Busan, Korea, 19–24 June 2016.
16. Yang, S.H.; Wang, Y.Q.; He, H.Z.; Zhang, J.; Chen, H. Dynamic properties and energy conversion efficiency of a floating multi-body wave energy converter. *China Ocean Eng.* **2018**, *32*, 347–357. [CrossRef]

17. Wavebob. Available online: http://www.wavebob.com (accessed on 3 April 2020).
18. ANSYS, A.W., Inc. *AQWA Manual Release 15.0*; ANSYS A.W., Inc.: Canonsburg, PA, USA, 2013.
19. Li, Y.C.; Teng, B. *The Effect of Waves on Marine Buildings*, 3rd ed.; Ocean Publication: Beijing, China, 2015; pp. 270–273.
20. Van Rijn, L.C. *Principles of Fluid Flow and Surface Waves in Rivers, Estuaries, Seas and Oceans*, 3rd ed.; Aqua Publications: Amsterdam, The Netherlands, 1990.
21. Morison, J.R.; Johnson, J.W.; Schaaf, S.A. The force exerted by surface waves on piles. *J. Pet. Technol.* **1950**, *2*, 149–154. [CrossRef]
22. Gao, W.; Dong, L.; Huang, J. *ANSYS AQWA Software Introduction and Improvement*, 1st ed.; China Water Resources and Hydropower Press: Beijing, China, 2018.
23. Cao, H.; Tang, Y.G.; Tao, H.C.; Qin, Y. Design and frequency domain analysis of semi-submersible floating foundation for offshore wind turbine. *Ocean Eng.* **2013**, *31*, 61–67.
24. Bai, J.; Li, Y.; Qu, Z.S.; Tang, Y.G. Optimization design of heave suppression structure for new cylindrical FPSO. *Ocean Eng.* **2020**, *38*, 20–29.
25. Wang, Y.P.; Shi, W.; Zhang, L.X.; Michailides, C.; Zhou, L. Hydrodynamic analysis of a floating hybrid renewable energy system. In Proceedings of the 30th International Society of Offshore and Polar Engineers, Shanghai, China, 14–19 June 2020.
26. Fang, M.C.; Kim, C.H. Hydrodynamically coupled motions of two ships advancing in oblique waves. *J. Ship Res.* **1986**, *30*, 159–171.
27. Oortmerssen, G.V. Hydrodynamic interaction between two structures, floating in waves. In Proceedings of the 2nd International Conference on Behaviour of Offshore Structures, BOSS'79, London, UK, 28–31 August 1979.
28. Zhao, Z.X.; Li, X.; Wang, W.H.; Shi, W. Analysis of dynamic characteristics of an ultra-large semi-submersible floating wind turbine. *Mar. Sci. Eng.* **2019**, *7*, 169. [CrossRef]
29. Tom, N.M. Design and Control of a Floating Wave-Energy Converter Utilizing a Permanent Magnet linear Generator. Ph.D. Thesis, UC Berkeley, Berkeley, CA, USA, 2013.
30. Falnes, J. A review of wave-energy extraction. *Mar. Struct.* **2007**, *20*, 185–201. [CrossRef]
31. Gao, Z.; Moan, T.; Wan, L.; Michailides, C. Comparative numerical and experimental study of two combined wind and wave energy concepts. *J. Ocean Eng. Sci.* **2016**, *1*, 36–51. [CrossRef]
32. Ren, N.X.; Zhu, Y.; Ma, Z.; Wu, H.B. Comparative study of hydrodynamic responses of two combined wind turbine and wave energy converter systems under typical operational sea cases. In Proceedings of the 37th International Conference on Ocean, Offshore and Arctic Engineering (OMAE), Madrid, Spain, 17–22 June 2018.

Journal of
Marine Science and Engineering

MDPI

Article

Local Enhancements of the Mean Drift Wave Force on a Vertical Column Shielded by an Exterior Thin Porous Shell

Peiwen Cong [1] and Yingyi Liu [2,*]

[1] State Key Laboratory of Coastal and Offshore Engineering, Dalian University of Technology, Dalian 116024, China; pwcong@dlut.edu.cn

[2] Research Institute for Applied Mechanics, Kyushu University, Fukuoka 8168580, Japan

* Correspondence: liuyingyi@riam.kyushu-u.ac.jp

Received: 12 April 2020; Accepted: 12 May 2020; Published: 14 May 2020

Abstract: The wave interaction with a vertical column shielded by an exterior porous shell is studied within the framework of potential flow theory. The structures are fixed rigidly at the sea bottom. The interior cylinder is impermeable, and the exterior shell is slightly porous and thin. Additionally, the exterior shell is assumed to have fine pores, and a linear pressure drop is adopted at the porous geometry. The mean drift wave force on the system is thereby formulated by two alternative ways, based respectively on the direct pressure integration, i.e., the near-field formulation, and the application of the momentum conservation theorem in the fluid domain, i.e., the far-field formulation. The consistency of the two formulations in calculating the mean drift wave force is assessed for the present problem. Numerical results illustrate that the existence of the porous shell can substantially reduce the mean drift wave force on the interior column. It also appears that the far-field formulation consists of a conventional part as well as an additional part caused by the energy dissipation through the porous geometry. The mean drift wave force on the system is dominated by the first part, which resembles that on an impermeable body. Local enhancements of the mean drift wave force are found at some specific wave frequencies at which certain propagation modes of the fluid satisfy a no-flow condition at the porous shell.

Keywords: concentric cylinders; porous structure; thin shell; linear resistance; mean drift wave force

1. Introduction

A geometry with slots or pores can be used to enhance energy dissipation and in turn reduce the environmental impact. Currently, porous structures have been widely constructed in the coastal and offshore industries for the purpose of shore production or reduction of the wave forces. To achieve a good understanding of the hydrodynamic properties related to porous structures, the behavior of a porous body in waves has attracted considerable interest among researchers.

So far, various studies have been conducted on this subject due to its importance in practical engineering. The porous elements are included in many offshore structures, such as fish cages, wave breakwaters, and offshore platforms equipped with damping devices. The porous geometry was commonly assumed to be with fine pores and thin in thickness. A linear or quadratic resistance law was used to relate the pressure drop to the crossflow velocity. Linear relations between the pressure drop across the porous geometry and the traversing velocity were derived by researchers, e.g., by using Darcy's law [1] or by using the convection neglected and porous effect modelled Euler equation [2]. The linear laws were adopted in many studies in assessing the functional performance of porous breakwaters of different types, such as vertical or submerged horizontal porous plates, perforated caissons, and porous columns [3–8]. Based on a linear resistance law, Zhao et al. [9] examined the various hydrodynamic identities for porous structures. In addition, Dokken et al. [10] formulated the wave diffraction/radiation problem of a porous geometry of an arbitrary shape by a set of integral equations. On the other hand, to model the flow separation through porous materials, quadratic laws were proposed in some studies, such as Molin [11] and An and Faltinsen [12]. The quadratic pressure drops were considered by researchers in prediction of the added mass and the damping coefficients for a porous stabilizer, plate, or disc undergoing forced motions as well as the wave force on thin porous sheets. Examples include Molin and Remy [13] and Mackay et al. [14].

The application of a porous material can lead to a low level of wave transmission and reflection. In the meantime, it can also reduce the wave impact on the enclosed structure. A vertical column has been an essential component of various offshore structures. Previous studies indicated that the linear wave force on an impermeable column can be reduced significantly by surrounding it with porous geometries (see Wang and Ren [15] for example). Besides the linear wave force, the nonlinear wave force is also closely relevant to the proper design of offshore structures. The nonlinear hydrodynamic properties related to a vertical column or a column array in open seas has been investigated in some studies spanning from analytical studies to model tests and numerical modelling, such as references [16–20]. However, the nonlinear wave interaction with a vertical column shielded by an exterior porous shell has been rarely investigated so far. The present study intends to contribute in this direction.

In this study, the mean drift wave force, which is of the second order from the perspective of the wave steepness, on a system consisting of an interior impermeable cylinder and an exterior porous shell is investigated. Definition of the concentric porous cylinder system is given in Figure 1. The potential flow theory is adopted, and a linear resistance law is assumed at the porous shell. Within the framework of the potential flow theory, a non-linear analysis can be achieved by applying a perturbation procedure, in which the velocity potential, wave force, and other physical quantities of interest are expressed in linear and higher-order components through some small parameters, and the wave steepness is normally used as the perturbation parameter. Based on the idea of the perturbation expansion, the mean drift wave force normally represents a time-independent force component which is proportional to the square of the wave steepness. In this study, semi-analytical solutions to the mean drift wave force on the system shown in Figure 1 are developed by two alternative ways. One is based on the direct pressure integration, while the other is based on the application of the momentum conservation theorem in the fluid domain. Unfortunately, solutions to the mean drift wave force on a vertical column shielded by an exterior porous shell have been rarely developed in the previous studies, and the research achievement in this study can help fill this gap. Based on the developed solutions, detailed numerical studies are performed. The effect of the exterior porous shell on the mean drift wave force on the interior column is examined, and the characteristics of the mean drift wave force on such a system are explored.

Following the introduction, the mathematical model and the solution of the velocity potential are introduced in Section 2, followed by a calculation of the mean drift wave force in Section 3. The parametric study is carried out thereafter in Section 4, with conclusions drawn in Section 5.

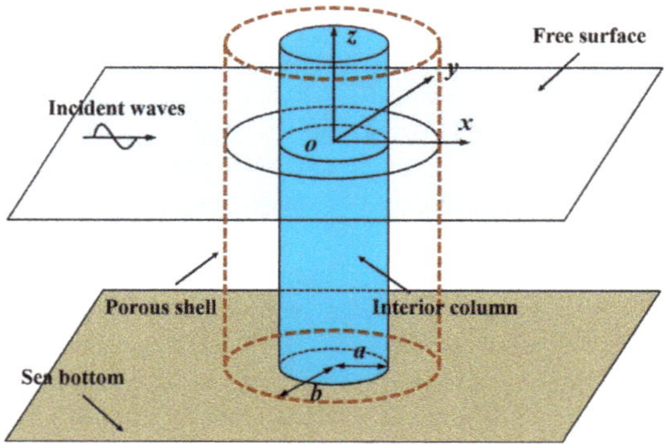

Figure 1. Definition sketch for a concentric porous cylinder system.

2. Mathematical Model and Solutions to the Velocity Potential

Let us consider a bottom-mounted, surface-piercing, and impermeable cylinder of radius a situated in a water of a finite depth d. This cylinder is surrounded by an exterior cylindrical shell of radius b. Both of them are fixed rigidly at the horizontal sea bottom. The exterior shell is porous and considered to be thin in thickness. As shown in Figure 1, a cylindrical coordinate system is employed with its origin located at the center of the interior cylinder and at the still free surface. The z-axis directs vertically upward and coincides with the vertical axis of the cylinders.

Within the framework of the potential flow theory, the flow is assumed to be inviscid, incompressible and irrotational. Thus, the flow field can be described by a scalar velocity potential Φ, satisfying Laplace's equation,

$$\nabla^2\Phi(\mathbf{x},\ t) = 0. \tag{1}$$

Regular incident waves are concerned in this study. The incident wave is time-harmonic with an angular frequency ω. Then, the time factor can be separated out, and the velocity potential is expressed as

$$\Phi(\mathbf{x},\ t) = \text{Re}\left[\phi(r,\ \theta,\ z)e^{-i\omega t}\right], \tag{2}$$

where "Re" denotes the real part of a complex expression; $i = \sqrt{-1}$.

The entire fluid domain is then divided into two sub-domains: the interior region Ω_1 ($a \le r \le b$), and the exterior region Ω_2 ($r \ge b$). Hereinafter, the velocity potentials in the interior and exterior regions are denoted by ϕ_1 and ϕ_2, respectively. Besides Laplace's equation, appropriate boundary conditions on the free surface and the seabed are also required, given by

$$\frac{\partial\phi_j}{\partial z} = \frac{\omega^2}{g}\phi_j, \quad \text{on } z = 0, \quad j = 1,\ 2; \tag{3}$$

$$\frac{\partial\phi_j}{\partial z} = 0, \quad \text{on } z = -d, \quad j = 1,\ 2, \tag{4}$$

where g is the acceleration due to gravity.

The fine-pore assumption is adopted, and a linear pressure drop assumption (see references [1,2] for example) is applied as in many previous studies. Then, the boundary condition on the porous shell can be expressed as

$$\frac{\partial\phi_1}{\partial r} = \frac{\partial\phi_2}{\partial r} = ikG_0(\phi_1 - \phi_2), \quad \text{on } r = b, \tag{5}$$

where k is the wave number satisfying the dispersion relation $\omega^2 = gk\tanh kd$. G_0 is a linearized porous effect parameter. On the surface of the interior impermeable cylinder, it is required that

$$\frac{\partial\phi_1}{\partial r} = 0, \quad \text{on } r = a. \tag{6}$$

A train of incident waves, of an amplitude A and an angular frequency ω, propagates in the direction of the positive x-axis to encounter the system. The incident velocity potential is then expressed as

$$\phi_I = -\frac{iAg}{\omega} Z(kz) \sum_{m=-\infty}^{+\infty} J_m(kr) i^m e^{im\theta}, \tag{7}$$

in which $J_m(kr)$ stands for the Bessel function of order m; $Z(kz)$ is an orthonormal function given at the interval $[-d, 0]$ and defined by

$$Z(kz) = \frac{\cosh k(z+d)}{\cosh kd}. \tag{8}$$

The presence of the system can result in diffraction of the incident waves in the exterior region. Then, the velocity potential in the exterior region can be decomposed as

$$\phi_2 = \phi_I + \phi_D, \tag{9}$$

in which ϕ_D denotes the diffraction potential in the exterior region. In addition to the boundary conditions in Equations (3), (4), and (5), ϕ_D also needs to satisfy the Sommerfeld radiation condition in the far field. That is,

$$\lim_{r\to\infty} \sqrt{r}\left(\frac{\partial \phi_D}{\partial r} - ik\phi_D\right) = 0. \tag{10}$$

The solution of the velocity potential is then obtained following the way shown in [15]. The separation of variables method is used, and the expressions of ϕ_1 and ϕ_2 can be written as

$$\phi_1(r,\ \theta,\ z) = \sum_{m=-\infty}^{+\infty} R_{1,\ m}(r) Z(kz) e^{im\theta}; \tag{11a}$$

$$\phi_2(r,\ \theta,\ z) = \phi_I(r,\ \theta,\ z) + \phi_D(r,\ \theta,\ z) = \sum_{m=-\infty}^{+\infty} R_{2,\ m}(r) Z(kz) e^{im\theta}, \tag{11b}$$

in which

$$R_{1,\ m}(r) = -\frac{iAg}{\omega} i^m [B_m J_m(kr) + C_m H_m(kr)]; \tag{12a}$$

$$R_{2,\ m}(r) = -\frac{iAg}{\omega} i^m [J_m(kr) + A_m H_m(kr)]. \tag{12b}$$

In Equation (12), A_m, B_m and C_m are unknown coefficients; $H_m(kr)$ is the Hankel function of the first kind of order m. The velocity potential given in Equation (11) satisfies Laplace's equation and Equations (3) and (4). The unknown coefficients in these expressions can be determined by imposing the boundary condition at $r = a$ and $r = b$ (see Equations (5) and (6)). Then, A_m, B_m and C_m are derived as

$$A_m = -\frac{\beta_m J'_m(kb) + 2G_0 J'_m(ka)}{\beta_m H'_m(kb) + 2G_0 H'_m(ka)}; \tag{13a}$$

$$B_m = \frac{2G_0 H'_m(ka)}{\beta_m H'_m(kb) + 2G_0 H'_m(ka)}; \tag{13b}$$

$$C_m = -\frac{2G_0 J'_m(ka)}{\beta_m H'_m(kb) + 2G_0 H'_m(ka)}, \tag{13c}$$

in which

$$\beta_m = \pi\kappa b[J'_m(kb) H'_m(ka) - J'_m(ka) H'_m(kb)]. \tag{14}$$

In Equations (13) and (14), the prime appearing in the superscript denotes the derivative with respect to the argument.

3. Calculation of the Mean Drift Wave Force Based on Direct Pressure Integration

Based on the derived velocity potential, various quantities of engineering interest can be determined. The mean drift wave force can be computed from the quadratic products of the quantities derived from the linear wave theory. The calculation of the mean drift wave force based on the direct pressure integration, i.e., the near-field formulation, is quite straightforward. Referring to the well-established second-order theory, the mean drift wave fore on the interior column, $f_c^{(2)-}$, can be determined according to

$$f_c^{(2)-} = -\frac{\rho}{4}\iint_{S_c} \nabla\phi_1 \cdot \nabla\phi_1^* n_x ds + \frac{\rho\omega^2}{4g}\oint_{\Gamma_c} \phi_1\phi_1^* n_x dl. \tag{15}$$

In Equation (15), S_c stands for the undisturbed wetted surface of the column; Γ_c denotes the intersection of S_c with mean free surface ($z = 0$); $\mathbf{n} = (n_x, n_y, n_z)^T$ is the normal vector on the boundary surface which is positive when pointing out of the fluid domain. The analysis of the mean drift force on a porous geometry is similar to that of an impermeable one. This was done by some researchers, such as the authors of references [9,10]. As for the present system, the mean drift wave force on the porous shell, $f_s^{(2)-}$, is due to the difference between the fluid pressure on its outer and inner surfaces. Then, we have

$$f_s^{(2)-} = -\frac{\rho}{4}\left(\iint_{S_s^-} \nabla\phi_1 \cdot \nabla\phi_1^* n_x ds + \iint_{S_s^+} \nabla\phi_2 \cdot \nabla\phi_2^* n_x ds\right) + \frac{\rho\omega^2}{4g}\left(\oint_{\Gamma_s^-} \phi_1\phi_1^* n_x dl + \oint_{\Gamma_s^+} \phi_2\phi_2^* n_x dl\right), \tag{16}$$

in which an asterisk in the superscript represents the complex conjugate; S_s^+ and S_s^- represent the outer and inner undisturbed wetted surfaces of the porous shell; Γ_s^+ and Γ_s^- stands for the intersection of S_s^+ and S_s^- with the mean free surface ($z = 0$). For vertically axisymmetric bodies, the surface integrals in Equations (15) and (16) can be simplified by integrating in θ and applying the orthogonality. Making use of Equations (5), (6), and (11), we can have

$$f_c^{(2)-} = \frac{\rho\pi a}{2}\mathrm{Re}\left\{\sum_{m=-\infty}^{+\infty} R_{1,\,m}(a)R_{1,\,m+1}^*(a)\left[\frac{1}{a^2}m(m+1) - k^2\right]N(kd)\right\}, \tag{17}$$

and

$$f_s^{(2)-} = \frac{\rho\pi b}{2}\mathrm{Re}\left\{\sum_{m=-\infty}^{+\infty}\left[R_{2,\,m}(b)R_{2,\,m+1}^*(b) - R_{1,\,m}(b)R_{1,\,m+1}^*(b)\right]\left[\frac{1}{b^2}m(m+1) - k^2\right]N(kd)\right\}. \tag{18}$$

In Equations (17) and (18), $N(kd)$ is defined by

$$N(kd) = \int_{-d}^{0} [Z(kz)]^2 dz = \frac{1}{\cosh^2 kd}\frac{d}{2}\left(\frac{\sinh 2kd}{2kd} + 1\right). \tag{19}$$

Then, the total mean drift wave force on the system, $f^{(2)-}$, can be expressed as a sum of $f_c^{(2)-}$ and $f_s^{(2)-}$, i.e.,

$$f^{(2)-} = f_c^{(2)-} + f_s^{(2)-}. \tag{20}$$

4. Calculation of the Mean Drift Wave Force Based on the Application of the Momentum Conservation Theorem

Besides the approach of direct pressure integration, the mean drift wave force can also be obtained by applying the momentum conservation theorem in the entire fluid domain, namely the far-field formulation. By using the kinematic transport theorem and Gauss' theorem, the total linear momentum along the direction of wave propagation can be expressed as

$$\frac{d\widetilde{M}}{dt} = -\iint_{S_b \cup S_\infty \cup S_d \cup S_f} [Pn_x + \rho u(\mathbf{u} \cdot \mathbf{n} - U_n)]ds, \tag{21}$$

in which $\widetilde{M}$ is the momentum; P is the fluid pressure; S_b represents a summation of the wetted surface of the column as well as the inner and outer wetted surface of the porous shell, respectively, i.e., $S_b = S_c + S_s^+ + S_s^-$; S_∞ is a circular cylindrical control surface at a large distance from the system; S_f and S_d represent the overall free surface and the overall sea bed in the entire fluid domain, respectively; $\mathbf{u} = \nabla\phi = (u, v, w)^T$ is the vector of the fluid particle velocity; U_n denotes the normal velocity of the boundary surface. On S_f and S_d, it is required that $\mathbf{u}\cdot\mathbf{n} - U_n = 0$. In addition, the entire system and the control surface at infinity are fixed, yielding $U_n = 0$ on S_b and S_∞. On S_d, we have $n_x = 0$. In the meantime, on the free surface, the fluid pressure P is equivalent to the atmospheric pressure which is assumed zero. Then, based on Equation (21), the wave force on the system, F, can be related to the total linear momentum in the fluid domain. That is,

$$F = \iint_{S_b} Pn_x ds = -\iint_{S_b} \rho u(\mathbf{u} \cdot \mathbf{n})ds - \iint_{S_\infty} [Pn_x + \rho u(\mathbf{u} \cdot \mathbf{n})]ds - \frac{d\widetilde{M}}{dt}. \tag{22}$$

When the time average is taken, and the periodicity is invoked, the last term on the right-hand side of Equation (22) gives no contribution. Then, the mean drift wave force on the system is

$$f^{(2)-} = f_f^{(2)-} + f_n^{(2)-}, \tag{23}$$

in which

$$f_f^{(2)-} = \overline{-\iint_{S_\infty} [Pn_x + \rho u(\mathbf{u} \cdot \mathbf{n})]ds}; \tag{24a}$$

$$f_n^{(2)-} = \overline{-\iint_{S_s^+ + S_s^-} \rho u(\mathbf{u} \cdot \mathbf{n})ds}. \tag{24b}$$

In Equation (24), an over bar indicates averaging over a wave period. In the far field, using asymptotic expressions for Hankel functions, the diffraction potential for large r can be expressed in an asymptotic form. Then, we have

$$\phi_2 = -\frac{iAg}{\omega}\left[e^{ikr\cos\theta} + \sqrt{\frac{2}{2kr}}e^{i(kr-\frac{\pi}{4})}\sum_{m=-\infty}^{+\infty} \Lambda_m e^{im\theta}\right]Z(kz), \quad \text{as } r \to +\infty, \tag{25}$$

in which

$$\Lambda_m = i^m A_m e^{-\frac{m\pi}{2}i}. \tag{26}$$

Inserting Equation (25) into Equation (24a) and applying the stationary phase method to the double integral, Equation (24a) can be rewritten as

$$f_f^{(2)-} = -\frac{\rho g A^2}{k}\frac{2kd + \sinh 2kd}{2\sinh 2kd}\mathrm{Re}\left[\sum_{m=-\infty}^{+\infty} 2\left(\Lambda_m \Lambda_{m+1}^* + \Lambda_m\right)\right]. \tag{27}$$

Combining Equations (5), (6), and (11), Equation (24b) can be rewritten as

$$f_n^{(2)-} = \frac{\rho}{2}\mathrm{Re}\left\{N(kd)i\pi kG_0 \sum_{m=-\infty}^{+\infty} [(2m+1)\delta_m^*\delta_{m+1}]\right\}, \tag{28}$$

in which

$$\delta_m = R_{1,\,m}(b) - R_{2,\,m}(b) = -\frac{2Ag}{\pi kb\omega}\frac{i^m\beta_m}{\beta_m H'_m(kb) + 2G_0 H'_m(ka)}. \tag{29}$$

From the derivation of Equations (23), (27), and (28), it is noted that when a vertical column is shielded by an exterior porous shell, the far-field formulation consists of a conventional part similar to that on an impermeable body, i.e., $f_f^{(2)-}$, as well as an additional part caused by the energy dissipation through the external porous shell, i.e., $f_n^{(2)-}$.

In this section, solutions to the mean drift wave force on the system are derived by applying the momentum conservation theorem in the entire fluid domain. Actually, if the momentum conservation theorem is applied in a finite fluid domain surrounding the structure, solutions can also be derived. Then, the quantities on a control surface, which is a distance away from the structure, are involved in the calculation. The derivation of the semi-analytical solutions to the mean drift wave force on structures using control surfaces can be found in some previous studies, such as Cong et al. [21].

5. Numerical Results and Discussion

In the previous sections, two different formulations have been proposed for the mean drift wave force on a concentric porous cylinder system. When the near-field formulation is used, the mean drift wave force on the interior column and the exterior shell can be evaluated based on Equations (17) and (18), respectively. The force components involved in the far field formulation can be evaluated according to Equations (27) and (28), respectively. Hereinafter, the factor $\rho gaAA^*$ is introduced for the normalization, and the symbols f_c^-, f_s^-, f_f^-, f_n^- and f^- are used to denote the normalized mean drift wave force. The convergence of the wave force based on the two proposed formulations both depends on the number of Fourier modes. In the numerical algorithm, in total, $2M + 1$ Fourier modes (from the order $(-M)$ to the order M) have been included. To examine the convergence of the present solution with respect to M, calculations are performed for the case of $b = 2a$, $d = 3a$ and $G_0 = 0.1$. Tables 1 and 2 list the normalized mean drift wave force on the interior column, i.e., f_c^-, and the exterior porous shell, i.e., f_s^-, as a function of M for a set of wave frequencies. The variation of f_f^- and f_n^-, which are the force components involved in the far-field formulation, with respect to M are given in Tables 3 and 4. In these tables, "NF" and "FF" refer to the results evaluated according to the near-field and the far-field formulations, respectively. Moreover, the term ka represents the product by the wave number k and the radius a. In addition, in the tables, positive and negative signs mean that the mean drift wave force is along the positive and negative x-directions, respectively. Inspecting the results listed in these tables, it appears that the two models both possess good convergence characteristics. Thirty-one Fourier modes ($M = 15$) is sufficient to achieve a convergent result, and therefore, $M = 15$ is adopted in the subsequent computations. Meanwhile, in order to confirm the validity of the present solution, a comparison between the results based on the two formulations is made (see Table 5). Comparison confirms the good agreement between the convergent results based on the two formulations.

Table 1. Variation of the normalized mean drift wave force on the interior column, f_c^-, with respect to M ($d = 3a$, $b = 2a$, and $G_0 = 0.1$).

ka = / M =	1.0	1.5	2.0	2.5	3.0
	NF	NF	NF	NF	NF
5	0.294×10^{-1}	0.214×10^{-1}	0.291×10^{-1}	0.218×10^{-1}	0.554×10^{-1}
10	0.294×10^{-1}	0.214×10^{-1}	0.291×10^{-1}	0.225×10^{-1}	0.585×10^{-1}
15	0.294×10^{-1}	0.214×10^{-1}	0.291×10^{-1}	0.225×10^{-1}	0.585×10^{-1}
20	0.294×10^{-1}	0.214×10^{-1}	0.291×10^{-1}	0.225×10^{-1}	0.585×10^{-1}
30	0.294×10^{-1}	0.214×10^{-1}	0.291×10^{-1}	0.225×10^{-1}	0.585×10^{-1}

Table 2. Variation of the normalized mean drift wave force on the exterior porous shell, f_s^-, with respect to M ($d = 3a$, $b = 2a$, and $G_0 = 0.1$).

ka = / M =	1.0	1.5	2.0	2.5	3.0
	NF	NF	NF	NF	NF
5	1.083	1.057	1.091	1.031	0.940
10	1.083	1.068	1.116	1.075	1.035
15	1.083	1.068	1.116	1.075	1.035
20	1.083	1.068	1.116	1.075	1.035
30	1.083	1.068	1.116	1.075	1.035

Table 3. Variation of the normalized mean drift wave force component involved in the far-field formulation, f_f^-, with respect to M ($d = 3a$, $b = 2a$, and $G_0 = 0.1$).

ka = / M =	1.0	1.5	2.0	2.5	3.0
	FF	FF	FF	FF	FF
5	1.328	1.297	1.218	1.068	1.068
10	1.328	1.301	1.274	1.272	1.294
15	1.328	1.301	1.274	1.272	1.294
20	1.328	1.301	1.274	1.272	1.294
30	1.328	1.301	1.274	1.272	1.294

Table 4. Variation of the normalized mean drift wave force component involved in the far-field formulation, f_n^-, with respect to M ($d = 3a$, $b = 2a$, and $G_0 = 0.1$).

ka = / M =	1.0	1.5	2.0	2.5	3.0
	FF	FF	FF	FF	FF
5	−0.214	−0.197	-0.612×10^{-1}	-0.518×10^{-1}	-0.494×10^{-1}
10	−0.216	−0.212	−0.129	−0.174	−0.201
15	−0.216	−0.212	−0.129	−0.174	−0.201
20	−0.216	−0.212	−0.129	−0.174	−0.201
30	−0.216	−0.212	−0.129	−0.174	−0.201

Table 5. Comparison of the normalized mean drift wave force on the concentric porous cylinder system, f^-, based on different formulations ($d = 3a$, $b = 2a$, and $G_0 = 0.1$).

ka = / M =	1.0		1.5		2.0		2.5		3.0	
	NF	FF	NF	FF	NF	FF	NF	FF	NF	FF
5	1.112	1.113	1.078	1.100	1.120	1.157	1.053	1.016	0.995	1.018
10	1.112	1.112	1.089	1.089	1.145	1.145	1.098	1.098	1.094	1.094
15	1.112	1.112	1.089	1.089	1.145	1.145	1.098	1.098	1.094	1.094
20	1.112	1.112	1.089	1.089	1.145	1.145	1.098	1.098	1.094	1.094
30	1.112	1.112	1.089	1.089	1.145	1.145	1.098	1.098	1.094	1.094

With the validation of the developed model, a parametric study is then performed. The far-field formulation is adopted firstly. As shown in Equation (23), if the far-field formulation is used, the total mean drift wave force on the system consists of a component associated with the quantities in the far-field, i.e., f_f^-, as well as an additional component caused by the energy dissipation through the porous shell, i.e., f_n^-. The effect of the porous parameter G_0 on f_f^-, f_n^-, and f^- is shown in Figure 2 with $b = 2a$ and $d = 3a$. In Figure 2, $G_0 = 0$ corresponds to the situation when the exterior shell becomes impermeable. When $G_0 = 0$, f_n^- remains zero, and f_f^- coincides with f^-. As shown in Figure 2a,b, when $G_0 > 0$, f_f^- and f_n^- in general make opposite contributions to the total force on the system in which the dominant part is f_f^-. In the meantime, from Figure 2c, it is found that, when the exterior shell becomes porous, i.e., $G_0 > 0$, the trend of the mean drift wave force on the system is different from that when $G_0 = 0$, as there are a series of small peaks appearing around ka = 0.678, 1.341, 1.979, 2.588, etc. The occurrence of these small peaks is explained as below. The continuity of the fluid velocity between the inner and the outer regions is fulfilled by Equation (5). Based on Equation (5), the normal fluid velocity across the porous shell can be expressed as

$$\left.\frac{\partial \phi_1}{\partial r}\right|_{r=b} = \left.\frac{\partial \phi_2}{\partial r}\right|_{r=b} = \sum_{m=-\infty}^{+\infty} u_{b,\,m} = -\frac{2Ag}{\pi b\omega} Z(kz) \sum_{m=-\infty}^{+\infty} \frac{i^{m+1}\beta_m G_0}{\beta_m H'_m(kb) + 2G_0 H'_m(ka)} e^{im\theta}, \tag{30}$$

in which

$$u_{b,\,m} = \left.\frac{\partial \varphi_{1,\,m}}{\partial r}\right|_{r=b} = \left.\frac{\partial \varphi_{2,\,m}}{\partial r}\right|_{r=b} = ikG_0 \left(\varphi_{1,\,m} - \varphi_{2,\,m}\right)\Big|_{r=b}. \tag{31}$$

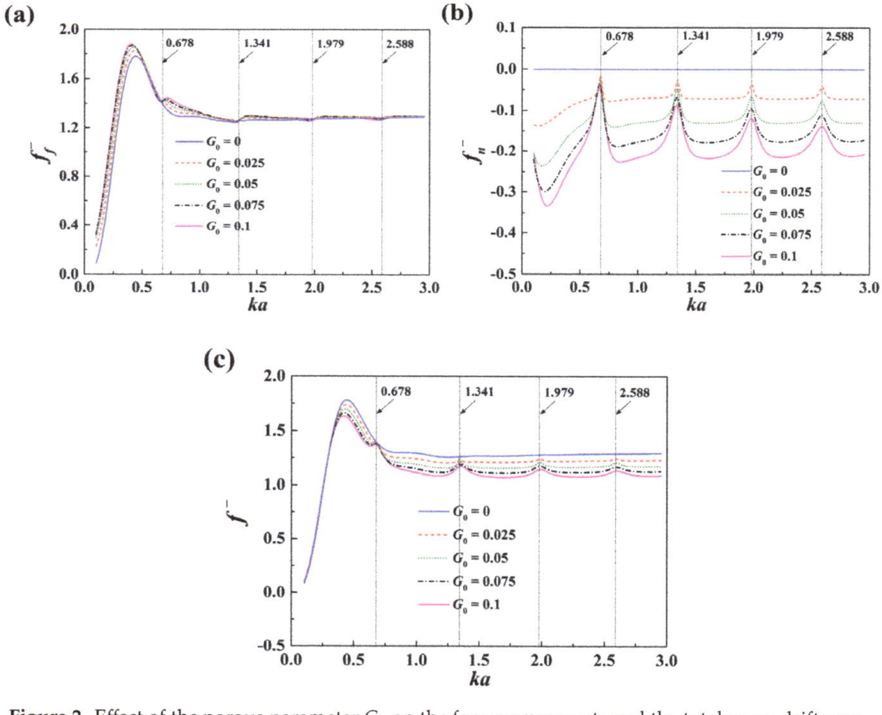

Figure 2. Effect of the porous parameter G_0 on the force components and the total mean drift wave force on the system using a far-field formulation ($b = 2a$ and $d = 3a$): (**a**) f_f^- (**b**) f_n^- (**c**) f^-.

In Equations (30) and (31), $u_{b,\ m}$ is the mth order Fourier component of the normal fluid velocity across the porous shell; $\varphi_{1,\ m}$ and $\varphi_{2,\ m}$ are the Fourier components of the velocity potential in the interior and the exterior domains, respectively. It is interesting to find that as ka gets close to 0.678, 1.341, 1.979 and 2.588 (when $b = 2a$ and $d = 3a$), $u_{b,\pm1}$, $u_{b,\pm2}$, $u_{b,\pm3}$, and $u_{b,\pm4}$ approach zero, respectively. Equation (5) ensures the continuity of the fluid velocity between the inner and outer regions. From Equation (31), it is found that at these wave frequencies, Fourier components of the velocity potential of the order ±1, ±2, ±3, ±4, etc. are continuous as well between the inner and outer regions. It means that to these components of the incoming waves, the porous shell is no longer a barrier but, on the contrary, thoroughly "transparent," permitting them to transmit across without any diffraction or dissipation. In this mechanism, the porous shell acts as a "wave filter" for special frequencies, thereby certain components of the incident waves cannot be dissipated by the shell. Then, the dissipation effect of the porous shell is significantly weakened at these wave frequencies, triggering an apparent diminishment of f_n^- in magnitude (see Figure 2b). In the meantime, as not all the components of the incoming waves can be diffracted by the shell around these special wave frequencies, the diffracted waves in the far filed would be affected, leading to the small oscillations of f_f^- around these wave frequencies (see Figure 2a). As a result, due to the aforementioned large diminishment of f_n^-, local enhancements of the total mean drift wave force on the system arise at these wave frequencies.

The near-field formulation is then adopted. The effect of the porous parameter G_0 on the mean drift wave force on the interior column and the exterior shell, i.e., f_c^- and f_s^-, is shown in Figure 3 with $b = 2a$ and $d = 3a$. In Figure 3a, "SC" corresponds to the situation when the exterior shell is removed and the interior column is exposed directly to the action of the incoming waves. Figure 3a illustrates that with the shielding of a slightly porous shell, the mean drift wave force on the interior column is significantly reduced in comparison to that without the shell. In addition, in Figure 3a, peaks can be observed around ka = 1.341, 1.979, and 2.588. As mentioned before, this is due to the reason that around these wave frequencies, certain components of the incoming waves can completely transmit into the inner region. Then, the wave action upon the interior column is enhanced, as demonstrated by the peaks. From Figure 3a, it is also interesting to find that around ka = 0.678, f_c^- is close to zero, which can be explained as below. Around ka = 0.678, the waves transmitting into the inner region are dominated by the Fourier components of the order ±1, resulting in a phase difference of π between the lee side and the weather side of the interior column. In addition, at ka = 0.678, the wave length is long enough when compared to the radius of the interior column, and the wave diffraction by the column is relatively negligible in comparison to the incident waves. Therefore, at this wave frequency, the wave run-up along the column distributes almost anti-symmetric with respect to the y-axis. Then, the quadratic fluid pressure, which is proportional to the square of the fluid velocity, on the lee side of the column is almost the same as that on the weather side, leading to that the mean drift wave on the column is close to zero. In Figure 3b, it appears that the mean drift wave force on the shell is much larger in magnitude than that on the column, and the results of f_s^- are close to those of f^-. It suggests that if the near-filed formulation is adopted, the total force on the system is dominated by f_s^-. The effect of G_0 on f_s^- is not apparent for long incident waves. However, for short incident waves, the effect is notable since f_s^- in general decreases as G_0 increases.

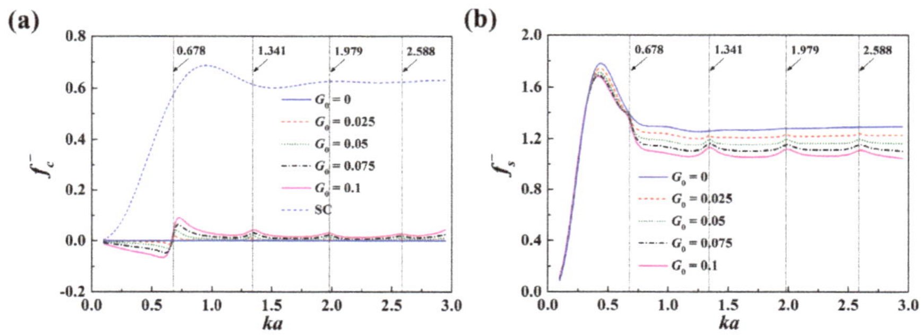

Figure 3. Effect of the porous parameter G_0 on the mean drift wave force on the interior column and the exterior porous shell using the near-field formulation ($b = 2a$ and $d = 3a$): (**a**) f_c^- (**b**) f_s^-.

The effect of the radius of the exterior porous shell and the effect of the water depth on the mean drift wave force acting on the impermeable column, the porous shell and the system as a whole are shown in Figures 4 and 5, respectively. As shown in Figure 4, the total mean drift wave force on the system increases with the increase of b. This is due to that the system interacts with more incoming waves as the radius of the exterior porous shell increases. It is also found that the frequencies of the small peaks move to the high-frequency region as the shell gets closer to the enclosed column. In Figure 5, the effect of the water depth gets less apparent as ka increases. This is due to the fact that in short incident waves, the fluid velocity decays quickly downward along the gravity direction.

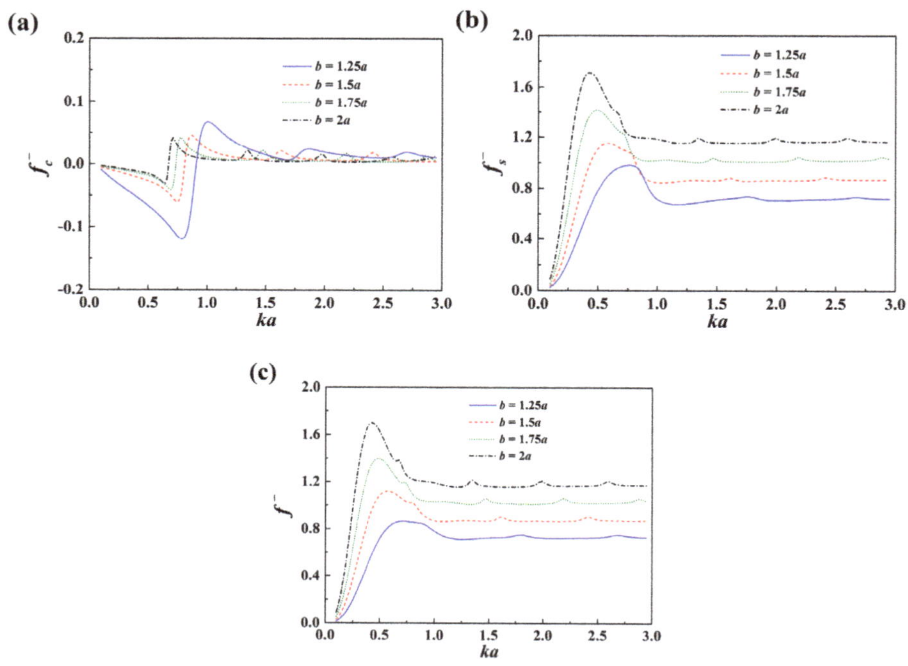

Figure 4. Effect of the radius of the exterior porous shell on the mean drift wave force on the interior column, the exterior porous shell, and the system ($G_0 = 0.05$ and $d = 3a$): (**a**) f_c^- (**b**) f_s^- (**c**) f^-.

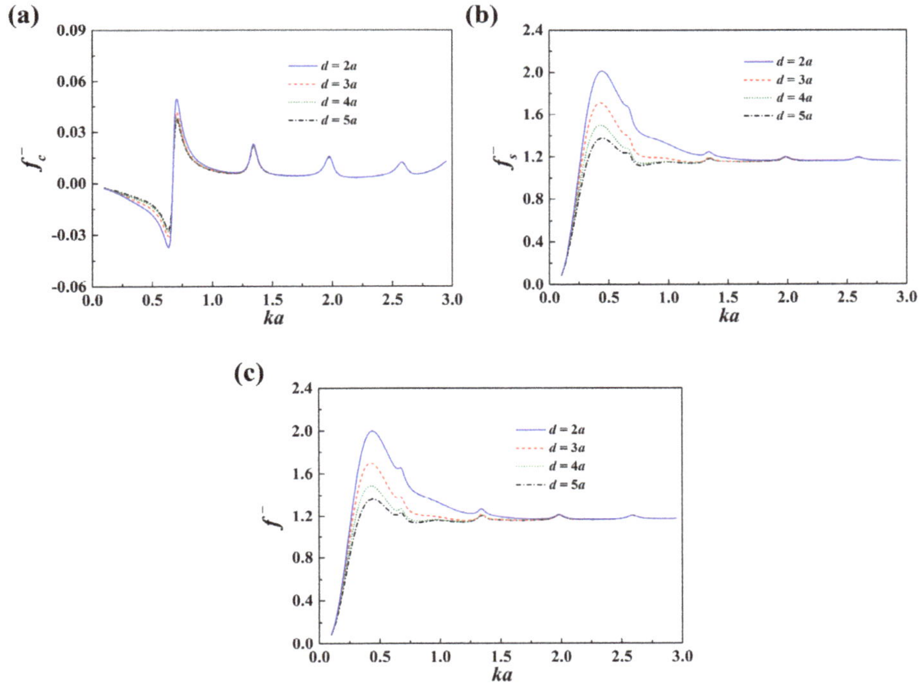

Figure 5. Effect of the water depth on the mean drift wave force on the interior column, the exterior porous shell, and the system ($G_0 = 0.05$ and $b = 2a$): (**a**) f_c^- (**b**) f_s^- (**c**) f^-.

6. Conclusions

The wave interaction with a concentric porous cylinder system, consisting of an interior impermeable column and an exterior slightly porous thin shell, is described. The exterior shell is assumed to have fine pores, and a linear resistance law is used to relate the pressure drop across the porous shell with the normal velocity. This study aims to investigate the effect of the exterior porous shell on the nonlinear mean drift wave force on the interior column. The main conclusions of this study are summarized as follows:

(1) Two different formulations have been derived for the calculation of the mean drift wave force on such a system. One is based on the direct pressure integration, i.e., the near-field formulation, and the other is based on the application of the momentum conservation theorem in the fluid domain, i.e., the far-field formulation. It is found that the results based on the two formulations agree exactly with each other.

(2) The far-field formulation of the mean drift wave force for the present problem consists of two parts, i.e., a conventional part, which is contributed by the quantities in the far field (similar to that on impermeable bodies), as well as an additional part caused by the energy dissipation across the porous shell. The former part dominates the total mean drift wave force on the system.

(3) The mean drift wave force on the interior column shielded by an exterior porous shell can be largely reduced when compared to that without the exterior porous shell. With the exterior shell being porous, the trend of the mean drift wave force on the system is different from that when the exterior shell is impermeable, as there is a series of small peaks riding on the main trend. It is found that these small peaks occur when certain propagation modes of the fluid velocity satisfy a no-flow condition at the porous shell. Then, the porous shell cannot dissipate energy from these modes, enhancing the wave impact upon the system.

The characteristics of the mean drift wave force on concentric porous cylinders are explored in this study. This reveals that, due to the presence of the porous shell, the interior column does not experience any mean drift wave force at specific conditions. The conclusions achieved in the present study are important references to subsequent studies concerning floating systems, which can be further explored as a future work.

Author Contributions: Conceptualization, P.C. and Y.L.; methodology, P.C.; validation, P.C.and Y.L.; investigation, P.C. and Y.L.; writing—original draft preparation, P.C.; writing—review and editing, Y.L. All authors have read and agreed to the published version of the manuscript.

Funding: This research was funded by the National Natural Science Foundation of China (Grant Number 51809037), and the Fundamental Research Funds for the Central Universities (DUT18RC(4)048)

Conflicts of Interest: The authors declare no conflict of interest. The funders had no role in the design of the study; in the collection, analyses, or interpretation of data; in the writing of the manuscript, or in the decision to publish the results.

Nomenclature

a	radius of the interior column
A	complex incident wave amplitude
b	radius of the exterior porous shell
d	water depth
$\mathbf{f}^{-}$	mean drift wave force on the total system
$\mathbf{f}_c^{-}$	mean drift wave force on the interior column
$\mathbf{f}_s^{-}$	mean drift wave force on the exterior porous shell
$\mathbf{f}_n^{-}$	force component in the far-field formulation caused by the energy dissipation through the external porous shell
$\mathbf{f}_f^{-}$	force component in the far-field formulation contributed by the quantities at the far field
ϕ_1, ϕ_2	velocity potential in the interior and exterior regions
ϕ_I	incident velocity potential
ϕ_D	diffraction velocity potential in the exterior region
g	gravitational acceleration
G_0	porous effect parameter
H_m	Hankel function of the first kind of order m
i	imaginary unit, $i^2 = -1$
J_m	Bessel function of order m
k	wave number
P	fluid pressure
r, θ, z	cylindrical coordinates
$R_{1,\,m}(r)$, $R_{2,\,m}(r)$	radial functions for the velocity potentials
$Z(kz)$	vertical functions for the velocity potential
ρ	density of water
ω	wave frequency
x, y, z	Cartesian coordinates

References

1. Chwang, A.T. A porous-wavemaker theory. *J. Fluid. Mech.* **1983**, *132*, 395–406. [CrossRef]
2. Yu, X.P. Diffraction of water waves by porous breakwaters. *J. Waterway Port Coastal Ocean Eng.* **1995**, *121*, 275–282. [CrossRef]
3. Teng, B.; Zhang, X.T.; Ning, D.Z. Interaction of oblique waves with infinite number of perforated caissons. *Ocean Eng.* **2004**, *31*, 615–632. [CrossRef]
4. Liu, Y.; Li, Y.C.; Teng, B. The reflection of oblique waves by an infinite number of partially perforated caissons. *Ocean Eng.* **2007**, *34*, 1965–1976. [CrossRef]

5. Chen, J.T.; Lin, Y.J.; Lee, Y.T.; Wu, C.F. Water wave interaction with surface-piercing porous cylinders using the null-field integral equations. *Ocean Eng.* **2011**, *38*, 409–418. [CrossRef]
6. Liu, Y.; Li, H.J.; Li, Y.C.; He, S.Y. A new approximate analytic solution for water wave scattering by a submerged horizontal porous disk. *Appl. Ocean Res.* **2011**, *33*, 286–296. [CrossRef]
7. Cho, I.H.; Kim, M.H. Transmission of oblique incident waves by a submerged horizontal porous plate. *Ocean Eng.* **2013**, *61*, 56–65. [CrossRef]
8. Cong, P.W.; Bai, W.; Teng, B. Analytical modeling of water wave interaction with a bottom-mounted surface-piercing porous cylinder in front of a vertical wall. *J. Fluid. Struct.* **2019**, *88*, 292–314. [CrossRef]
9. Zhao, F.F.; Kinoshita, T.; Bao, W.G.; Wan, R.; Liang, Z.L.; Huang, L.Y. Hydrodynamics identities and wave-drift force of a porous body. *Appl. Ocean Res.* **2011**, *33*, 169–177. [CrossRef]
10. Dokken, J.; Grue, J.; Karstensen, L.P. Wave analysis of porous geometry with linear resistance law. *J. Mar. Sci. Appl.* **2017**, *16*, 480–489. [CrossRef]
11. Molin, B. Hydrodynamic modeling of perforated structures. *Appl. Ocean Res.* **2011**, *33*, 1–11. [CrossRef]
12. An, S.; Faltinsen, O.M. An experimental and numerical study of heave added mass and damping of horizontally submerged and perforated rectangular plates. *J. Fluid. Struct.* **2013**, *39*, 87–101. [CrossRef]
13. Molin, B.; Remy, F. Experimental and numerical study of the sloshing motion in a rectangular tank with a perforated screen. *J. Fluid. Struct.* **2013**, *43*, 463–480. [CrossRef]
14. Mackay, E.; Johanning, L.; Ning, D.Z.; Qiao, D.S. Numerical and experimental modelling of wave loads on thin porous sheets. In Proceedings of the 38th International Conference on Ocean, Offshore and Arctic Engineering, Glasgow, UK, 9–14 June 2019; pp. 1–10.
15. Wang, K.H.; Ren, X. Wave interaction with a concentric porous cylinder system. *Ocean Eng.* **1994**, *21*, 343–360. [CrossRef]
16. Chau, F.P.; Eatock Taylor, R. Second-order wave diffraction by a vertical cylinder. *J. Fluid. Mech.* **1992**, *240*, 571–599. [CrossRef]
17. Malenica, Š.; Eatock Taylor, R.; Huang, J.B. Second-order water wave diffraction by an array of vertical cylinders. *J. Fluid. Mech.* **1999**, *390*, 349–373. [CrossRef]
18. Cong, P.W.; Gou, Y.; Teng, B.; Zhang, K.; Huang, Y.F. Model experiments on wave elevation around a four-cylinder structure. *Ocean Eng.* **2015**, *96*, 40–55. [CrossRef]
19. Cong, P.W.; Bai, W.; Teng, B.; Gou, Y. Semi-analytical solution to the second-order wave loads on a vertical cylinder in bi-chromatic bi-directional waves. *Ocean Eng.* **2018**, *161*, 205–220. [CrossRef]
20. He, G.H.; Zhang, Z.G.; Wang, W.; Wang, Z.K.; Jing, P.L. Near-Trapping on a Four-Column Structure and the Reduction of Wave Drift Forces Using Optimized Method. *J. Mar. Sci. Eng.* **2020**, *8*, 174. [CrossRef]
21. Cong, P.W.; Liu, Y.Y.; Gou, Y.; Teng, B. Theoretical modeling of hydrodynamic characteristics of a compound column-plate structure based on a novel derivation of mean drift force formulation. *P I MECH ENG C-J MEC* **2019**, *233*, 1022–1036. [CrossRef]

Journal of
Marine Science and Engineering

MDPI

Article

Two-Dimensional Numerical Modelling of a Moored Floating Body under Sloping Seabed Conditions

Aichun Feng [1], Hooi Siang Kang [2], Binbin Zhao [3] and Zhiyu Jiang [4,*]

1 Department of Civil and Environmental Engineering, National University of Singapore, Singapore 117576, Singapore; fengaichun2018@163.com
2 Marine Technology Center, Faculty of Engineering, Universiti Teknologi Malaysia, Johor Bahru 81310, Malaysia; kanghs@utm.my
3 College of Shipbuilding Engineering, Harbin Engineering University, Harbin 150001, China; 15045628379@163.com
4 Department of Engineering Sciences, University of Agder, N-4879 Grimstad, Norway
* Correspondence: zhiyu.jiang@uia.no

Received: 16 April 2020; Accepted: 14 May 2020; Published: 28 May 2020

Abstract: A coupled floating body-mooring line model is developed by combining a boundary element model for a two-dimensional floating body and a catenary mooring line model. The boundary element model is formulated in the time domain by a continuous Rankine source, and a reflection potential is introduced to account for the wave reflection due to sloping seabed. This newly developed model is validated by comparisons against available data. Then, dynamic response analyses are performed for the moored body in various seabed conditions. Compared with a flat seabed, a sloping seabed causes unsymmetrical mooring line configuration and generates noticeable effects in the motion responses of the floating body.

Keywords: coupled model; boundary element method; unsymmetrical mooring lines; sloping seabed; linear potential flow

---

1. Introduction

Many types of offshore structures, including floating islands [1], floating shelters [2], floating wind turbines [3], floating wave energy converters [4] and floating offshore fish farms [5], are moored in complicated coastal environments. These structures are exposed to ocean waves, with the possibility of wave reflection and wave shoaling from coastal seabeds. Dynamic responses of these floating bodies are the combined effects of ocean environments and mooring systems. Accurate predictions of dynamic responses are of practical importance for the design and manufacture of these offshore structures.

Compared with that of deep water offshore structures, design and construction of floating structures in coastal environments face unique challenges and the dynamics of these structures could be more complicated because of the seabed effects. Previous researches suggest that shallow waters can excite larger responses of offshore structures under hydrodynamic loading because of the flat seabed effects [6–8]. As numerical approaches for flat seabeds are not directly applicable to sloping seabed profiles, refs. [9,10] developed second-body models to account for the sloping seabed effect within the boundary element model frame and found that a sloping seabed significantly influences the cross coupling hydrodynamic coefficients. Refs. [11,12] developed multi-domain approaches that divide the fluid domain into an interior domain of variable water depth and an exterior domain of constant depth. An extra term accounting for the sloping-bottom effects is introduced to correct the incident wave potential so that the sloping seabed condition is satisfied. They found that the sloping seabed significantly affects the body motion response amplitude operator (RAO). Ref. [13] coupled a Rankine source model to the Boussinesq equation, which supplies all relevant information concerning the fluid domain. They reported that the peak frequency of the exciting forces and motion responses are shifted ahead due to the sloping seabed effects. Refs. [14,15] introduced a reflection velocity potential to account for the sloping seabed effect. Numerical results demonstrated that the sloping seabed alters the symmetrical profile of the fluid domain and the coupled effects between different motion modes become important. However, the mooring lines are not considered in these numerical models.

Mooring systems serve as station-keeping devices used to maintain a floating body in acceptable positions. Mooring systems can be categorised based on the restoring mechanisms, weathervaning characteristics and so forth. For a taut-leg mooring, the elastic stiffness due to line stretch is the dominant source of restoring, while for a catenary mooring configuration, the mooring restoring comes primarily from the geometry stiffness of the lines in normal sea states as the floating structure moves within certain offset ranges [16]. As a result of the relatively cheap anchoring costs and convenient offshore installation, the catenary mooring configuration has been extensively applied in various water depths [17,18] and with different component compositions [19]. For instance, Ref. [20] found that proper application of clump weights and buoys can increase the mooring restoring force and floating body capacity, but this will add extra difficulty for practical operation.

Numerical modelling of catenary mooring lines has different levels of fidelity. The quasi-static approach based on the catenary equations does not consider mooring dynamics and facilitates frequency-domain analysis for preliminary design purposes [21]. For deep-water applications, the dynamic mooring line tension becomes more important, and semi-analytical methods accounting for simplified dynamics were also proposed [22]. Later, finite element models [23,24] and lumped-mass models [25] were proposed for time-domain simulations, and these models provide accurate representation of the mooring line dynamics at increased computational costs.

Concerning the modelling of floating body and mooring lines, both uncoupled and coupled analysis have been applied. An uncoupled model essentially ignores the interactions between the floating body and treats a mooring line as a simple massless spring [26,27]. Such an approach benefits from less computational cost but sacrifices result accuracy. In comparison, a coupled model addresses the interaction between the floating body and mooring lines and can be particularly important if accurate mooring responses are needed. For a coupled model, frequency-domain and time-domain panel models have been developed to describe the wave phenomena and wave forces. As shown by [28], both frequency- and time-domain models can be used to deal with nonlinear waves around the floating body, although the latter is less ambiguous for the scattered waves.

Many offshore renewable energy facilities such as wave energy converters are moored in coastal areas with shallow water. Depending on the site conditions, the mooring lines may lie on a sloping seabed, which will change the fluid dynamics and eventually affect the mooring dynamics and power production performance. To examine the influence of seabed conditions without loss of generality, we only consider a floating body with mooring lines in this work. A simplified catenary model is adopted together with a boundary integral method to develop a time-domain coupled numerical model. The boundary integral model comprises three boundary element equations accounting for wave diffraction, wave radiation and wave reflection, respectively. This coupled model is validated by comparison against published data for the static offset, free decay and regular wave tests. Furthermore, numerical simulations are performed for various seabed profiles and mooring line conditions. The numerical study shows that the sloping seabed significantly changes the fluid domain and mooring line profile and therefore results in noticeable effects on the dynamic responses of the coupled floating body-mooring line system. These effects also vary under different sloping seabed with asymmetrical mooring lines conditions compared with the flat seabed case.

It should be noticed that in real ocean environments, inclined seabeds can lead to formation of vortices of various scales which affect the mooring line mechanism as well as the floating body motion characteristics. The present fluid domain is described by potential flow and flow viscosity is essentially ignored. The investigation of vortices on the mooring line and body motion is out of the scope of the present study.

2. Problem Formulation

2.1. Coupled Floating Body-Mooring Line Model

For a free-floating body moored in coastal environment, the position and velocity of the floating body vary at each time step and mooring line supplies restoring force to keep the body in acceptable positions. The coupled motion equation for a floating body with mooring lines can be expressed in the form:

$$\sum_{j=1}^{3} M_{ij}\ddot{X}_j(t) = F_i^{wave}(t) + F_i^{mooring}(t) + F_i^{static}(t) \ (i = 1,2,3), \tag{1}$$

where M_{ij} and $\ddot{X}_j(t)$ respectively denote the body mass matrix and body acceleration. $F_i^{wave}(t)$, $F_i^{mooring}(t)$ and $F_i^{static}(t)$ denote wave force, mooring line force and hydrostatic force, respectively. $F_i^{wave}(t)$ consists of wave radiation and wave diffraction forces. $i, j = 1,2,3$ stand for surge, heave and pitch motion, respectively. In the present study, the coupling effects between different motion modes are ignored for simplification and therefore Equation (1) can be further expressed as:

$$M_{11}\ddot{X}_1(t) = F_1^{wave}(t) + F_1^{mooring}(t) + F_1^{static}(t), \tag{2}$$

$$M_{22}\ddot{X}_2(t) = F_2^{wave}(t) + F_2^{mooring}(t) + F_2^{static}(t), \tag{3}$$

$$M_{33}\ddot{X}_3(t) = F_3^{wave}(t) + F_3^{mooring}(t) + F_3^{static}(t). \tag{4}$$

2.2. Catenary Mooring Line Model

The boundary conditions applied on the mooring lines are the constraints at the two ends of the line: the upper end is attached to the floating platform and is forced to move with the platform, and the lower end is connected to an anchor point at seafloor. For a catenary mooring configuration, a portion of catenary line should lie on the seabed to avoid large vertical forces on the anchors.

The following key assumptions are applied during the analysis of a catenary model: (1) The mooring line moves very slowly such that the drag forces and inertial forces are negligible. (2) The environmental loads on the mooring line are insignificant and can be excluded in the model. (3) The mooring line is inelastic. (4) The anchor point does not move in any directions. (5) The mooring line is lying on a horizontal seabed. In the present study, the sloping seabed inclination is small and the sloping seabed effects for the mooring line mechanism are therefore ignored. This simplification justifies the application of catenary mooring line model to horizontal seabed profiles. For a catenary mooring line with length L, under its own weight of W per unit length in water, the governing equations can be analytically expressed as [29] (pp. 258–263):

$$L - L_z(1 + 2\frac{T_x}{L_zW})^{\frac{1}{2}} + \frac{T_x}{W}\cosh^{-1}(1 + \frac{L_zW}{T_x}) - L_x = 0, \tag{5}$$

$$T_z = W(L_x^2 + 2L_xW)^{\frac{1}{2}}, \tag{6}$$

where L_x and L_z are the horizontal and vertical distances between the two ends points of a mooring line, respectively. T_x and T_z are the horizontal and vertical tension, respectively. A sketch of the catenary mooring line model is presented in Figure 1.

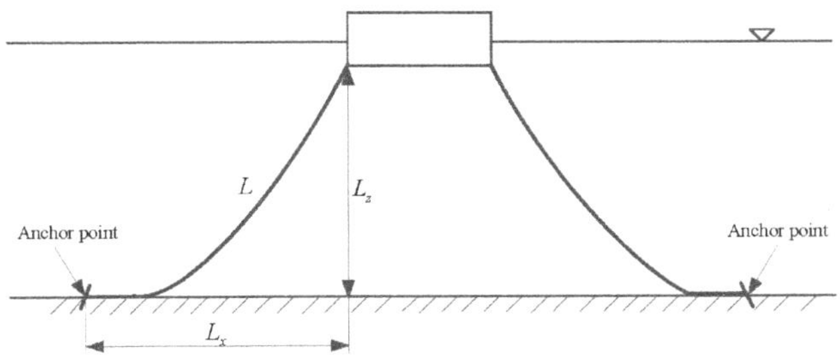

Figure 1. Sketch of a catenary mooring line model.

At an initial time $t = 0$, a mooring line pretension T_{pre} is applied to keep the floating body at a static equilibrium position. At each time step, an iterative scheme is applied to calculate the mooring line forces based on Equation (5). This force is used as the exciting mooring line force on the right-hand side of Equation (1) and then the equation is solved using a fourth-order Runge-Kutta method. The floating body displacement works as boundary conditions for the upper end of mooring line. Then, at the next time step, Equation (5) is solved again to start another loop until the end of the numerical simulation.

2.3. *Fluid Domain Formulation*

2.3.1. Fluid Motion Equations

Figure 2 illustrates a coordinate frame OXZ with its origin O at the centre of a two-dimensional floating body-mooring line system in a fluid domain of finite water depth. The fluid domain Ω is bounded by a free surface S_f, a body surface S_b, a seabed surface S_0 and an enclosing surface at infinity, S_∞. The fluid flow is assumed to be irrotational and can be described by the velocity potential, ϕ, which satisfies the Laplace equation as follows,

$$\frac{\partial^2 \phi}{\partial x^2} + \frac{\partial^2 \phi}{\partial z^2} = 0 \text{ in the fluid domain } \Omega. \tag{7}$$

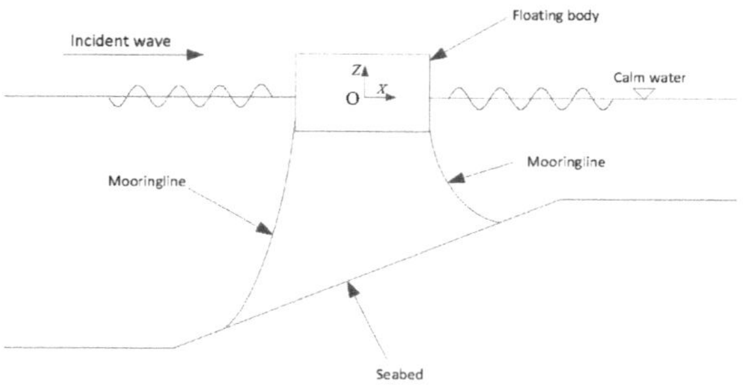

Figure 2. Illustration of a coupled floating body-mooring line model with a seabed condition.

Based on linear theory, the velocity potential ϕ can be expressed as:

$$\phi(x,z,t) = \phi_I(x,z,t) + \phi_D(x,z,t) + \phi_R(x,z,t) + \phi_{re}(x,z,t), \tag{8}$$

where ϕ_I, ϕ_D, ϕ_R and ϕ_{re} denote the incident wave potential, diffraction potential, radiation potential and reflection potential, respectively. The radiation potential is used to describe the velocity potential generated by floating body motions. For two-dimensional problems, a floating body can undergo surge, heave and pitch motions. For different degrees of freedom, the radiation potential essentially has different values.

2.3.2. Diffraction, Radiation and Reflection Potential Problems

The velocity potentials ϕ_I, ϕ_D, ϕ_R and ϕ_{re} satisfy the Laplace equation, and ϕ_D, ϕ_R and ϕ_{re} can be investigated by formulating the boundary integral equations accordingly based on the work of [8,14,30]. The diffraction potential, $\phi_D(x,z,t)$, describes the fluid field potential generated by the incoming wave within the seabed environment. For an incoming progressive linear wave with a wave amplitude, A, in a finite water depth, h, the velocity potential, ϕ_I and the wave elevation, η_I, can be expressed as

$$\phi_I = \frac{gA}{\omega}\frac{\cosh k(z+h)}{\cosh kh}\sin(kx-\omega t), \tag{9}$$

$$\eta_I = A\cos(kx-\omega t), \tag{10}$$

where g, k and ω denote the gravitational acceleration, wave number and wave frequency, respectively.

As no fluid penetrates the rigid body and seabed surfaces, the boundary conditions applied to the body and flat seabed are:

$$\frac{\partial \phi_D}{\partial \mathbf{n}_b} = -\frac{\partial \phi_I}{\partial \mathbf{n}_b} \quad \text{on } S_b, \tag{11}$$

$$\frac{\partial \phi_D}{\partial \mathbf{n_0}} = 0 \quad \text{on } S_0, \tag{12}$$

where $\mathbf{n}_b$ and $\mathbf{n}_0$ denote normal vector on body surface and seabed surface, respectively.

The linearised kinematic condition and the dynamic free surface boundary condition are given as

$$\frac{\partial \phi_D}{\partial z} - \frac{\partial \eta_D}{\partial t} = 0 \quad \text{on } S_f, \tag{13}$$

$$\frac{\partial \phi_D}{\partial t} + g\eta_D = 0 \quad \text{on } S_f, \tag{14}$$

where η_D refers to the wave elevation caused by wave diffraction. The body boundary condition (Equation (11)), seabed condition (Equation (12)) and free surface boundary conditions (Equations (13) and (14)) create a closed boundary value formulation to the diffraction potential problem.

The radiation potential, $\phi_R(x, z, t)$, accounts for the fluid field potential caused by body motions in the fluid domain. Nonpermeable boundary condition is applied to the body and seabed conditions. The free surface boundary condition is simplified as a linear form:

$$\frac{\partial \phi_R}{\partial \mathbf{n}_b} = \mathbf{v}_b \cdot \mathbf{n}_b \quad \text{on } S_b, \tag{15}$$

$$\frac{\partial \phi_R}{\partial \mathbf{n}_0} = 0 \quad \text{on } S_0, \tag{16}$$

$$\frac{\partial \phi_R}{\partial z} - \frac{\partial \eta_R}{\partial t} = 0 \quad \text{on } S_f, \tag{17}$$

$$\frac{\partial \phi_R}{\partial t} + g\eta_R = 0 \quad \text{on } S_f, \tag{18}$$

where $\mathbf{v}_b$ denotes the body motion velocity. For a floating body experiencing forced oscillatory motion with an amplitude a, the body motion velocity $\mathbf{v}_b$ can be described as:

$$\mathbf{v}_b = (a\omega cos\omega t,\ 0) \quad \text{for surge motion}, \tag{19}$$

and

$$\mathbf{v}_b = (0,\ a\omega cos\omega t) \quad \text{for heave motion}. \tag{20}$$

In the proposed model, the free surface is divided into inner and outer domains, which are discussed in the next section. In the outer domain, the source panels are arranged in an exponentially increasing manner. Using this distribution method, a very large free surface can be covered by the source panels and therefore the generated waves can move towards the far field in this enlarged computational domain before the end of numerical simulation. The radiation condition is satisfied without applying any artificial damping on the free surface.

The sloping seabed provides a source of wave reflections into the fluid domain. In this investigation, the wave reflection potential, ϕ_{re}, is introduced to satisfy the Laplace Equation (7) whose solution is determined by solving the following boundary value equations

$$\frac{\partial \phi_{re}}{\partial \mathbf{n}_b} = 0 \quad \text{on } S_b, \tag{21}$$

$$\frac{\partial \phi_{re}}{\partial z} - \frac{\partial \eta_{re}}{\partial t} = 0 \quad \text{on } S_f, \tag{22}$$

$$\frac{\partial \phi_{re}}{\partial t} + g\eta_{re} = 0 \quad \text{on } S_f, \tag{23}$$

$$\frac{\partial \phi_{re}}{\partial \mathbf{n}_0} = -\frac{\partial \phi_I}{\partial \mathbf{n}_0} \quad \text{on } S_0, z = h(x). \tag{24}$$

Notice that this boundary value equation is analogous to the solution of the wave diffraction potential. In this regard, the potential ϕ can be considered as the potential caused by the interaction of the floating body (the first body) and the sloping seabed (the second body). Equation (24) describes the sloping seabed boundary condition whereas, in Equation (12), the constant water depth refers to the deeper water depth on the left hand side of the sloping seabed. If the seabed is completely flat, the reflection potential ϕ_{re} disappears. In this scenario, the reflected wave moves towards the left hand side of the fluid domain and travels into the far field. As the proposed free surface source distribution method covers a very large free surface area on both sides of the body, the reflected wave will not reflect backwards the fluid domain.

2.3.3. Numerical Discretisation of Velocity Potential ϕ_D, ϕ_R and ϕ_{re}

The velocity potential ϕ can be expressed as a boundary integral of Rankine sources continuously distributed on the fluid boundary surface $S = S_b \cup S_f \cup S_0$ as

$$\phi(\mathbf{x}',t) = \int_{S_b} \sigma^b \ln|\mathbf{x}' - \mathbf{x}| ds_{\mathbf{x}} + \int_{S_f} \sigma^f \ln|\mathbf{x}' - \mathbf{x}| ds_{\mathbf{x}} + \int_{S_0} \sigma^0 \ln|\mathbf{x}' - \mathbf{x}| ds_{\mathbf{x}}, \tag{25}$$

where $\mathbf{x}' = (x', z')$ and $\mathbf{x} = (x, z)$ are the control point and the source point, respectively. In this paper, a continuous panel is applied to replace the isolated source point method. The control point is the point where the boundary condition is satisfied and it is located in the middle of a continuous panel. The source point is the end point defining the continuous panel. Each continuous panel has two source points and one control point. σ^b, σ^f and σ^0 denote the source strength of the body, free surface and seabed, respectively.

Let the integral domain boundary surfaces S_b, S_f and S_0 be approximated as the sums of N_b, N_f and N_0 panels, respectively such that $S_b = \cup_{i=1}^{N_b} S_i^b$, $S_f = \cup_{j=1}^{N_f} S_j^f$ and $S_0 = \cup_{k=1}^{N_0} S_k^0$. Let $\mathbf{x'}_i^b \in S_b$, $\mathbf{x'}_j^f \in S_f$ and $\mathbf{x'}_k^0 \in S_0$ be the control points for $i = 1, ..., N^b$, $j = 1, ..., N^f$ and $k = 1, ..., N^0$. The discretised form of the boundary integral (Equation (25)) describing the diffraction problem is given as:

$$\begin{aligned}\frac{\partial \phi_D(\mathbf{x'}_i^b)}{\partial \mathbf{n}_b} &= \sum_{i=1}^{N^b} \sigma_i^b \frac{\partial}{\partial \mathbf{n}_b^i} \int_{S_i^b} \ln|\mathbf{x'}_i^b - \mathbf{x}_i| ds_{\mathbf{x}} \\ &+ \sum_{j=1}^{N^f} \sigma_j^f \frac{\partial}{\partial \mathbf{n}_b^i} \int_{S_j^f} \ln|\mathbf{x'}_i^b - \mathbf{x}_j| ds_{\mathbf{x}} + \sum_{k=1}^{N^0} \sigma_k^0 \frac{\partial}{\partial \mathbf{n}_b^i} \int_{S_k^0} \ln|\mathbf{x'}_i^b - \mathbf{x}_k| ds_{\mathbf{x}},\end{aligned} \tag{26}$$

$$\begin{aligned}\frac{\partial \phi_D(\mathbf{x'}_j^f)}{\partial \mathbf{n}_f} &= \sum_{i=1}^{N^b} \sigma_i^b \frac{\partial}{\partial \mathbf{n}_f^j} \int_{S_i^b} \ln|\mathbf{x'}_j^f - \mathbf{x}_i| ds_{\mathbf{x}} \\ &+ \sum_{j=1}^{N^f} \sigma_j^f \frac{\partial}{\partial \mathbf{n}_f^j} \int_{S_j^f} \ln|\mathbf{x'}_j^f - \mathbf{x}_j| ds_{\mathbf{x}} + \sum_{k=1}^{N^0} \sigma_k^0 \frac{\partial}{\partial \mathbf{n}_f^j} \int_{S_k^0} \ln|\mathbf{x'}_j^f - \mathbf{x}_k| ds_{\mathbf{x}},\end{aligned} \tag{27}$$

$$\begin{aligned}\frac{\partial \phi_D(\mathbf{x'}_k^0)}{\partial \mathbf{n}_0} &= \sum_{i=1}^{N^b} \sigma_i^b \frac{\partial}{\partial \mathbf{n}_0^k} \int_{S_i^b} \ln|\mathbf{x'}_k^0 - \mathbf{x}_i| ds_{\mathbf{x}} \\ &+ \sum_{j=1}^{N^f} \sigma_j^f \frac{\partial}{\partial \mathbf{n}_0^k} \int_{S_j^f} \ln|\mathbf{x'}_k^0 - \mathbf{x}_j| ds_{\mathbf{x}} + \sum_{k=1}^{N^0} \sigma_k^0 \frac{\partial}{\partial \mathbf{n}_0^k} \int_{S_k^0} \ln|\mathbf{x'}_k^0 - \mathbf{x}_k| ds_{\mathbf{x}}.\end{aligned} \tag{28}$$

At any time step t during the numerical simulation, the body and the seabed boundary conditions are described by their boundary value equations. The free surface boundary conditions for diffraction potential ϕ_D, radiation potential ϕ_R and reflection potential ϕ_{re} are all updated by a 3rd-order Adams-Bashforth scheme. The time interval $\Delta t = T_0/100$ is applied in the numerical simulations. Here, T_0 is the wave period.

A matrix equation is formulated by discretising Equations (26)–(28). At each time step t, the boundary conditions work as inputs to this matrix equation and the source strength σ_i^b, σ_j^f and σ_k^0 are the solutions of this equation. Once the source strengths are obtained, the velocity potential ϕ can be investigated based on Equation (25).

The discretisation for the wave radiation, ϕ_R, and the reflection potential, ϕ_{re}, is exactly the same as that for the diffraction potential, ϕ_D and is therefore omitted herein. Calculation of the Rankine source lnr integral in the boundary integral equations is computationally expensive. In this study, a continuous Rankine source panel integral is developed instead of an isolated Rankine source approximation to reduce numerical errors and increase numerical efficiency.

Note that, physically, the free surface, S_f, is of infinite size and the integration over this infinite dimension is not possible by numerical techniques. For this reason, the computational domain is truncated in the numerical simulations and the far-field boundary condition is satisfied on this truncated free surface. In the present study, the Rankine source panels are distributed in an exponential manner, allowing for the coverage of a very large area of free surface. The numerical computations are completed before the incident wave and reflected wave reaches the truncated boundary.

2.3.4. Boundary Panel Distribution

In the present study, the sloping seabed $z = h(x)$ is described mathematically as:

$$h(x) = \begin{cases} -H - \frac{L_v}{2} & x < -\frac{L_h}{2} \\ -H + x\tan\theta & -\frac{L_h}{2} \le x \le \frac{L_h}{2} \\ -H + \frac{L_v}{2} & x > \frac{L_h}{2} \end{cases} \tag{29}$$

where H is the mean water depth and θ denotes the sloping seabed angle which is defined as:

$$\theta = arctan(\frac{L_h}{L_v}), \tag{30}$$

where L_h and L_v are the projected length of the slope seabed in horizontal and vertical directions, respectively. Figure 3 illustrates the definitions of these parameters in the coupled floating body-mooring line problem under investigation. The sloping seabed is divided into three regions, h_1, $h_2(x)$ and h_3. The constants h_1 and h_3 correspond to the constant depth flat portions on the left and right hand sides of the sloping seabed whereas the variable $h_2(x)$ represents the depth of water in the sloping region.

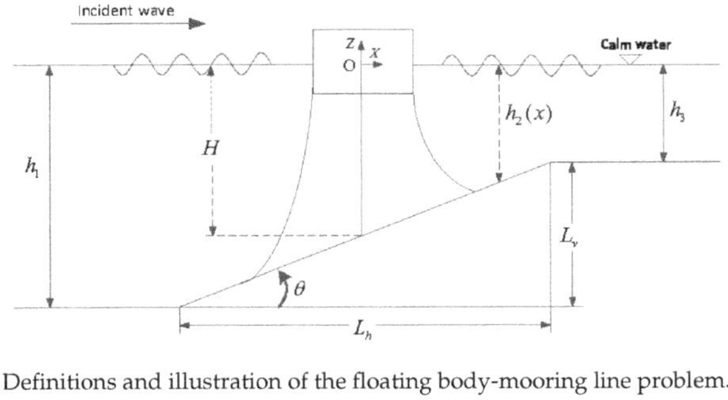

Figure 3. Definitions and illustration of the floating body-mooring line problem.

Both the free surface and seabed surface are divided into inner and outer domains. To approximate the inner domain, the source panel distribution method discussed by [8] is adopted here. A parameter $\gamma = 1 + \tan\theta$ is introduced to deal with this asymmetric characteristic of the surface. For free surface and flat seabed, the angle $\theta = 0$ and therefore $\gamma = 1$. On the right side of S_0, starting from $x = 0$, the horizontal distances between neighbouring control points are expressed as:

$$|\mathbf{x}'^0_k - \mathbf{x}'^0_{k-1}| = \frac{L_b \alpha_k}{\gamma \omega^2} \text{ in the inner domain,} \tag{31}$$

$$|\mathbf{x}'^0_k - \mathbf{x}'^0_{k-1}| = \frac{L_b \alpha_k}{\gamma \omega^2} 1.05^{k(k-1)/2} \text{ in the outer domain,} \tag{32}$$

and on the left hand side of S_0:

$$|\mathbf{x}'^0_k - \mathbf{x}'^0_{k-1}| = \frac{\gamma L_b \alpha_k}{\omega^2} \text{ in the inner domain,} \tag{33}$$

$$|\mathbf{x}'^0_k - \mathbf{x}'^0_{k-1}| = \frac{\gamma L_b \alpha_k}{\omega^2} 1.05^{k(k-1)/2} \text{ in the outer domain,} \tag{34}$$

where L_b is the body surface panel length. The parameters α_k define the separation distances and are decided according to the method discussed by [8]. In the outer domain the distances between the centres of neighbouring source panels are displaced in an exponentially increasing manner as aforementioned.

3. Model Validation

In order to validate the proposed coupled floating body-mooring line model, numerical simulations are performed for the static offset test, the free decay test without mooring line, the free decay test with mooring line and the regular wave simulation. All numerical simulations are conducted in the time domain and converted into the frequency domain for direct comparison against available published data.

3.1. Static Offset Test

The stiffness characteristics of catenary mooring lines are investigated preliminarily by the static offset tests. For a mooring line with a length of L = 100 m and a unit weight of W = 828 N/m, moored in water depth h = 25 m, Figure 4 shows the comparison of this mooring line stiffness curves with respect to horizontal force T_x (a) and vertical force T_z (b) between the proposed numerical method and [29] (pp. 258–263). Results from the proposed method show excellent agreement with those of [29]. It is noticed that both the horizontal force T_x and vertical force T_z demonstrate nonlinear characteristics as the mooring line horizontal distance L_x increases and this phenomenon is also called nonlinear "hardening" behaviour by [26].

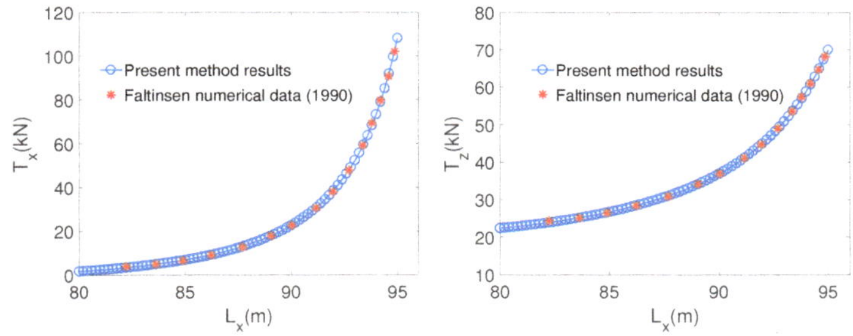

(**a**)Horizontal force T_x (**b**)Vertical force T_z

Figure 4. Comparison of mooring line stiffness with respect to horizontal force T_x (a) and vertical force T_z (b) between proposed numerical method and numerical data of [29].

3.2. Free Decay Test

The free decay test works as a measurement to determine the natural period T and the damping ratio β of the system. In the free decay test, the floating system was offset an initial displacement and allowed to damp out until reaching the equilibrium position. A heave decay test was performed for a semi-circular cylinder with radius 10 m excluding mooring line as a preliminary study. Figure 5 shows comparison of the heave decay test for the semi-circular cylinder between the proposed method, the [31] numerical simulation and the [32] experiment. As shown, the proposed model yields a good agreement compared with the numerical prediction of [31] and experimental data of [32]. The natural period for this cylinder in heave motion is 7.18 s.

Numerical simulations are further performed for the semi-circular cylinder with different mooring line stiffness K_i. The mooring line system stiffness K_i is defined as:

$$K_i = \frac{F_i^{mooring}}{X_i}. \tag{35}$$

where X_i denotes the body offset and is corresponding to the body displacement in Equation (1). Figure 6 shows the surge decay test for the cylinder with mooring line stiffness $K_1 = 90{,}000$ N/m, 135,000 N/m, 180,000 N/m and 225,000 N/m, respectively. It is noticed that the decay period and motion amplitude decrease as stiffness K_i increases. The coupled system free decay period can be analytically evaluated by the following equation of [33] (pp. 275–278):

$$T = 2\pi/(1-\beta^2)/\sqrt{K/(M+a)}, \tag{36}$$

with

$$\beta = \frac{b}{2\sqrt{(M+a)K}}, \tag{37}$$

where subscripts are omitted for simple notation. The body added mass a and damping b can be obtained from the work of [8]. Table 1 shows a comparison of the surge decay period T between the analytical method by [33] and the proposed numerical method for the mooring line stiffness of $K_1 = 90{,}000$ N/m, 135,000 N/m, 180,000 N/m and 225,000 N/m, respectively. It is observed that the proposed method agrees well with the analytical expression of [33]. The natural period of this floating body-mooring system is around 10 s. Therefore, the wave frequencies dominate the system responses and low-frequency responses are expected to be small.

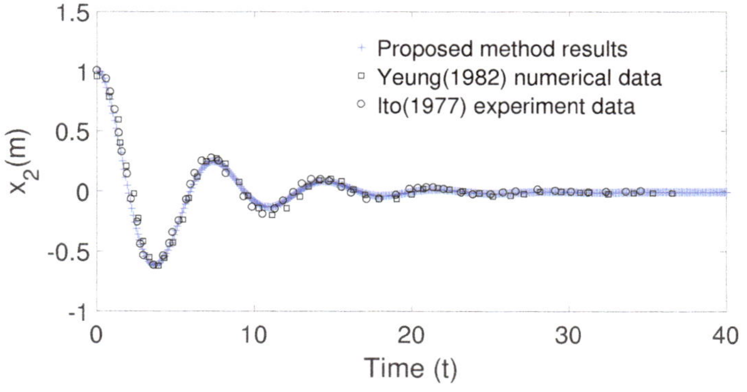

Figure 5. Comparison of heave decay test for the semi circular cylinder between proposed method results, [31] numerical results and [32] experimental data.

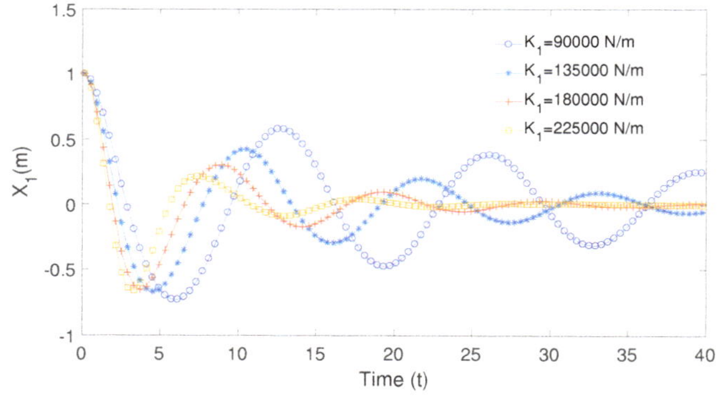

Figure 6. Surge decay test for cylinder with mooring line stiffness $K_1 = 90{,}000$ N/m, 135,000 N/m, 180,000 N/m and 225,000 N/m, respectively.

Table 1. Comparison of surge decay period T between Patel (1989) analytical method and proposed numerical method data for mooring line stiffness $K_1 = 90{,}000$ N/m, 135,000 N/m, 180,000 N/m and 225,000 N/m, respectively.

	Surge Decay Period T		
Stiffness K_1 (N/m)	Patel (1989) analytical method	Proposed numerical method	Error
90,000	12.77	13.05	2.1%
135,000	10.80	10.65	1.4%
180,000	9.35	9.15	2.0%
225,000	7.15	7.30	2.1%

3.3. Regular Wave Simulation

Numerical simulations are further carried out for the floating body-mooring system subject to regular waves. As a preliminary study, numerical simulations are conducted for a rectangular cylinder with breadth $B = 0.30$ m and draft $d = 0.20$ m, floating in water depth $h = 0.40$ m without mooring lines attached. Figure 7 shows a comparison of the time records of surge motion $X_1(t)$ (a) and heave motion $X_2(t)$ (b) between the proposed method, the experimental data by [34] and the numerical data by [35]. The floating rectangular cylinder is subject to a regular wave with wave period $T_0 = 1.2$ s and wave amplitude $A = 0.02$ m. It is shown that the proposed model results agree favourably well with the model test data of [34] and the numerical prediction of [35]. This floating body is moored by two symmetrical mooring lines, with $L = 1.6$ m and a weight of $W = 39.2$ N/m in water. Figure 8 demonstrates the comparison of time records of surge motion $X_1(t)$ (a) and heave motion $X_2(t)$ (b) between proposed method results and [35] numerical data for the rectangular cylinder with the mooring effect included. The numerical data of [35] is shifted with $T = 0$ as the starting time for better comparison. It is noticed that the proposed model data agrees reasonably well with the data of [35] for both surge and heave motions. The discrepancy can be explained by the fact that the viscous effects are included in the work of [35] but excluded in the present model.

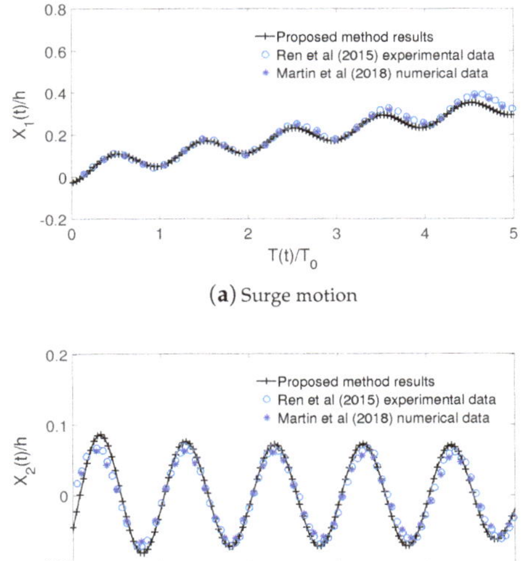

(**a**) Surge motion

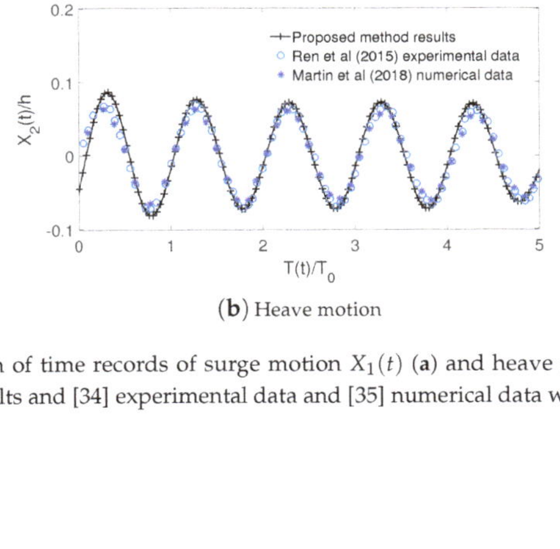

(**b**) Heave motion

Figure 7. Comparison of time records of surge motion $X_1(t)$ (**a**) and heave motion $X_2(t)$ (**b**) between proposed method results and [34] experimental data and [35] numerical data without mooring effect.

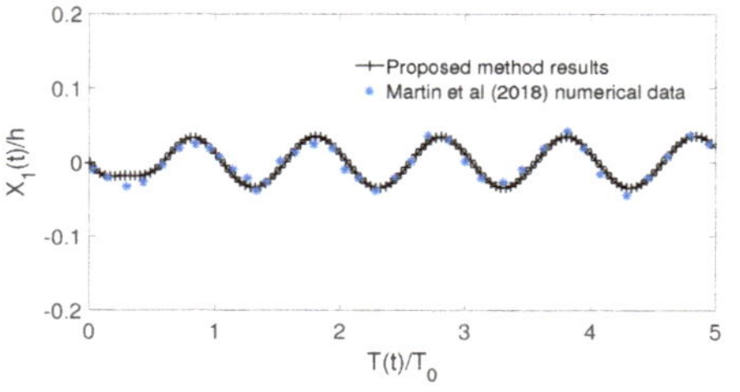

(**a**) Surge motion

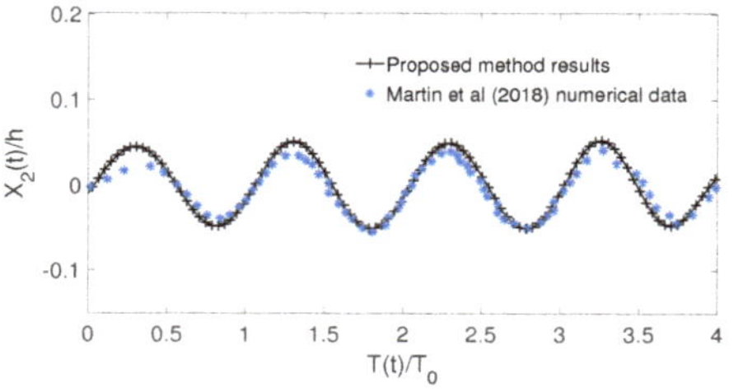

(**b**) Heave motion

Figure 8. Comparison of time records of surge motion $X_1(t)$ (**a**) and heave motion $X_2(t)$ (**b**) between proposed method results and [35] numerical data with mooring effect included.

4. Case Study

For a floating body moored in coastal areas, if the mooring line anchors are located in different water depths, then the mooring line profiles are not symmetrical. The asymmetry of seabed and mooring line profiles significantly affects fluid domain and hence the body response characteristics. By comparing the asymmetrical seabed and moorings with the flat seabed and symmetrical mooring lines, this section investigates the floating body response characteristics in sloping seabed and asymmetrical mooring lines conditions.

A rectangular cylinder with breadth B = 20 m and draft d = 10 m is selected as the floating body. Such a body size is typical of floating bodies in ocean and offshore engineering. An average water depth of H = 200 m is considered for the first numerical case. Two types of seabed and mooring line configurations are used in the numerical simulations. The first type is a flat seabed with two identical mooring lines, and the second one is a sloping seabed with two different mooring line arrangement. Figure 9 shows the profiles of symmetrical and asymmetrical seabeds with mooring lines. For the flat seabed condition, the left-side mooring line length L_1 and anchor horizontal distance AL_1 equal the right-side values L_2 and AL_2. The average water depths H for the sloping seabeds are same as the flat ones for direct comparisons. Each mooring line has a weight of W = 5 kN/m in water and two mooring lines are used on each side in the present study unless otherwise stated.

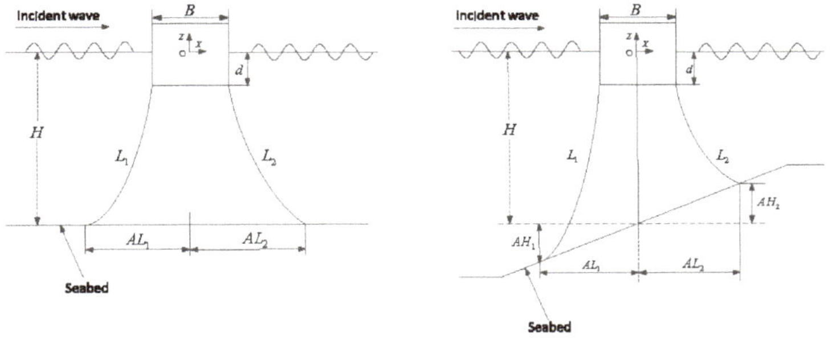

Figure 9. Profiles of symmetrical and asymmetrical seabeds with mooring lines.

Table 2 shows parameters of the four sets of mooring lines for the symmetrical seabed condition. In the present study, 10% of the mooring line length are lying on the seabed for all the numerical simulations unless stated otherwise. The incident wave has a wave amplitude $A = 1$ m and the mooring line is kept in a catenary profile in all numerical simulations. Free decay test shows the heave nature period for this coupled body and mooring line system are 8.78 s. Figure 10 shows a comparison of floating body surge motion $RAO_{11}(\omega)$ between without mooring line condition, mooring line set 1, mooring line set 2, mooring line set 3 and mooring line set 4. It is observed that the mooring line reduces the surge motion $RAO_{11}(\omega)$ noticeably and the $RAO_{11}(\omega)$ becomes smaller as the mooring line length increases. Figure 11 shows a comparison of the heave motion $RAO_{22}(\omega)$ of the floating body between without mooring line condition, mooring line set 1, mooring line set 2, mooring line set 3 and mooring line set 4. It is noticed that the mooring line reduces the heave motion amplitude significantly. Longer mooring line generates larger restoring force for the floating body and therefore the heave motion amplitude reduces more. The peak value of motion amplitude is shifted towards the high wave frequency as the mooring line length increases. This finding is in line with that of [36].

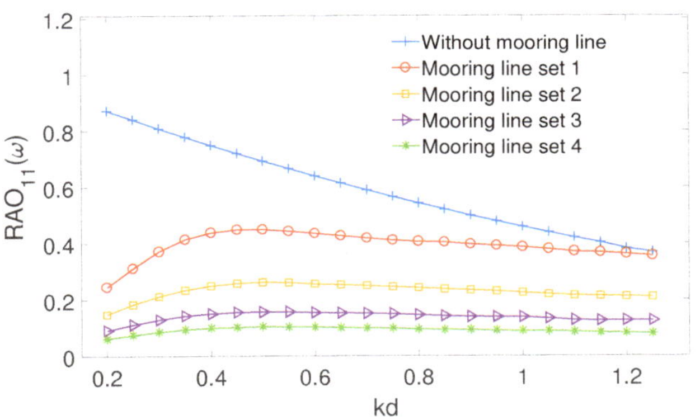

Figure 10. Comparison of the floating body surge motion $RAO_{11}(\omega)$ between without mooring line condition, mooring line set 1, mooring line set 2, mooring line set 3 and mooring line set 4.

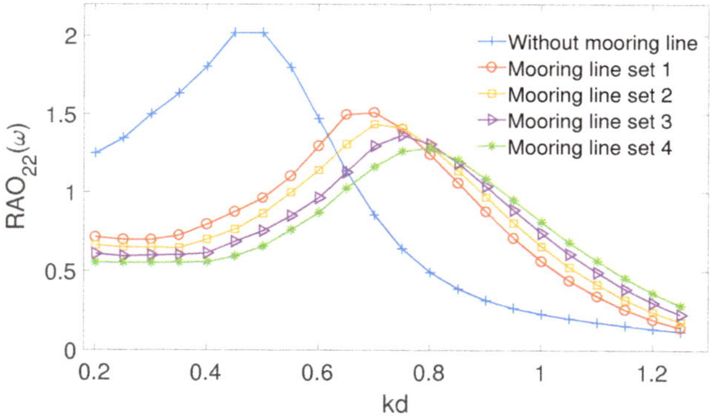

Figure 11. Comparison of the floating body heave motion $RAO_{22}(\omega)$ between without mooring line condition, mooring line set 1, mooring line set 2, mooring line set 3 and mooring line set 4.

Table 2. Parameters of four sets of mooring lines for symmetrical seabed condition with water depth $h = 200$ m.

Mooring Line Type	L_1	L_2	AL_1	AL_2
Set 1	600 m	600 m	549 m	549 m
Set 2	700 m	700 m	656 m	656 m
Set 3	800 m	800 m	762 m	762 m
Set 4	900 m	900 m	766 m	766 m

In order to investigate the hydrodynamic effects of asymmetrical seabed and mooring line, a symmetrical mooring line set $L_1 = L_2 = 700$ m is taken as the basis. Two asymmetrical mooring line sets are used in the numerical simulations and the parameters for these two sets of mooring lines are presented in Table 3. For direct comparison, the total mooring line lengths are same for both symmetrical and asymmetrical mooring line sets. It should be emphasised that the floating body has to be at static equilibrium. This requirement is naturally satisfied for symmetrical mooring line set. For asymmetrical seabed and mooring line set, the left side mooring line is arranged similarly as the symmetrical case, i.e., 10% of the mooring line length are laid on the seabed. The right-side mooring line has to be arranged to balance the horizontal mooring line force generated by the left side. An iterative procedure is developed to calculate the anchor horizontal distance AL_2 and the anchor vertical distance AH_2 is readily available according to the seabed slope once AL_2 is obtained. In such arrangement, the difference of vertical mooring line forces generated by the left and right sides is less than 3% and has marginal effects on the pitch motion of the floating body.

Table 3. Parameters of two sets of mooring lines for asymmetrical seabed condition with water depth $h = 200$ m.

Mooring Line Type	L_1	L_2	AL_1	AL_2	AH_1	AH_2
Set 1	720 m	680 m	674 m	640 m	9.7 m	9.2 m
Set 2	740 m	660 m	695 m	620 m	10.0 m	8.8 m

Figure 12 shows a comparison of the surge motion $RAO_{11}(\omega)$ of the floating body between symmetrical mooring line condition, asymmetrical mooring line set 1 and set 2. It is noticed that the surge motion RAO_{11} reduces appreciably as the asymmetry level increases across the whole wave frequency range. Compared with the symmetrical case, the left-side length of asymmetrical mooring line set 2 increases by 6%, but it reduces the RAO_{11} by approximately 15%. If less mooring line length is allowed to lie on the seabed, effects due to the mooring line asymmetry are expected to be more significant. It should be noticed that the difference between the left-and right-side mooring lines is in a reasonable range, otherwise the mooring line set can not keep the floating body at a static equilibrium position. The heave motion $RAO_{22}(\omega)$ is marginally affected by the asymmetrical mooring line setup and therefore the data are not presented here.

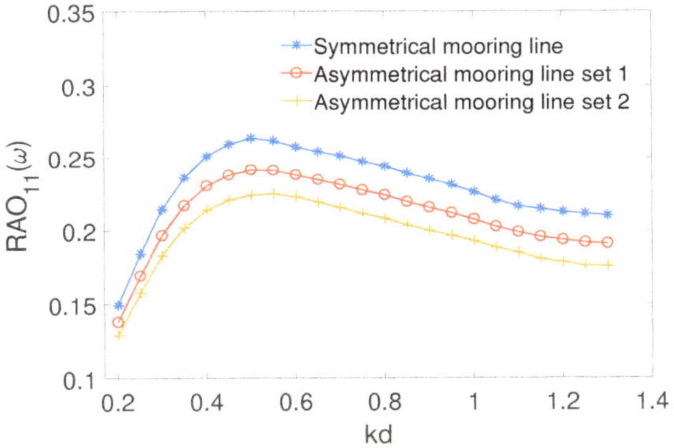

Figure 12. Comparison of the floating body surge motion $RAO_{11}(\omega)$ between symmetrical mooring line condition, asymmetrical mooring line set 1 and set 2.

Numerical simulations are further performed for the floating body positioned in a water depth of $H = 20$ m. Both flat seabed with symmetrical mooring lines and sloping seabed with asymmetrical mooring lines are considered. The mooring lines on the sloping seabed have length $L_1 = L_2$ but are arranged in asymmetrical position. These mooring lines are lying on flat seabed portion whereas sloping seabed portion is located within two anchor points. Both the symmetrical and asymmetrical mooring line configurations have the same length $L_1 = L_2 = 60$ m for direct comparison. The 15% mooring line length is allowed to lie on the seabed for both cases. Figure 13 shows the profile of an asymmetrical seabed with mooring lines positioned in the water depth $H = 20$ m.

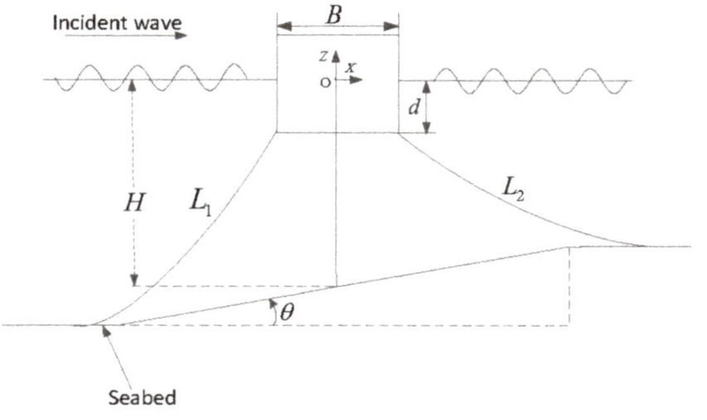

Figure 13. Profile of asymmetrical seabed with mooring lines positioned in water depth $H = 20$ m.

The sloping seabed effect for the incident wave amplitude is first investigated, and no floating body or mooring lines are included in the fluid domain. A linear incident wave with a wave amplitude of A_{inc} comes from the left-hand side and moves towards the right-hand side as illustrated in Figure 13. Figure 14 shows the incident wave amplitude A on the right- and left-hand sides over a sloping seabed with sloping angle $\tan\theta = 1/8$. The left-hand side has a wave amplitude of $A = A_{inc}$ as the incident wave comes from the left. The wave amplitude changes gradually on the sloping part due to the sloping seabed effects. The wave becomes steady when it moves into the flat part on the right-hand side. It is noticed that the incident wave amplitude increases due to the sloping seabed and the sloping seabed has largest effect for the incident wave when the wave frequency is approximately $kd = 0.8$.

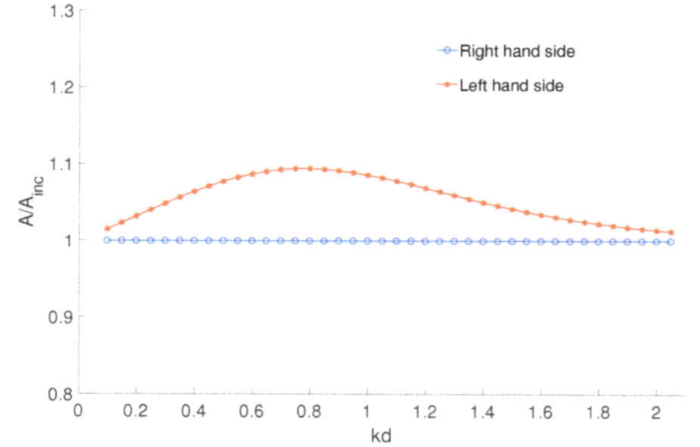

Figure 14. Incident wave amplitude A on the right hand and left hand sides over a sloping seabed with sloping angle $\tan\theta = 1/8$.

The left-side mooring line weight has a weight of $W = 5$ kN/m in water for the sloping seabed case, and the floating body is at a static equilibrium. A preliminary iterative procedure is developed to calculate the right-side mooring line weight per meter to reach static equilibrium. Figure 15 shows a comparison of the surge motion $RAO_{11}(\omega)$ of the floating body between a flat seabed and a sloping seabed with

$\tan\theta = 1/20$, $\tan\theta = 1/10$ and $\tan\theta = 1/8$, respectively. It is noticed that the surge motion $RAO_{11}(\omega)$ is smaller for the flat seabed than the sloping seabed conditions and $RAO_{11}(\omega)$ increases as the sloping seabed angle θ increases across the frequency range.

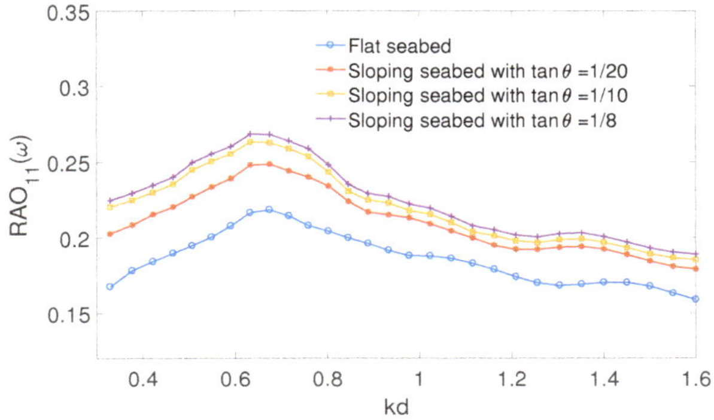

Figure 15. Comparison of the floating body surge motion $RAO_{11}(\omega)$ between flat seabed, sloping seabed with $\tan\theta = 1/20$, $\tan\theta = 1/10$ and $\tan\theta = 1/8$, respectively with mooring line length $L_1 = L_2 = 60$ m.

Figure 16 shows a comparison of the heave motion $RAO_{22}(\omega)$ of the floating body between flat seabed, sloping seabed with $\tan\theta = 1/20$, $\tan\theta = 1/10$ and $\tan\theta = 1/8$, respectively. It is noticed that the peak responses of the heave motion $RAO_{22}(\omega)$ are shifted towards lower wave frequency and the peak response values increase along with the sloping seabed angle. The phenomenon presented in Figures 15 and 16 implies that the sloping seabed with asymmetrical mooring line arrangement produces less mooring stiffness than does the symmetrical case. In the preliminary study to investigate the right-side mooring weight, numerical simulations show that less weight is required compared with the left side.

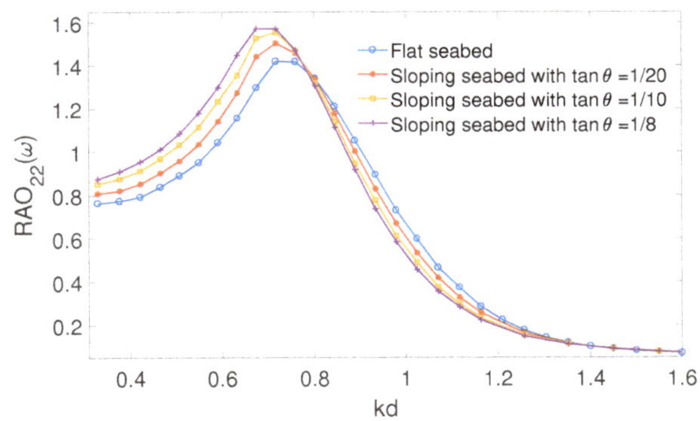

Figure 16. Comparison of the floating body heave motion $RAO_{22}(\omega)$ between flat seabed, sloping seabed with $\tan\theta = 1/20$, $\tan\theta = 1/10$ and $\tan\theta = 1/8$, respectively with mooring line length $L_1 = L_2 = 60$ m.

Figure 17 shows a comparison of the time record of the floating-body heave motion $X_2(t)$ for wave frequency $kd = 1.0$ between flat seabed, sloping seabed with $\tan\theta = 1/20$, $\tan\theta = 1/10$ and $\tan\theta = 1/8$, respectively. The heave motion $X_2(t)$ has the largest amplitude for flat seabed condition and the $X_2(t)$ motion amplitudes decrease as the sloping angles increase. These time-domain results are aligned with those frequency-domain results presented in Figure 16. In this study, the fluid domain is described by a linear boundary element model. The peak value for heave motion $X_2(t)$ is around 0.7 whereas the trough value is about 0.5. This nonharmonic body motion track demonstrates the nonlinear characteristics of the coupled floating body-mooring line system.

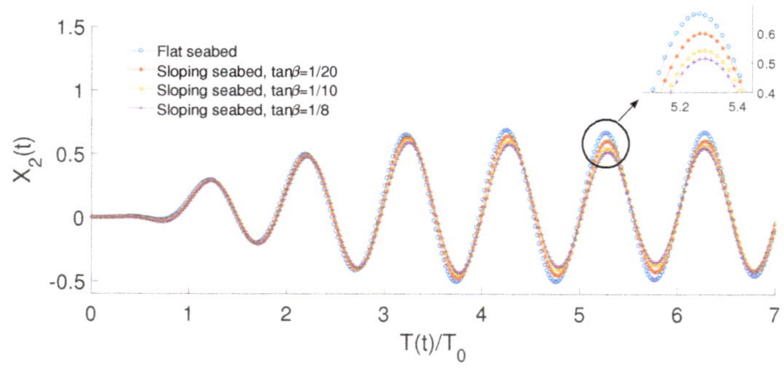

Figure 17. Comparison of time record of floating body heave motion $X_2(t)$ for wave frequency $kd = 1.0$ between flat seabed, sloping seabed with $\tan\theta = 1/20$, $\tan\theta = 1/10$ and $\tan\theta = 1/8$, respectively.

Figure 18 shows a comparison of mooring line force $F_2^{mooring}(t)$ for wave frequency $kd = 1.0$ between flat seabed, sloping seabed with $\tan\theta = 1/20$, $\tan\theta = 1/10$ and $\tan\theta = 1/8$, respectively. The mooring line pretension is excluded in the mooring line force $F_2(t)$. The mooring line force $F_2^{mooring}(t)$ has the largest amplitude for flat seabed condition and $F_2^{mooring}(t)$ amplitudes decrease as the sloping angles increase. It is also noticed that the force peak values are smaller than the trough values and this phenomenon is due to the fact that the heave motion $X_2(t)$ peak values are bigger than the trough values. Figure 19 demonstrates a comparison of time record of floating body surge motion $X_1(t)$ for wave frequency $kd = 1.0$ between flat seabed, sloping seabed with $\tan\theta = 1/20$, $\tan\theta = 1/10$ and $\tan\theta = 1/8$, respectively. The surge motion $X_1(t)$ peak values are almost same for these different seabed conditions but the $X_1(t)$ has the least value for flat seabed condition. It is also noticed that the nonlinear characteristics become more obvious as the seabed inclination increases.

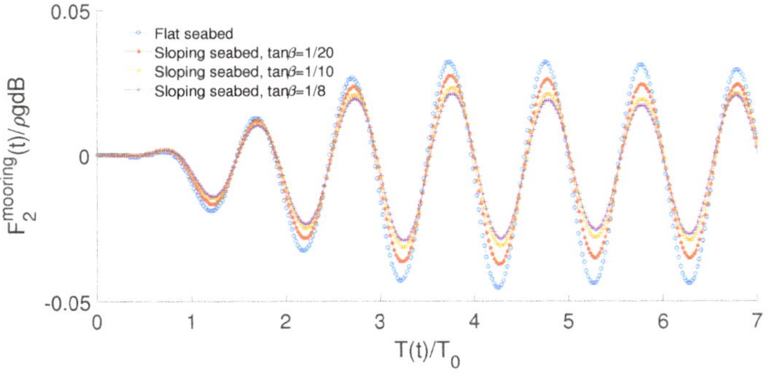

Figure 18. Comparison of time record of mooring line force $F_2(t)$ for wave frequency $kd = 1.0$ between flat seabed, sloping seabed with $\tan\theta = 1/20$, $\tan\theta = 1/10$ and $\tan\theta = 1/8$, respectively.

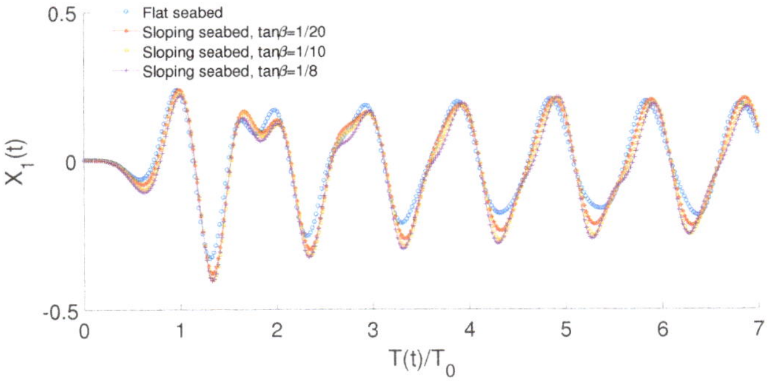

Figure 19. Comparison of time record of floating body surge motion $X_1(t)$ for wave frequency $kd = 1.0$ between flat seabed, sloping seabed with $\tan\theta = 1/20$, $\tan\theta = 1/10$ and $\tan\theta = 1/8$, respectively.

Figure 20 shows the wave force $F_2^{wave}(\omega)$ on the moored body excited by an incident wave for the slope seabed angle of $\tan\theta = 1/8$. It is noticeable that the wave force $F_2^{wave}(\omega)$ increases with the wave frequency for $kd < 0.6$. This wave force $F_2^{wave}(\omega)$ reaches its peak in the period $0.6 < kd < 0.7$ and decreases gradually with the wave frequency for $kd > 0.7$. This variation trend is in line with the moored body motion as presented in Figure 16.

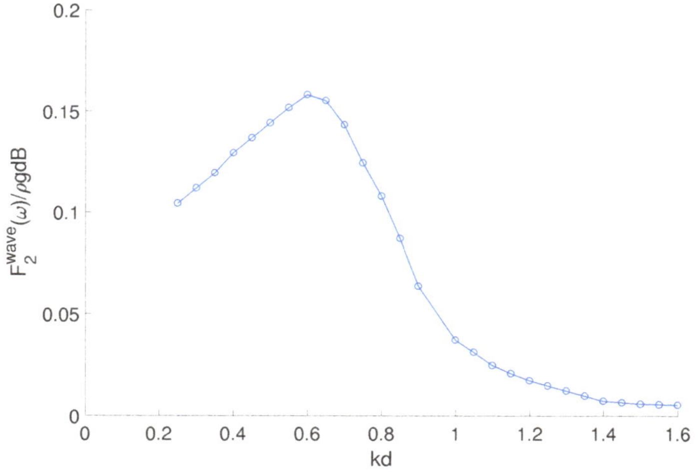

Figure 20. Wave force $F_2^{wave}(\omega)$ on the moored body excited by an incident wave.

To demonstrate the slope seabed inclination effects for the mooring line force. Numerical simulations are performed for this floating body positioned above a sloping seabed with various inclinations. Figure 21 shows a comparison of the mooring line force $F_2^{mooring}$ amplitude for different slope seabed angle tanθ for the floating body experiencing an incident wave with frequency $kd = 1.0$. It is noticed that the mooring line force $F_2^{mooring}$ reduces as seabed angle tanθ increases. In the present study, the mooring line and floating body are coupled into an integrated model. The mooring line force is excited mainly due to the floating body motion which is caused by the incident wave. The mooring line force amplitude conforms with the floating body motion amplitude.

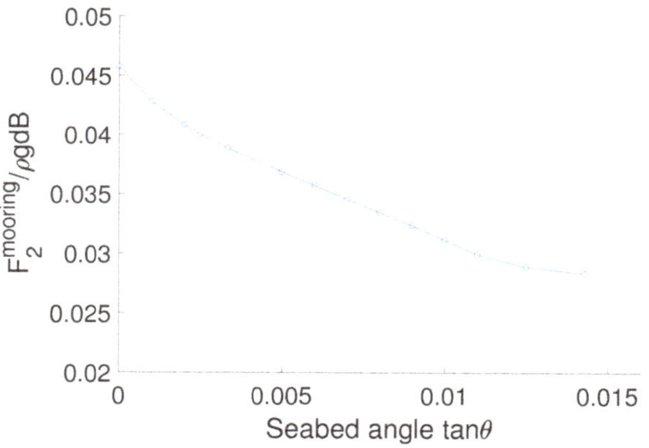

Figure 21. Comparison of the mooring line force $F_2^{mooring}$ amplitude for different slope seabed angle tanθ for floating body experiencing an incident wave with frequency $kd = 1.0$.

It should be noticed that the coupled floating body and mooring line system natural frequency is within the wave frequency range and only the wave-frequency responses are investigated in this paper. The low-frequency responses are very important for large floating bodies.

5. Conclusions

A two-dimensional coupled floating body-mooring line model is developed to study the freely floating body motion responses in both flat and sloping seabed conditions. A continuous Rankine source based boundary element model is established to describe the fluid domain. This model comprises three time-domain boundary integral equations accounting for diffraction problem, radiation problem and reflection problem, respectively. The mooring line is formulated by a catenary mooring line model and coupled with the boundary element model at each time step.

Numerical validations were carried out for the static offset test, free decay test without mooring line, free decay test with mooring line and regular wave simulation. Reasonable agreement with published data demonstrates the accuracy of proposed model.

Two numerical cases are investigated with an emphasis on the hydrodynamic effects of sloping seabed and asymmetrical mooring line. In the first numerical study, both the surge motion RAO_{11} and the heave motion RAO_{22} decrease as the mooring line length increases for flat seabed condition. For the sloping seabed case, the left and right sides of mooring line have different lengths and are laid on sloping seabed portion. The right-side mooring line is positioned to keep the floating body at static equilibrium. Numerical results show that the surge motion RAO_{11} decreases as the asymmetrical level increases across the whole frequency range but the heave motion RAO_{22} keeps constant regardless of the asymmetry level. These numerical findings indicate that in such sloping seabed conditions, the mooring line system shows better station-keeping capability than it does in flat seabed conditions.

For the second numerical case, both the left and right sides of mooring lines are laid on flat seabed portion with sloping portion between two anchor points. The left and right mooring line has same length but different weight to keep the floating body in static equilibrium position. Numerical study shows that both the surge motion RAO_{11} and heave motion RAO_{22} increase as the sloping seabed angle becomes larger. The coupled floating body-mooring line system demonstrates nonlinear characteristics for the body motion and mooring line force responses in the time domain. The surge motion has clearer nonlinear performance than the heave motion. An asymmetrical mooring line configuration demonstrates less station-keeping capability than an symmetrical configuration, but it also requires less mooring line weight and steel consumption. From an economical viewpoint, such a mooring line configuration could be beneficial. In actual industrial projects, the design of asymmetrical mooring line is a trade-off between station-keeping capability and economical benefit.

This paper deals with two-dimensional cases and therefore has academic values, while practical engineering problems are three-dimensional. The study of three-dimensional coupled floating body-mooring line problem and an extention to wave energy converters will be part of our future work.

Author Contributions: Conceptualization, A.F. and H.S.K.; methodology, A.F., H.S.K., B.Z. and Z.J.; programming, A.F.; validation, A.F.; investigation, A.F., H.S.K. and B.Z.; writing—original draft preparation, A.F.; writing—review and editing, H.S.K., B.Z. and Z.J.; project administration, Z.J. All authors have read and agreed to the published version of the manuscript.

Funding: The APC was funded by University of Agder.

Conflicts of Interest: The authors declare no conflict of interest.

References

1. Lefebvre, S.; Collu, M. Preliminary design of a floating support structure for a 5 MW offshore wind turbine. *Ocean Eng.* **2012**, *40*, 15–26. [CrossRef]
2. Ardianti, A.; Mutsuda, H.; Kawawaki, K.; Doi, Y. Fluid structure interactions between floating debris and tsunami shelter with elastic mooring caused by run-up tsunami. *Coast. Eng.* **2018**, *137*, 120–132. [CrossRef]
3. Jiang, Z.; Moan, T.; Gao, Z. A comparative study of shutdown procedures on the dynamic responses of wind turbines. *J. Offshore Mech. Arctic Eng.* **2015**, *137*, 011904. [CrossRef]
4. Zanuttigh, B.; Angelelli, E. Experimental investigation of floating wave energy converters for coastal protection purpose. *Coast. Eng.* **2013**, *80*, 148–159. [CrossRef]
5. Li, L.; Jiang, Z.; Høiland, A.V.; Ong, M.C. Numerical analysis of a vessel-shaped offshore fish farm. *J. Offshore Mech. Arctic Eng.* **2018**, *140*. [CrossRef]
6. Bai, K. The added mass of two-dimensional cylinders heaving in water of finite depth. *J. Fluid Mech.* **1977**, *81*, 85–105. [CrossRef]
7. Xiong, L.; Lu, H.; Yang, J.; Zhao, W. Motion responses of a moored barge in shallow water. *Ocean Eng.* **2015**, *97*, 207–217. [CrossRef]
8. Feng, A.; Bai, W.; You, Y.; Chen, Z.; Price, W. A Rankine source method solution of a finite depth, wave-body interaction problem. *J. Fluids Struct.* **2016**, *62*, 14–32. [CrossRef]
9. Buchner, B. The motions of a ship on a sloped seabed. In Proceedings of the 25th International Conference on Offshore Mechanics and Arctic Engineering, Hamburg, Germany, 4–9 June 2006; pp. 339–347.
10. Ferreira, M.; Newman, J. Diffraction effects and ship motions on an artificial seabed. In Proceedings of the 24th International Workshop on Water Waves and Floating Bodies, Zelenogorsk, Russia, 19–22 April, 2009.
11. Belibassakis, K.; Athanassoulis, G. Three-dimensional Green's function for harmonic water waves over a bottom topography with different depths at infinity. *J. Fluid Mech.* **2004**, *510*, 267–302. [CrossRef]
12. Belibassakis, K. A boundary element method for the hydrodynamic analysis of floating bodies in variable bathymetry regions. *Eng. Anal. Bound. Elements* **2008**, *32*, 796–810. [CrossRef]
13. Kim, T.; Kim, Y. Numerical analysis on floating-body motion responses in arbitrary bathymetry. *Ocean Eng.* **2013**, *62*, 123–139. [CrossRef]
14. Feng, A.; Cai, H.; You, Y. Freely floating body motion responses induced by wave and current in seabed conditions. *Ocean Eng.* **2019**, *172*, 396–406. [CrossRef]
15. Feng, A.; Magee, A.; Price, W. Two dimensional wave-current-structure interaction with flat or sloping seabed environment in a linearized framework. *Ocean Eng.* **2019**, *174*, 296–306. [CrossRef]
16. Li, L.; Jiang, Z.; Ong, M.C.; Hu, W. Design optimization of mooring system: an application to a vessel-shaped offshore fish farm. *Eng. Struct.* **2019**, *197*, 109363. [CrossRef]
17. Mavrakos, S.; Papazoglou, V.; Triantafyllou, M.; Hatjigeorgiou, J. Deep water mooring dynamics. *Mar. Struct.* **1996**, *9*, 181–209. [CrossRef]
18. Montasir, O.; Yenduri, A.; Kurian, V. Effect of mooring line configurations on the dynamic responses of truss spar platforms. *Ocean Eng.* **2015**, *96*, 161–172. [CrossRef]
19. Smith, R.; MacFarlane, C. Statics of a three component mooring line. *Ocean Eng.* **2001**, *28*, 899–914. [CrossRef]
20. Yuan, Z.; Incecik, A.; Ji, C. Numerical study on a hybrid mooring system with clump weights and buoys. *Ocean Eng.* **2014**, *88*, 1–11. [CrossRef]
21. Lie, H. *MIMOSA, a Computation Program for Mooring and Stationkeeping Analysis, User's Manual*; Technical Report; MARINTEK: Trondheim, Norway, 1990.
22. Larsen, K.; Sandvik, P. Efficient methods for the calculation of dynamic mooring line tension. In Proceedings of the 1st ISOPE European Offshore Mechanics Symposium, Trondheim, Norway, 20–22 August 1990.
23. Ormberg, H.; Passano, E. *RIFLEX Theory Manual*; Technical Report; MARINTEK: Trondheim, Norway, 2012.
24. Kim, M.; Koo, B.; Mercier, R.; Ward, E. Vessel/mooring/riser coupled dynamic analysis of a turret-moored FPSO compared with OTRC experiment. *Ocean Eng.* **2005**, *32*, 1780–1802. [CrossRef]

25. Hall, M.; Goupee, A. Validation of a lumped-mass mooring line model with DeepCwind semisubmersible model test data. *Ocean Eng.* **2015**, *104*, 590–603. [CrossRef]
26. Ran, Z.; Kim, M. Nonlinear coupled responses of a tethered spar platform in waves. *Int. J. Offshore Polar Eng.* **1997**, *7*, 1–8.
27. Girón, A.R.C.; Corrêa, F.N.; Hernández, A.O.V.; Jacob, B.P. An integrated methodology for the design of mooring systems and risers. *Mar. Struct.* **2014**, *39*, 395–423. [CrossRef]
28. Molen, W.; Wenneker, I. Time-domain calculation of moored ship motions in nonlinear waves. *Coast. Eng.* **2008**, *55*, 409–422. [CrossRef]
29. Faltinsen, O. *Sea Loads on Ships and Offshore Structures*; Cambridge University Press: Cambridge, UK, 1990.
30. Feng, A.; Price, W. Numerical simulations of the hydrodynamic responses of a body interacting with wave and current over a sloping seabed. *Appl. Ocean Res.* **2018**, *79*, 184–196. [CrossRef]
31. Yeung, R. The transient heaving motion of floating cylinders. *J. Eng. Math.* **1982**, *16*, 97–119. [CrossRef]
32. Ito, S. Study of the Transient Heave Oscillation of a Floating Cylinder. Master's Thesis, Massachusetts Institute of Technology, Cambridge, MA, USA, 1977.
33. Patel, M. *Dynamics of Offshore Structures*; Butterworth-Heinemann: Oxford, UK, 1989.
34. Ren, B.; He, M.; Dong, P.; Wen, H. Nonlinear simulations of wave-induced motions of a freely floating body using WCSPH method. *Appl. Ocean Res.* **2015**, *50*, 1–12. [CrossRef]
35. Martin, T.; Bihs, H.; Kamath, A.; Arntsen, Ø. Simulation of Floating Bodies in Waves and Mooring in a 3D Numerical Wave Tank Using REEF3D. In Proceedings of the 4th International Conference in Ocean Engineering, Chennai, India, 18–21 February 2018; pp. 673–683.
36. Bian, S.; Xiang, S. Manage tensioner stroke for dry tree semisubmersibles. In Proceedings of the 32nd International Conference on Ocean, Offshore and Arctic Engineering, Nantes, France, 9–14 June 2013; pp. 1–9.

Journal of
Marine Science and Engineering

MDPI

Article

Frequency-Based Performance Analysis of an Array of Wave Energy Converters around a Hybrid Wind–Wave Monopile Support Structure

Sofia Gkaraklova, Pavlos Chotzoglou and Eva Loukogeorgaki *

Department of Civil Engineering, Aristotle University of Thessaloniki, University Campus, 54124 Thessaloniki, Greece; sofentgka@civil.auth.gr (S.G.); pavchogeo@civil.auth.gr (P.C.)
* Correspondence: eloukog@civil.auth.gr; Tel.: +30-2310-99-5951

Abstract: In this paper, we investigate, in the frequency domain, the performance (hydrodynamic behavior and power absorption) of a circular array of four semi-immersed heaving Wave Energy Converters (WECs) around a hybrid wind–wave monopile (circular cylinder). The diffraction/radiation problem is solved by deploying the conventional boundary integral equation method. Oblate-spheroidal and hemispherical-shaped WECs are considered. For each geometry, we assess the effect of the array's net radial distance from the monopile and of the incident wave direction on the array's performance under regular waves. The results illustrate that by placing the oblate spheroidal WECs close to the monopile, the array's power absorption ability is enhanced in the low frequency range, while the opposite occurs for higher wave frequencies. For hemispherical-shaped WECs, the array's power absorption ability is improved when the devices are situated close to the monopile. The action of oblique waves, with respect to the WECs' arrangement, increases the absorbed power in the case of oblate spheroidal WECs, while these WECs show the best power absorption ability among the two examined geometries. Finally, for the most efficient array configuration, consisting of oblate spheroidal WECs situated close to the monopile, we utilize an "active" Power Take-Off (PTO) mechanism, facilitating the consideration of a variable with frequency PTO damping coefficient. By deploying this mechanism, the power absorption ability of the array is significantly enhanced under both regular and irregular waves.

Keywords: hybrid wind–wave system; wave energy converters; monopile; oblate spheroids; arrays; hydrodynamic analysis; power absorption; power take-off mechanism

1. Introduction

The increase of energy demand, the rising cost of fossil fuels, and the environmental problems derived from their overexploitation have put a spotlight on renewable energy resources. Offshore renewable energy, which includes both the ocean (wave and tidal) energy and the mature, nowadays, offshore wind energy, has shown a great potential for development [1,2] and it is anticipated to play a fundamental role in the EU energy policy, as identified by the European Strategic Energy Technology Plan (SET-Plan). The set target for 2050 corresponds to 188 GW and 460 GW of installed capacity for ocean energy and offshore wind respectively [3,4]. As for the contribution of solely wave and offshore wind power to the EU electricity mix, the goal is to satisfy 15% of the European electricity demand and, in some countries, up to 20% of the national demand by 2050 [3].

Aiming at exploiting simultaneously both offshore wind and wave energy, Offshore Wind Turbines (OWTs) can be integrated with Wave Energy Converters (WECs). This integration can be realized in a twofold way by (a) locating OWTs and WECs in the same marine area, leading to the formation of the so-called "co-located parks" [4–7] and/or (b) combining offshore wind and wave energy technologies into one structure, resulting to the deployment of the so-called "hybrid systems" or "combined systems". The latter systems can be further categorized into bottom-mounted and floating ones, depending upon the type of the support structure utilized (bottom-mounted or floating platform, respectively). The advantages of combining offshore wind and wave energy technologies into a hybrid system are numerous (e.g., [4,5]). First of all, the energy yield per unit of marine surface is enhanced and, hence, better use of natural resources is achieved. Moreover, a smooth and more predictable power output can be realized, since the exploitation of the less variable and more predictable waves enables the avoidance of sudden losses in power generation due to unpredictable wind variations. One of the most considerable benefits is the decrease of the levelized cost of energy, as costs related to infrastructure, transmission, grid connection, permissions, and operation and maintenance are shared. Furthermore, the ecological footprint of a hybrid wind and wave energy exploitation system is expected to be smaller than that of the separate alternative [8,9]. However, there are a few challenges to overcome as well. The main problem is the different level of maturity of the two specific technologies. OWTs correspond to a well-established technology with a total installed capacity in Europe by the end of 2019 equal to 22.072 GW [10]. The wave energy industry, on the other hand, is still at a nascent phase. This is primarily due to the wide range of WECs' types available and the fact that most of them are still at early stages of development. The above items in combination with lack of practical experience and data related to combined technologies lead to an additional economic risk associated with hybrid projects [11].

Up until now there have been many investigations that focus on floating hybrid systems. Michailides et al. [12–14] proposed a semi-submersible flap type combined system (SFC system) and performed an integrated time domain numerical analysis of this system as well as relevant experiments. Muliawan et al. [15,16] conducted an integrated numerical and experimental analysis for the case of a spar-type floating wind turbine and a coaxial floating torus WEC (STC system). The performance of the aforementioned floating hybrid systems (SFC and STC) was also compared by Gao et al. [17] using both numerical and experimental data. Inspired by the STC system, a 5 MW wind turbine on a semi-submersible platform combined with a heaving WEC was investigated in [18]. The performance of a floating hybrid system with a Tension Leg Platform (TLP) and three point absorbers was numerically studied and assessed by [19], while in [20–22], the WindWEC system, combining a floating wind turbine on a spar and an oscillating in heave and pitch WEC buoy, was proposed and it was numerically investigated in time domain. A floating hybrid system consisting of a 5 MW wind turbine and oscillating water column devices moored using tensioned tethers in a TLP concept was investigated numerically and experimentally by [23]. Finally, Lee et al. [24] studied in the frequency domain the performance of multiple heaving WECs placed on a floating semi-submersible platform that can be utilized as a support structure of more than one wind turbines.

Contrary to the floating hybrid systems, the case of deploying a bottom-mounted hybrid system has been considered by a few researchers, who mainly focused on the design and the development of a WEC concept that can be utilized on a hybrid wind–wave bottom-mounted support structure. More specifically, Perez-Collazo et al. [25] carried out an experimental investigation of an oscillating water column WEC attached to hybrid wind–wave jacket support structure, while the aforementioned WEC type attached to a hybrid wind–wave monopile support structure has been the subject of the experimental research of [26]. Inspired by the STC floating hybrid system, a heaving WEC attached on hybrid wind–wave monopile support structure using a pulley slide mechanism was studied both numerically and experimentally in [27].

In the present paper, we numerically investigate the performance (hydrodynamic behavior and power absorption) of a circular array of four semi-immersed heaving WECs distributed uniformly around a hybrid wind–wave monopile support structure. The array consists of either hemispherical-shaped or oblate-spheroidal WECs, while each WEC is assumed to absorb power through a linear Power Take-Off (PTO) mechanism, actuated from its heave motion. The monopile corresponds to a bottom-mounted, surface-piercing circular cylinder. The examined WEC array and the monopile could be considered as components of a new bottom-mounted hybrid offshore wind and wave energy exploitation system, which is proposed in this paper (Figure 1a). The system was inspired by Wave Treader [28] and Wave Star [29] and enables the motion of the WECs only along the vertical direction by attaching them on the monopile via arms, which move vertically along sliding guideways, i.e., sliders (Figure 1b). The analysis is implemented in the frequency domain under the action of regular, head, and oblique waves. The corresponding diffraction/radiation problem, taking into account the hydrodynamic interactions among the WECs and between the monopile and the WECs, is solved by utilizing the conventional Boundary Integral Equation (BIE) method. For each of the examined geometries and for a constant PTO damping coefficient, we initially present extended results in order to assess the effect of the net radial distance of the array from the monopile and of the incident wave direction on the hydrodynamic behavior and the power absorption ability of the array. Moreover, comparison with the case of an isolated array is performed in order to illustrate the effect of the existence of the monopile on the performance of the array. Finally, the most efficient array configuration (WECs' geometry and net radial distance) is chosen in order to further enhance its power absorption ability by appropriately adjusting the damping coefficient of the PTO mechanism along the examined frequency range (i.e., utilization of an "active" PTO mechanism). In this case, the power absorbed by the array is assessed for regular waves as well as for various sea states by utilizing the Jonswap spectrum.

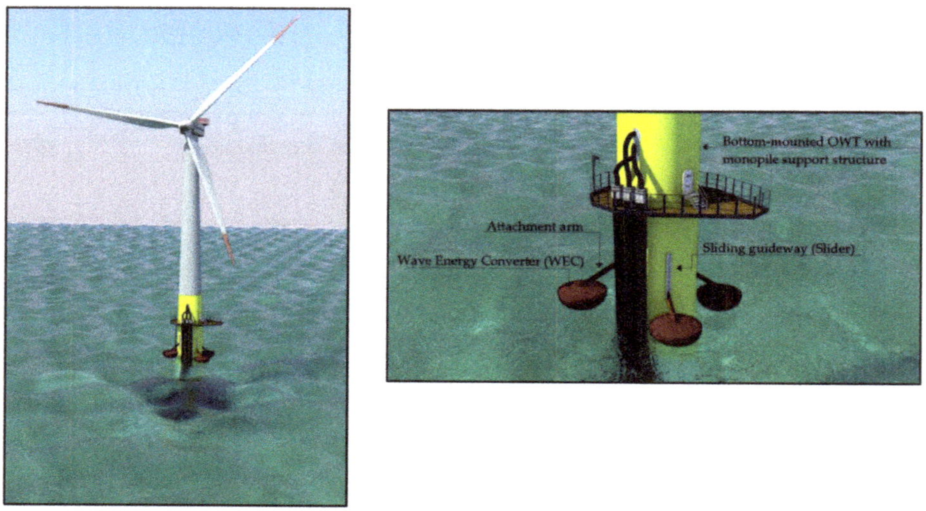

(**a**) (**b**)

Figure 1. Proposed hybrid offshore wind and wave energy exploitation system: (**a**) conceptual representation; (**b**) view of the WECs' attachment on the bottom-mounted OWT.

2. Numerical Modeling

A circular array of M hydrodynamically interacting, semi-immersed heaving WECs is placed around a bottom-mounted cylindrical monopile in a marine region of constant water depth d (Figure 2). The array consists of identical WECs, which have either an oblate spheroidal shape or a hemispherical shape (with a short cylindrical upper part close to the waterline), as shown in Figure 3. The radius and the total draft of the WECs are denoted, respectively, as r_k and h_k, with $k = 1$ corresponding to the dimensions of the oblate spheroidal WECs and $k = 2$ to those of the hemispherical-shaped devices. Each heaving WEC is assumed to absorb power through a linear PTO mechanism, which is schematically represented in Figure 2b as a linear damping system, with damping coefficient b_{PTOj}, $j = 1, \ldots, M$. All WECs are distributed uniformly in a circular arrangement around the monopile at a center-to-center radial distance equal to d_{radial} and net radial distance equal to d_{net} (Figure 2a). The monopile is modelled as a bottom-mounted, surface-piercing circular cylinder of radius r_0 and of draft d (Figure 3). All bodies are subjected to the action of monochromatic incident waves of circular frequency ω and linear amplitude A, which propagate at an angle β relative to the global horizontal X-axis (Figure 2a).

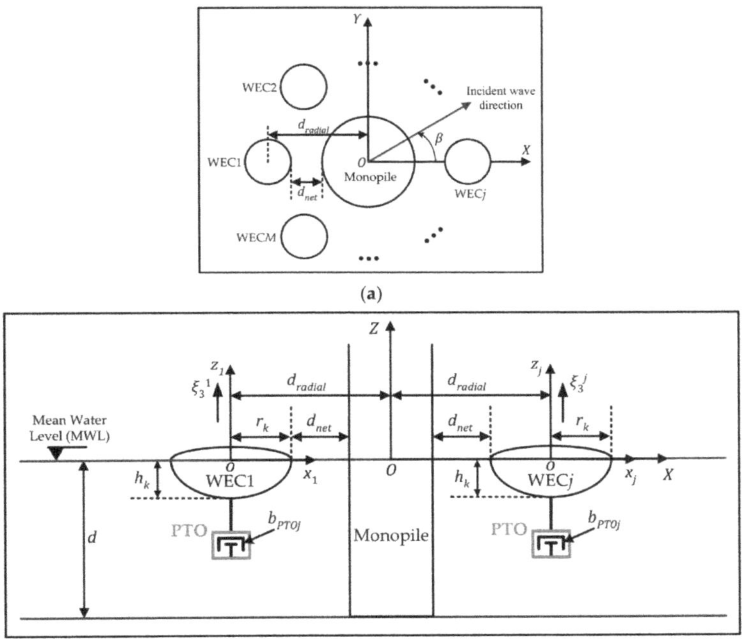

Figure 2. Geometry of the examined problem and definition of basic quantities: (**a**) $X - Y$ plane; (**b**) $X - Z$ plane.

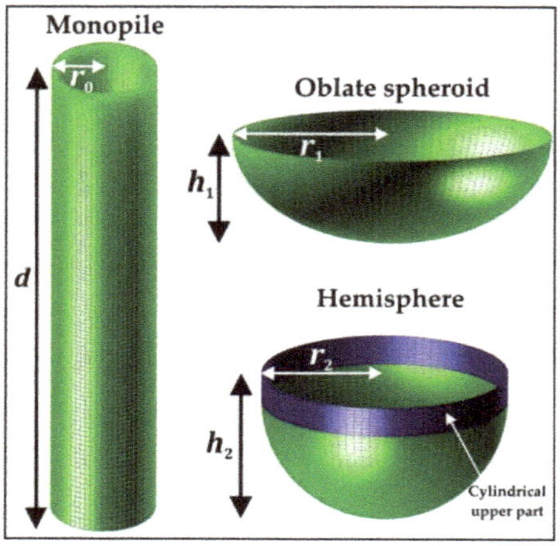

Figure 3. Submerged parts of all examined bodies.

The hydrodynamic analysis of the WEC circular array around the monopile, including the hydrodynamic interactions among all co-located bodies, is conducted in the frequency domain and it relies on the BIE method (e.g., [30–32]), which is numerically realized using WAMIT [33]. The analysis is based on a three-dimensional linear diffraction theory, where the monopile is considered to be fixed at its position, while all WECs are taken to undergo small amplitude oscillations only along the vertical z_j-axis, i.e., along their working direction (Figure 2b). Thus, for each WECj, $j = 1, \ldots, M$, all degrees of freedom, except the one corresponding to heave, are considered ideally restricted. Assuming inviscid and incompressible fluid with irrotational flow, the fluid motion is described by introducing the velocity potential. Its complex spatial part is defined as follows [33,34]:

$$\varphi = \underbrace{(\varphi_I + \varphi_S)}_{\varphi_D} + i\omega \sum_{j=1}^{M} \xi_3^j \varphi_j \tag{1}$$

$$\varphi_I = \frac{igA}{\omega} \frac{cosh[k(Z+d)]}{cosh(kd)} e^{-ik(Xcos\beta + Ysin\beta)} \tag{2}$$

where φ_I is the incident wave potential; φ_S is the scattered potential, related to the scattered disturbance of the incident waves from the WECs and the monopile; φ_D denotes the diffraction potential; and φ_j, $j = 1, \ldots, M$, correspond to the radiation potentials, associated with the waves radiated from the WECs due to their forced motion in heave, while ξ_3^j, $j = 1, \ldots, M$, denote the complex amplitudes of the heave motions of the WECs. Furthermore, g is the gravitational acceleration, k is the wave number, and $i^2 = -1$.

The velocity potentials φ_q ($q = D$ or $q = j$) satisfy the Laplace equation everywhere in the fluid domain, while, additionally, they are subjected to the following linearized boundary conditions corresponding to the combined kinematic and dynamic free-surface condition (Equation (3)), the bottom boundary condition (Equation (4)), and the Neumann boundary conditions on the wetted surface of the bodies (Equations (5) and (6)) [33,34]:

$$\frac{\partial \varphi_q}{\partial Z} - \frac{\omega^2}{g} \varphi_q = 0 \text{ on } Z = 0 \tag{3}$$

$$\frac{\partial \varphi_q}{\partial Z} = 0 \text{ on } Z = -d \tag{4}$$

$$\frac{\partial \varphi_D}{\partial n} = 0 \tag{5}$$

$$\frac{\partial \varphi_j}{\partial n} = n_3^j \text{ for } j = 1, \ldots, M \tag{6}$$

In Equation (6), n_3^j denotes the normal unit vector of WECj in the vertical direction.

Green's theorem is employed to form the boundary integral equations for the unknown diffraction and radiation potentials on the boundaries of all bodies (WECs and monopile) and of the WECs, respectively. The corresponding first order boundary value problem is then solved on the basis of the three dimensional low-order panel method [33,34].

Having solved the aforementioned boundary value problem, we calculate the first-order hydrodynamic forcing quantities as follows:

$$F_l = -i\omega\rho \iint_{S_b^m} n_l \varphi_D ds, \qquad l = 1,\ 2 \tag{7}$$

$$F_3^i = -i\omega\rho \iint_{S_b^i} n_3^i \varphi_D ds, \qquad i = 1, \ldots, M \tag{8}$$

$$A_{ij} - \frac{i}{\omega} B_{ij} = \rho \iint_{S_b^i} n_3^i \varphi_j ds, \quad i,\ j = 1, \ldots, M \tag{9}$$

where F_l, $l = 1,\ 2$, is the surge and the sway exciting force respectively applied on the monopile; F_3^i, $i = 1, \ldots, M$, is the heave exciting force applied on each WECi; A_{ij} and B_{ij} correspond to the added mass and radiation damping coefficients, respectively; S_b^m is the wetted surface of the monopile; n_l, $l = 1,\ 2$, denote the normal unit vector on S_b^m in the x and y directions, respectively; S_b^i is the wetted surface of the ith WEC; and ρ is the water density.

The complex amplitudes of the WECs' heave motions, ξ_3^j, $j = 1, \ldots, M$, are, then, obtained by solving the following linear system of equations:

$$\sum_{j=1}^{M} \left[-\omega^2 (M_{ij} + A_{ij}) + i\omega \left(B_{ij} + B_{ij}^{PTO} \right) + C_{ij} \right] \xi_3^j = F_3^i \quad i = 1, \ldots, M \tag{10}$$

where M_{ij} are the mass matrix coefficients; C_{ij} are the hydrostatic-gravitational stiffness coefficients; A_{ij} and B_{ij} are the added mass and the radiation damping coefficients, respectively, as defined in Equation (9); and B_{ij}^{PTO} correspond to the damping coefficients originating from the PTO mechanism. For a jth WEC of the array, this mechanism is modeled as a linear damping system (Figure 2b), which is actuated by the heave motion of the corresponding WEC and it has a damping coefficient b_{PTOj}. Thus, in Equation (10), $B_{ij}^{PTO} = b_{PTOj}$ for $i = j = 1, \ldots,\ M$, while $B_{ij}^{PTO} = 0$ for $i \neq j$.

The heave response of each WEC is expressed in terms of the response amplitude operator as

$$RAO_3^j = \frac{\left| \xi_3^j \right|}{A} \tag{11}$$

where $\left| \xi_3^j \right|$ denotes the amplitude of the complex quantity ξ_3^j.

The mean power, $p(\omega)$, absorbed by the whole array of the WECs at a given ω is obtained as

$$p(\omega) = \sum_{j=1}^{M} p_j(\omega) \tag{12}$$

where $p_j(\omega)$, $j = 1, \ldots,\ M$, correspond to the power absorbed by the jth WEC, calculated using the following equation:

$$p_j(\omega) = 0.5 b_{PTOj} \omega^2 \left| \xi_3^j \right|^2 \tag{13}$$

3. Characteristics of the Physical Problem Examined

The BIE-based numerical model described in Section 2 is applied for the case of a monopile of radius $r_0 = 4.0$ m and an array of $M = 4$ identical, semi-immersed WECs with the layout shown in Figure 4. All bodies are placed in an area of constant water depth $d = 40.0$ m. The oblate spheroidal WECs (Figure 3) have radius $r_1 = 2.5$ m and total draft $h_1 = 1.7$ m. The corresponding dimensions of the hemispherical-shaped WECs (Figure 3) are defined as equal to $r_2 = 2.0$ m and $h_2 = 2.5$ m (0.5 m correspond to the cylindrical upper part), aiming at comparing different shaped WECs that have similar submerged volumes, V, as well as similar heave natural frequencies, ω_{n3}^{ISO} (Table 1). For a given geometry, all WECs within the array are considered to have the same PTO characteristics; thus, in Equation (10), $B_{ij}^{PTO} = b_{PTO}$ for $i = j = 1, \ldots, 4$. Furthermore, the constant PTO damping coefficient, b_{PTO}, is appropriately tuned in terms of maximizing energy absorption at the natural frequency of the corresponding single, isolated WEC. On the basis of this and along the lines of [35], b_{PTO} is taken to be equal to the heave radiation damping coefficient of a single, isolated oblate spheroidal or hemispherical-shaped WEC at its heave natural frequency, ω_{n3}^{ISO}, namely, $b_{PTO} = B_{33}^{ISO}(\omega = \omega_{n3}^{ISO})$. The values of ω_{n3}^{ISO} and b_{PTO} for both examined geometries obtained by the hydrodynamic analysis of the corresponding single, isolated body are presented in Table 1. It is evident that the existence of different hydrodynamic properties among the examined geometries lead to different b_{PTO} values.

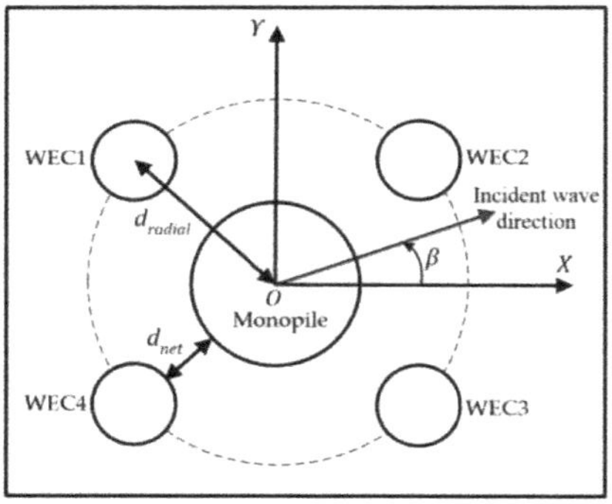

Figure 4. Examined layout of the array with $M = 4$ WECs.

Table 1. Geometrical and hydrodynamic properties of WEC geometries examined.

Geometry	r_k (m)	h_k (m)	V	C_{33} (kN/m)	ω_{n3}^{ISO} (rad/s)	b_{PTO} (kNs/m)
Oblate spheroid	2.5	1.7	22.150	197.076	2.282	20.615
Hemisphere	2.0	2.5 [1]	22.935	125.969	2.008	6.673

[1] 0.5 m correspond to the cylindrical upper part.

For each WEC geometry, we, initially examine and assess the effect of the net radial distance of the array from the monopile, d_{net}, and of the incident wave direction, β (Figure 4) on the performance of the array under the action of regular waves, with ω varying between 0.02 and 4.5 rad/s. Regarding the net radial distance, seven different values of d_{net} are examined (Table 2) for head incident waves (i.e., $\beta = 0^o$, Figure 4), while furthermore, representative results are compared with those of isolated arrays (i.e., without the monopile), aiming at illustrating more clearly the effect of the presence of the monopile on the power absorption ability of the WECs. With regard to the incident wave direction, we take three different values of β, representing head and oblique waves, into account (Table 2) for $d_{net} = 0.125r_0$ (oblate spheroidal WECs) and $d_{net} = 0.25r_0$ (hemispherical-shaped WECs). Finally, the most efficient array configuration (WECs' geometry and net radial distance) is chosen in order to further enhance its power absorption ability by adjusting appropriately the damping coefficient of the linear PTO mechanism along the examined frequency range (i.e., utilization of an "active" PTO mechanism). In this case, the power absorbed by the array is assessed for both regular and irregular waves with $\beta = 45^o$. More details about this "active" PTO mechanism are cited in Section 4.4, where the corresponding results are presented.

Table 2. Examined cases for the array with $M = 4$ WECs.

Design Parameter	WEC Geometry	d_{net} (m)	β (°)
Net radial distance (d_{net})	Oblate spheroid, Hemisphere	$0.125r_0$, $0.25r_0$, $0.5r_0$, $0.75r_0$, $1.0r_0$, $1.25r_0$, $1.5r_0$	0
Incident wave direction (β)	Oblate spheroid Hemisphere	$0.125r_0$ $0.25r_0$	0, 22.5, 45

4. Results and Discussion

4.1. Effect of Net Radial Distance on the Performance of the Array with Oblate Spheroidal WECs

The effect of the net radial distance of the oblate spheroidal WECs from the monopile on the WECs' non-dimensional (in terms of $\rho g A \pi r_1{}^2$) heave exciting forces is shown in Figure 5, where the variation of F_3 for all WECs of the circular array is presented as a function of ω for all examined d_{net} values. The symmetry of the array with respect to $\beta = 0^o$ (Figure 4) leads to the same values of F_3 for the WECs situated in the seaward side of the monopile (i.e., WEC1 and WEC4, Figure 4) as well as for those situated in the leeward side of the bottom-mounted support structure (i.e., WEC2 and WEC3, Figure 4).

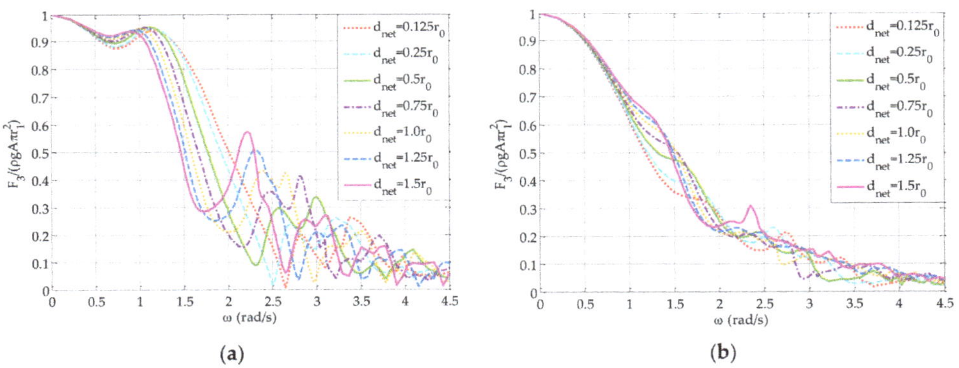

Figure 5. Effect of d_{net} on the non-dimensional heave exciting forces applied on the oblate spheroidal WECs placed around the monopile for $\beta = 0^o$: (**a**) WEC1 and WEC4; (**b**) WEC2 and WEC3.

Considering the WECs placed in the seaward side of the monopile (Figure 5a), the variation of the non-dimensional heave exciting force for the smallest examined net radial distance, $d_{net} = 0.125r_0$, starts from the limiting value of ≈ 1.00 at $\omega = 0.02$ rad/s and it is characterized by an almost smooth plateau up to $\omega \approx 1.2$ rad/s, followed by a rapid decrease up to $\omega \approx 2.6$ rad/s, where F_3 obtains its first local minimum with a value almost equal to zero. Then, F_3 varies quite intensively and obtains two local maxima at $\omega \approx 2.8$ rad/s and $\omega \approx 3.3$ rad/s. A similar variation pattern is observed for the rest of the d_{net} values examined. However, by successively increasing the WECs' net radial distance from the monopile, the extent of the aforementioned plateau is gradually reduced, while the first local minimum obtains non-zero values and it is shifted at lower frequencies. This, in turn, leads to larger values of the first F_3 local peak, which, moreover, occur at lower wave frequencies. Overall, it can be stated that the placement of the array close to the monopile (i.e., at $d_{net} \leq 0.5r_0$) leads to significant values of F_3 at 0.02 rad/s $< \omega <$ 1.4 rad/s; however, at the same time, it results in a substantial decrease of F_3 at 2.0 rad/s $< \omega <$ 2.5 rad/s, where resonance phenomena of the oblate spheroidal WECs are anticipated. The latter feature is not observed for $d_{net} \geq 0.75r_0$, where the relevant diffracted disturbances induced by the WECs and the monopile, as physically interpreted in the following paragraph, enable the formation of F_3 peaks at 2.0 rad/s $< \omega <$ 2.5 rad/s. Regarding WEC2 and WEC3 (Figure 5b), their placement in the leeward side of the monopile, where shadow effects are induced by both the monopile and the seaward WECs, results in a smoother variation of the heave exciting forces compared to those applied on WEC1 and WEC4, while, moreover, the change of d_{net} has a small effect on the values and the variation pattern of F_3.

Coming back to the characteristic F_3 local minima observed in the case of the seaward WECs (Figure 5a), their occurrence is attributed to the hydrodynamic interactions among the WECs as well as between the WECs and the monopile. For physically interpreting this, we take into account the spatial variation of the non-dimensional wave elevation, η_D/A, due to diffracted waves only. More specifically, Figure 6 shows the η_D/A spatial variation indicatively for $d_{net} = 0.125r_0$ and $1.0r_0$ calculated at -16 m $\leq X \leq 16$ m and -16 m $\leq Y \leq 16$ m and for $\omega \approx 2.6$ rad/s. At this frequency, F_3 of WEC1 and WEC4 obtains a characteristic local minimum for $d_{net} = 0.125r_0$, while the opposite holds true for $d_{net} = 1.0r_0$ (Figure 5a). In the case of the smallest examined d_{net} value (Figure 6a), the variation pattern of η_D/A in the seaward side of the monopile (i.e., at $X > 0$ m) is quite smooth and is characterized by the existence of well-formed lengthwise zones along the Y-axis, where η_D/A has values almost equal to zero. The positioning of WEC1 and WEC4 within one of these zones leads in turn to the existence of an almost symmetric, with respect to these WECs' local horizontal axes, diffracted wave field around them, with quite small η_D/A values. On the other hand, by placing WEC1 and WEC4 further from the monopile (Figure 6b), the diffraction disturbances induced by WEC1, WEC4, and the monopile are amplified in the seaward side of the monopile. As a result, a non-symmetric diffracted wave field with significant η_D/A values is formed around WEC1 and WEC4. All of the above advocate the existence of a very small F_3 value for $d_{net} = 0.125r_0$ at $\omega \approx 2.6$ rad/s contrary to $d_{net} = 1.0r_0$, in absolute accordance with the results of Figure 5a.

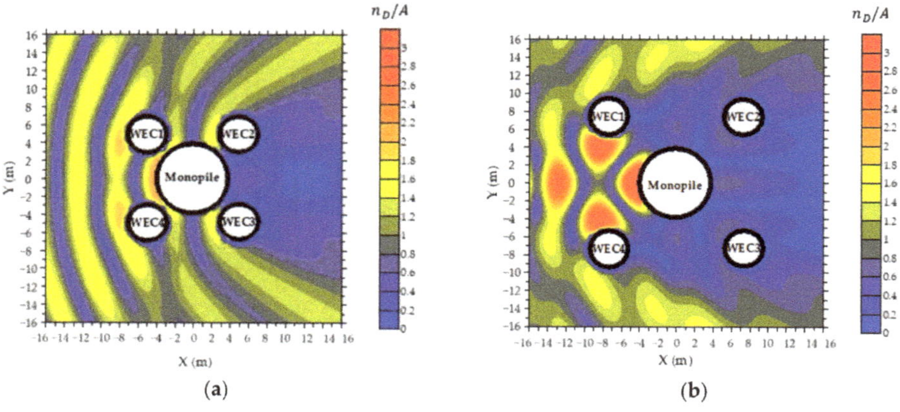

Figure 6. Spatial variation of the non-dimensional wave elevation, n_D/A, around the monopile and the oblate spheroidal WECs at $\omega \approx 2.6$ rad/s and for $\beta = 0^o$: (**a**) $d_{net} = 0.125r_0$; (**b**) $d_{net} = 1.0r_0$.

Continuing with the response of the oblate spheroidal WECs, Figure 7 shows the effect of d_{net} on RAO_3. In the case of WEC1 and WEC4 (Figure 7a), RAO_3 for all d_{net} values examined obtains a global peak at 1.0 rad/s< ω < 1.5 rad/s in absolute accordance with the variation of the corresponding heave exciting forces (Figure 5a). The largest peak value occurs for $d_{net} = 0.125r_0$, while by gradually increasing d_{net}, the peak values are successively decreased, following the relevant trend of F_3 (Figure 5a). For the four largest examined d_{net} values, a second RAO_3 peak occurs at 2.0 rad/s $< \omega <$ 2.5 rad/s due to resonance phenomena. However, this is not observed for $d_{net} \leq 0.5r_0$, due to the rapid decrease of F_3 in this frequency range (Figure 5a) for the reasons previously explained. For $\omega >$ 2.5 rad/s, RAO_3 for all d_{net} values shows an intense variation with successively decreasing values towards higher frequencies. As for the WECs situated in the leeward side of the monopile (Figure 7b), the change of d_{net} has a small effect on the values and the variation pattern of RAO_3, while for all d_{net} values investigated, RAO_3 varies quite smoothly. These trends are in absolute accordance with those observed for the case of the corresponding heave exciting forces (Figure 5b).

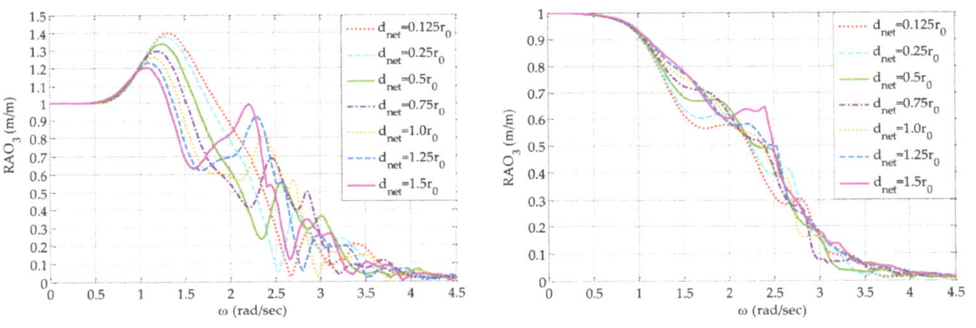

Figure 7. Effect of d_{net} on the heave response of the oblate spheroidal WECs placed around the monopile for $\beta = 0^o$: (**a**) WEC1 and WEC4; (**b**) WEC2 and WEC3.

Figure 8 shows the effect of d_{net} on the mean power absorbed by each oblate spheroidal WEC, $p_j(\omega)$ (Figure 8a,b), as well as on the mean power absorbed by the whole array, $p(\omega)$ (Figure 8c). When observing Figure 8a,b, one can easily realize the direct dependence of $p_j(\omega)$, $j = 1, \ldots, 4$, upon the heave responses (Figure 7), since for a given jth WEC, the variation patterns of p_j and RAO_3^j are very similar. Regarding WEC1 and WEC4, situated in the seaward side of the monopile (Figure 8a), it is evident that up to $\omega = 2.0$ rad/s, these WECs show better absorption ability for small net radial distances from the monopile (i.e., $d_{net} \leq 0.5r_0$), whereas at higher frequencies (i.e., 2.0 rad/s< $\omega <$ 2.5 rad/s), WEC1 and WEC4 are more efficient for d_{net} values larger than $0.5r_0$. From a physical point of view, the above conclusions can be explained as follows. For $d_{net} \leq 0.5r_0$, the power absorption ability of the seaward WECs is not driven by resonance phenomena, as large heave exciting forces and significant responses exist at wave frequencies outside the range where WECs' resonance occurs. However, the opposite holds true for $d_{net} \geq 0.75r_0$, where the hydrodynamic interactions among the WECs and between the WECs and the monopile do not impose any restrictions on the RAO_3 amplification of the seaward WECs due to resonance. The WECs placed in the leeward side of the monopile (Figure 8b) are not strongly affected by the change of d_{net}. However, for $d_{net} = 1.5r_0$, they demonstrate a better power absorption ability at 2.0 rad/s< $\omega <$ 2.5 rad/s, in absolute accordance with the variation of the corresponding heave response (Figure 7b). Moreover, WEC2 and WEC3 show a reduced energy absorption ability compared to WEC1 and WEC4; hence, the variation pattern of the mean power absorbed by the whole array, as shown in Figure 8c, is significantly affected by the features of the power absorption curves related to the WECs situated in the seaward side of the monopile. Consequently, regarding the effect of d_{net} on the power absorption ability of the whole array, similar conclusions to those derived for WEC1 and WEC4 hold true.

Having this in mind, the results of Figure 8c illustrate that the placement of the WECs at $d_{net} = 0.125r_0$ leads to an array that shows the best power absorption ability in the low examined frequency range among all examined d_{net} cases, with a $p(\omega)$ peak value approximately equal to 100 kW/m^2. Moreover, for this net radial distance, the frequency range where adequate amount of power is absorbed is quite wide. These features can be physically interpreted by using a similar rationale as in the case WEC1 and WEC4 for $d_{net} \leq 0.5r_0$. On the other hand, at higher wave frequencies, the array placed at $d_{net} = 1.5r_0$ from the monopile shows the best power ability among all examined d_{net} cases, having a sharp $p(\omega)$ peak with a value approximately equal to 150 kW/m^2. Although the latter value is larger compared to the corresponding one of 100 kW/m^2 obtained for $d_{net} = 0.125r_0$, the placement of the WECs at $d_{net} = 1.5r_0$ results in a more narrower frequency range, where adequate amount of power is absorbed. This is attributed to the fact that for $d_{net} = 1.5r_0$, the power absorption ability of the seaward WECs and, thus, of the whole array is mainly driven by resonance phenomena. Taking all the above into account, $d_{net} = 0.125r_0$ is chosen as the most appropriate net radial distance for placing the oblate spheroidal WECs around the monopile and it is considered for further investigation. This choice is also supported by the fact that most marine areas, including those in the North Sea, where the proposed hybrid offshore wind and wave energy exploitation system could be deployed, are mainly characterized by low-frequency sea waves (e.g., [36]).

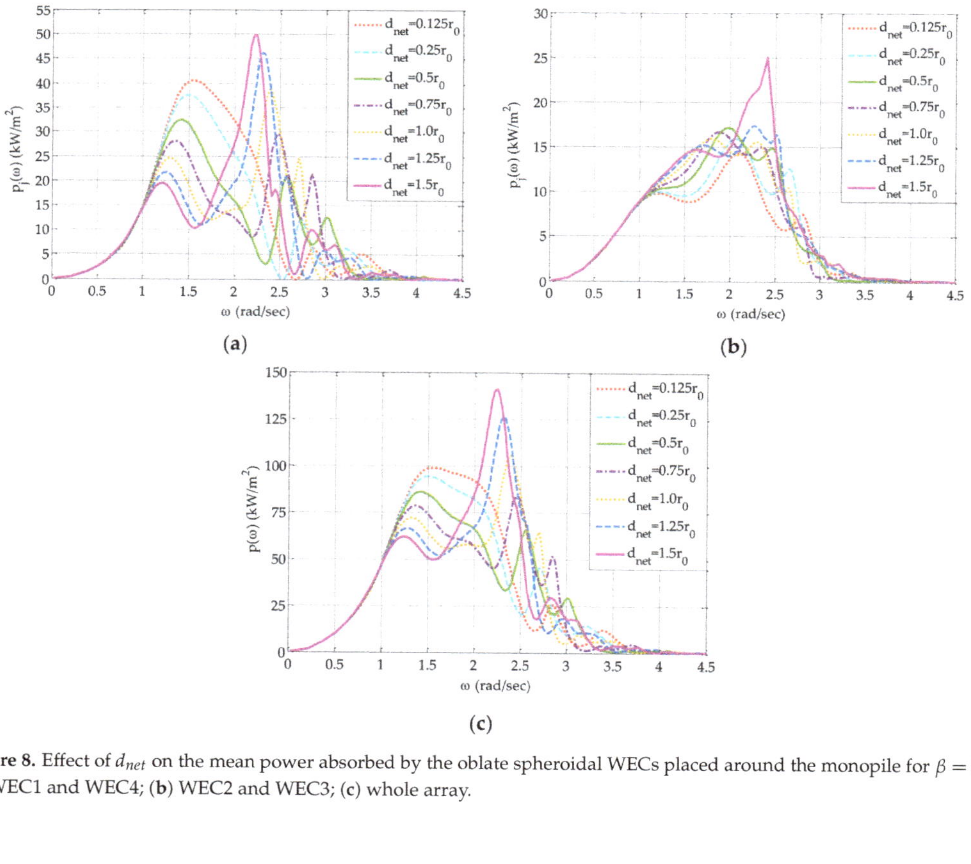

Figure 8. Effect of d_{net} on the mean power absorbed by the oblate spheroidal WECs placed around the monopile for $\beta = 0^o$: (**a**) WEC1 and WEC4; (**b**) WEC2 and WEC3; (**c**) whole array.

In order to illustrate more clearly the effect of the presence of the monopile on the power absorption ability of the oblate spheroidal WECs, we made a comparison of $p_j(\omega)$, $j = 1, \ldots, 4$, and $p(\omega)$ for $d_{net} = 0.125r_0$ with and without (i.e., isolated array) the monopile. The corresponding results are shown in Figure 9. For WEC1 and WEC4 situated in the seaward side of the monopile (Figure 9a), the existence of the latter structure positively affects their power absorption ability, since it leads to an $\approx$ 15% increase of the $p_j(\omega)$, $j = 1$ and 4, peak value, as well as to a much wider frequency range where adequate amount of power is absorbed. This is attributed to the existence of larger heave exciting forces of WEC1 and WEC4 at 1.0 rad/s $< \omega <$ 2.0 rad/s, due to the diffraction disturbances induced by the monopile, that, in turn, lead to larger heave responses of these WECs (relevant results are not included here due to space constraints). However, the shadow effects induced by the bottom-mounted cylinder reduce the power absorption ability of the WECs situated in the leeward side of the monopile (Figure 9b) at almost the whole examined frequency range. This in turn reduces at a small percentage ($\approx$ 8%) the peak value of the power absorbed by the whole array compared to the isolated one (Figure 9c). Nevertheless, the presence of the monopile enhances the power absorption ability of the whole array at a quite wide frequency range (i.e., 1.0 rad/s $< \omega <$ 1.9 rad/s).

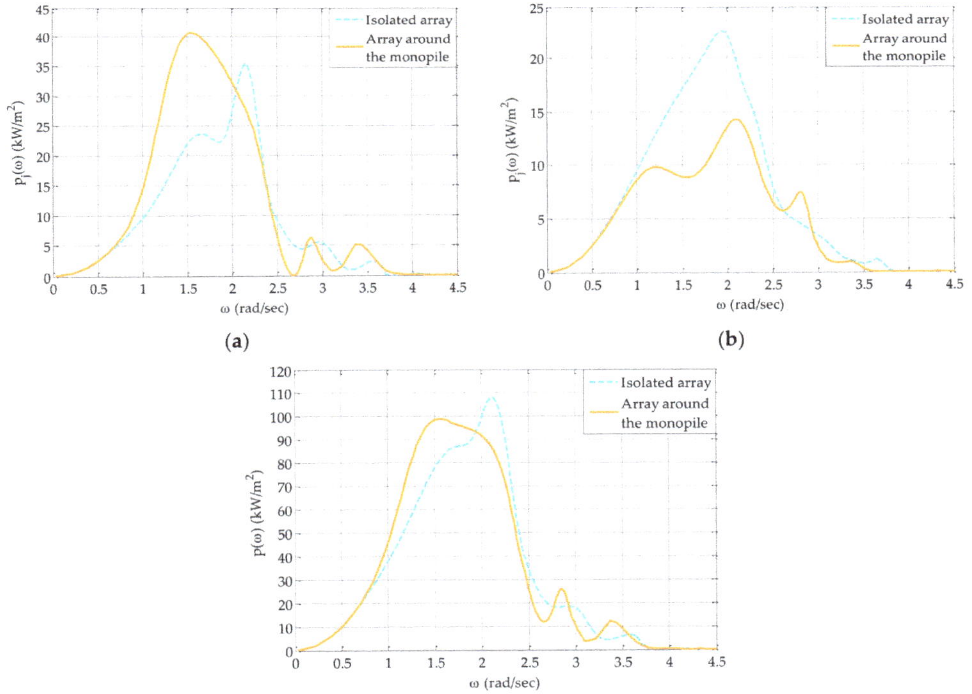

Figure 9. Comparisonof the mean power absorbed by the oblate spheroidal WECs with and without (isolated array) the monopile for $\beta = 0^{o}$ and $d_{net} = 0.125r_0$: (**a**) WEC1 and WEC4; (**b**) WEC2 and WEC3; (**c**) whole array.

Finally, it is interesting to demonstrate the effect of the presence of the oblate spheroidal WECs on the hydrodynamic loading applied on the monopile. This is realized with the aim of Figure 10, where the non-dimensional (in terms of $\rho g A r_0{}^2$) surge exciting forces applied on the monopile for all d_{net} values examined are compared with those applied on the isolated (i.e., without the WEC array) monopile. It is evident that the presence of the WECs, irrespective of their net radial distance from the monopile, affects the surge exciting force of the bottom-mounted cylinder at $\omega > 2.0$ rad/s, since in this frequency range, F_1 does not demonstrate a continuous smooth decrease as in the case of the isolated monopile. From a physical point of view, this can be related to the diffraction disturbances induced by the WECs, which are anticipated to be more pronounced at higher frequencies (i.e., for shorter wave lengths) due to WECs' geometrical characteristics.

4.2. Effect of Net Radial Distance on the Performance of the Array with Hemispherical-Shaped WECs

For the array consisting of hemispherical-shaped WECs, we focus on the effect of the net radial distance of the WECs from the monopile on the WECs' heave responses (Figure 11) and on the power absorbed by them (Figure 12). With regard to the heave exciting forces, similar conclusions can be drawn as in the case of the oblate spheroidal WECs and, thus, relevant results are not included due to space constraints.

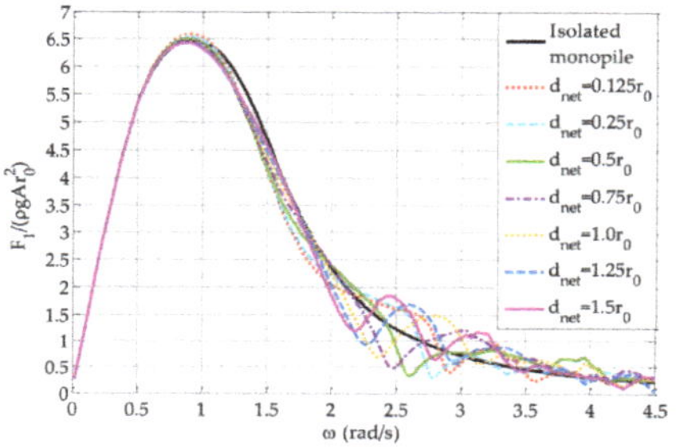

Figure 10. Comparisonof the non-dimensional surge exciting forces applied on the monopile with and without (isolated monopile) the oblate spheroidal WECs for $\beta = 0^o$.

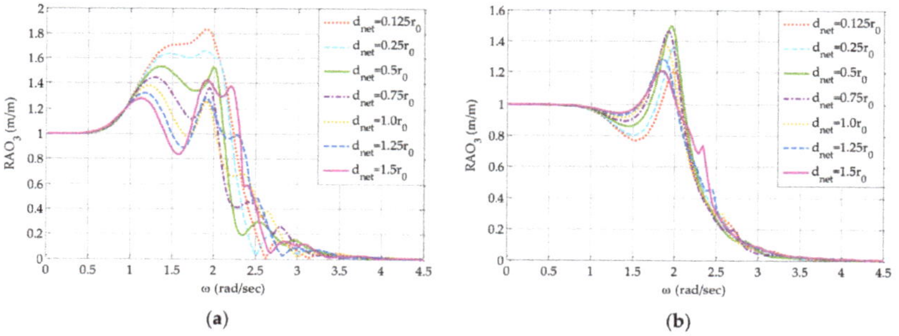

Figure 11. Effect of d_{net} on the heave response of the hemispherical-shaped WECs placed around the monopile for $\beta = 0^o$: (**a**) WEC1 and WEC4; (**b**) WEC2 and WEC3.

With regard to the effect of d_{net} on the WECs' heave responses (Figure 11), RAO_3 of WEC1 and WEC4 situated in the seaward side of the monopile (Figure 11a) obtains a peak at 1.7 rad/s $< \omega <$ 2.3 rad/s due to resonance phenomena for all d_{net} values examined. The largest peak value is observed for $d_{net} = 0.125r_0$, while the gradual increase of d_{net} leads successively to smaller RAO_3 peak values. For the arrays with $d_{net} \geq 0.5r_0$, a characteristic local minimum also occurs at 1.3 rad/s $< \omega <$ 1.9 rad/s in absolute accordance with the variation of the corresponding heave exciting forces. Finally, for $\omega > 2.3$ rad/s, RAO_3 for all d_{net} values shows an intense variation with continuously decreasing values towards higher frequencies. As for WEC2 and WEC3 situated in the leeward side of the monopile (Figure 11b), the change of d_{net} has a small effect on the values and the variation pattern of RAO_3. Compared to the oblate spheroidal WECs (Figure 7), the utilization of hemispherical-shaped WECs leads to larger RAO_3 peak values related to resonance phenomena. This is mainly attributed to the consideration of a smaller b_{PTO} value for the hemispherical-shaped WECs (Table 1), as a result of their intrinsic hydrodynamic characteristics (i.e., existence of smaller heave radiation damping coefficients).

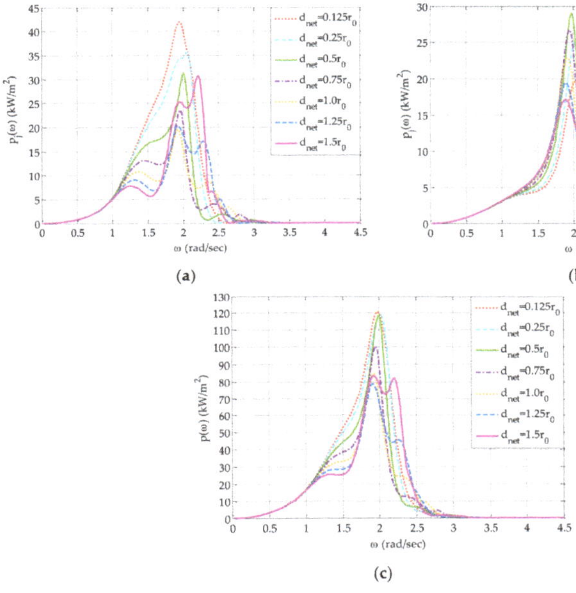

Figure 12. Effect of d_{net} on the mean power absorbed by the hemispherical-shaped WECs placed around the monopile for $\beta = 0^o$: (**a**) WEC1 and WEC4; (**b**) WEC2 and WEC3; (**c**) whole array.

As for the power absorption ability of the hemispherical-shaped WECs, Figure 12 shows the effect of d_{net} on the mean power absorbed by each WEC, $p_j(\omega)$ (Figure 12a,b), as well as on the mean power absorbed by the whole array, $p(\omega)$ (Figure 12c). In general, the variation of $p_j(\omega)$, $j = 1, \ldots, 4$, follows the variation pattern of the corresponding RAO_3^j (Figure 11). In the case of WEC1 and WEC4 (Figure 12a), it is evident that the placement of these WECs at the smallest examined net radial distance from the monopile (i.e., $d_{net} = 0.125r_0$) leads to the largest $p_j(\omega)$ peak, with a value approximately equal to 40 kW/m^2. By increasing d_{net} up to $1.0r_0$, a gradual decrease of the $p_j(\omega)$ peak values occurs, while the opposite trend is observed for the largest examined d_{net} values. Moreover, for $d_{net} > 0.5r_0$, an intense decrease of $p_j(\omega)$ occurs at 1.1 rad/s $< \omega <$ 1.8 rad/s. For the WECs situated in the leeward side of the monopile (Figure 12b) and contrary to WEC1 and WEC4, the increase of d_{net} up to $0.5r_0$ results successively in larger power peak values, while the opposite trend occurs for the rest examined d_{net} values. Hence, the maximum energy absorption for WEC2 and WEC3 is achieved for $d_{net} = 0.5r_0$, where the $p_j(\omega)$ peak is approximately equal to 30 kW/m^2. Moreover, for these two WECs, the frequency range, where adequate amount of power is absorbed, becomes narrower compared to the case of WEC1 and WEC4.

Regarding the power absorbed by the whole array (Figure 12c), the placement of the WECs at any of the three smallest examined net radial distances (i.e., $d_{net} \leq 0.5r_0$) does not introduce any significant differences on $p(\omega)$. Moreover, the aforementioned placement leads to arrays that show the best power absorption ability compared to the rest of the examined d_{net} cases, with a $p(\omega)$ peak value approximately equal to 120 kW/m^2. Taking all the above into account and aiming at forming arrays that enable an adequate power absorption ability for both the seaward and the leeward WECs, we chose $d_{net} = 0.25r_0$ as the most appropriate net radial distance for placing the hemispherical-shaped WECs around the monopile.

Finally, Figure 13 shows the effect of the presence of the monopile on the power absorption ability of the hemispherical-shaped WECs for the case of $d_{net} = 0.25r_0$. In the case of WEC1 and WEC4 (Figure 13a), the presence of the monopile in the leeward side of these WECs results in an $\approx 33\%$ decrease of the $p_j(\omega)$, $j = 1$ and 4, peak value; however, it enhances the power ability of WEC1 and WEC4 at 1.0 rad/s $< \omega <$ 1.9 rad/s. The shadow effects induced by the bottom-mounted cylinder reduce the power absorption ability of the WECs situated in the leeward side of the monopile (Figure 13b) at 1.1 rad/s $< \omega <$ 1.9 rad/s, while, moreover, they lead to an $\approx 11\%$ decrease of the $p_j(\omega)$, $j = 1$ and 4, peak value. As for the total power absorbed by the whole array (Figure 13c), it is evident that the presence of the monopile negatively affects the power absorption ability of the array, since it results in an $\approx 20\%$ decrease of the $p(\omega)$ peak value.

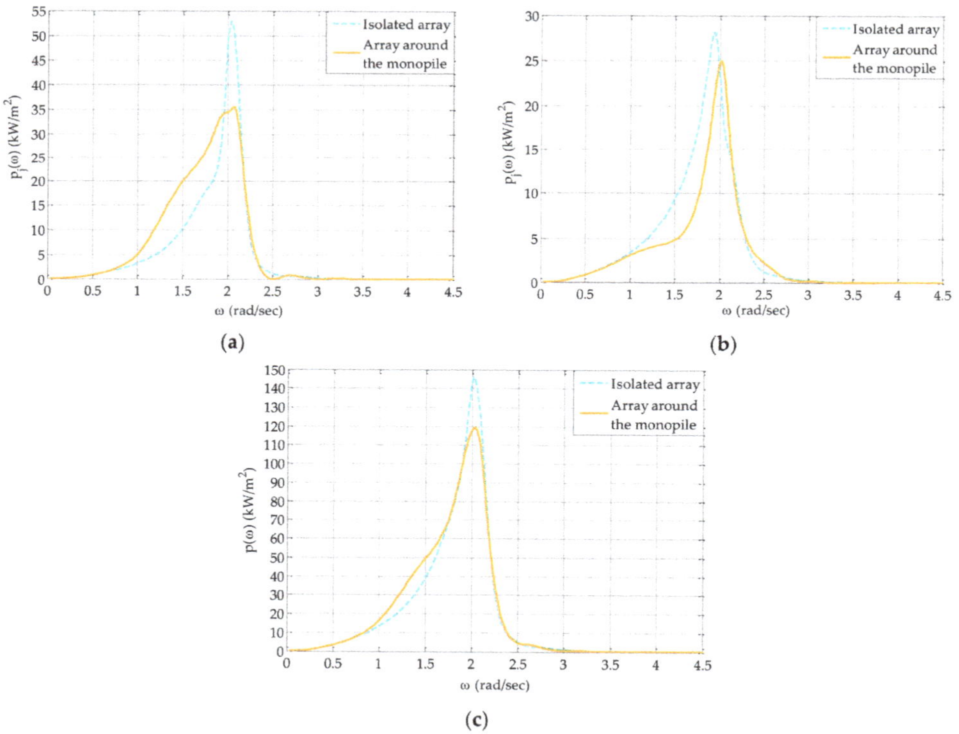

Figure 13. Comparisonof the mean power absorbed by the hemispherical-shaped WECs with and without (isolated array) the monopile for $\beta = 0^o$ and $d_{net} = 0.25r_0$: (**a**) WEC1 and WEC4; (**b**) WEC2 and WEC3; (**c**) whole array.

4.3. *Effect of the Incident Wave Direction on the Power Absorption Ability of the Array*

In this section, we examine the effect of the incident wave direction, β, on the mean power absorbed by the array consisting of oblate spheroidal WECs with $d_{net} = 0.125r_0$

and hemispherical-shaped WECs with $d_{net} = 0.25r_0$. It is recalled that for a given WEC geometry, the aforementioned d_{net} value corresponds to the best net radial distance for placing the WECs around the monopile in terms of power absorption ability. The corresponding results are shown in Figure 14, where the variation of $p(\omega)$ as a function of ω is presented for $\beta = 0^o$ (head waves), $\beta = 22.5^o$, and 45° (oblique waves). In the case of the oblate spheroidal WECs (Figure 14a), the increase of β enhances the power absorption ability of the array, while the exact opposite trend is observed for the array with the hemispherical-shaped WECs (Figure 14b). Hence, for the oblate spheroidal WECs, the action of oblique waves with $\beta = 45^o$ or "equivalently" the orientation of the array in a manner that enables only one of the WECs to be situated in the seaward side of the monopile leads to the best power absorption ability. On the other hand, the array consisting of hemispherical-shaped WECs perform better under the action of head incident waves (i.e., $\beta = 0^o$) or "equivalently" when two WECs of the array are situated in the seaward side of the monopile.

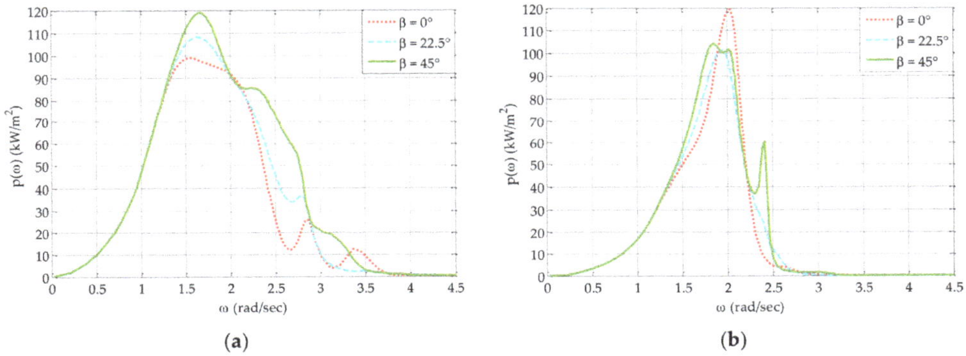

Figure 14. Effect of β on the mean power absorbed by the array with (**a**) oblate spheroidal WECs and $d_{net} = 0.125r_0$; (**b**) hemispherical-shaped WECs and $d_{net} = 0.25r_0$.

In addition to the above and aiming at determining the array configuration (WECs' geometry and net radial distance) that shows the best power absorption ability, we compare the power absorbed by the two aforementioned arrays (oblate spheroidal WECs with $d_{net} = 0.125r_0$ and hemispherical-shaped WECs with $d_{net} = 0.25r_0$) for a given incident wave direction. The relevant comparison is made with the aid of Figure 15. Under the action of waves with $\beta = 0^o$ (Figure 15a), the array consisting of oblate spheroidal WECs shows a smaller $p(\omega)$ peak value compared to the array with the hemispherical-shaped WECs. However, by deploying oblate spheroidal WECs, the frequency range, where adequate amount of power is absorbed, becomes much wider. Indicatively, the oblate spheroidal WECs absorb power larger than 80 kW/m^2 at 1.3 rad/s $< \omega <$ 2.2 rad/s, which is a frequency range almost three times wider compared to the corresponding one (1.8 rad/s $< \omega <$ 2.1 rad/s) observed for the hemispherical-shaped WECs. Under the action of oblique waves (Figure 15b,c), it is evident that the utilization of oblate spheroidal WECs around the monopile enhances the power absorption ability of the array, since compared to the hemispherical-shaped WECs, adequate power is absorbed at much wider frequency ranges, while, moreover, larger $p(\omega)$ peak values are observed. From a physical point of view, this trend can be attributed to a better contribution of WEC4 and WEC3 (Figure 4) to the array's power absorption ability in the case of the oblate spheroidal WECs. More specifically, under the action of oblique waves, the hydrodynamic interactions among the hemispherical-shaped WECs and between these WECs and the monopile affect the heave exciting forces and the responses of WEC4 in a manner that significantly reduces its power absorption ability (relevant results are not cited here due to space constraints). Hence, the contribution of WEC4 to the power absorption ability of the array is smaller in the case of the hemispherical-shaped WECs compared to the oblate spheroidal ones. Furthermore, it is evident that by increasing β, the power absorption ability of WEC3 increases for both examined geometries, since shadow effects induced by the monopile are mainly observed for WEC2. However, as in the case of $\beta = 0^o$, for the oblate spheroidal device, the frequency range, where adequate power is absorbed, is much wider compared to the hemispherical-shaped WEC.

Taking into account all the above, the array with the oblate spheroidal WECs situated at $d_{net} = 0.125r_0$ around the monopile could be considered as the most efficient array configuration in terms of power absorption ability among all cases examined in the present paper.

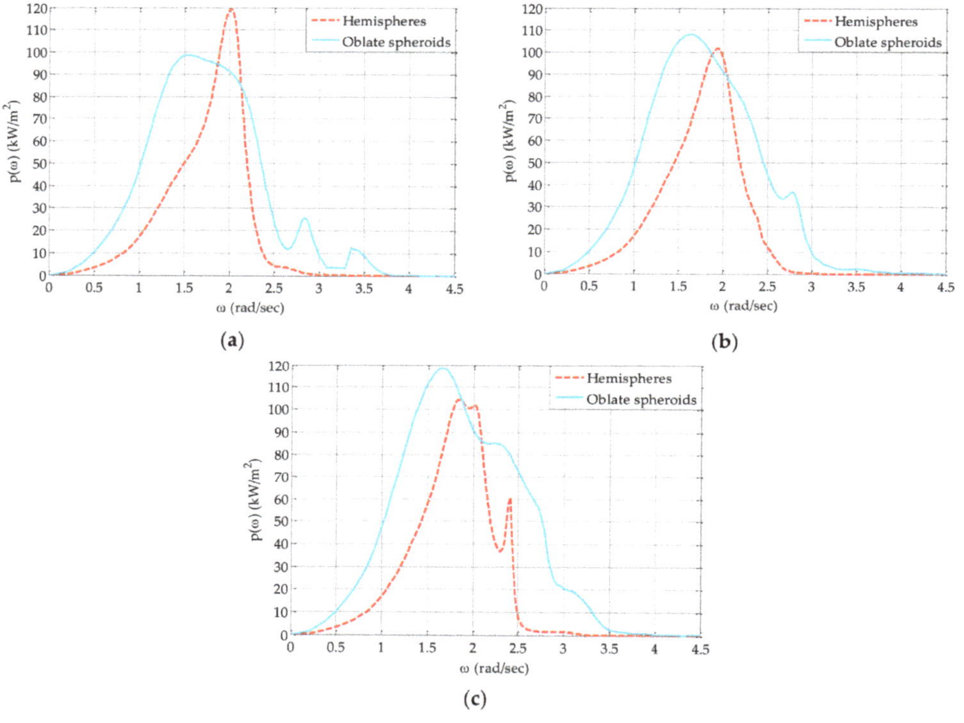

Figure 15. Comparisonof the mean power absorbed by the array with oblate spheroidal ($d_{net} = 0.125r_0$) and hemispherical-shaped ($d_{net} = 0.25r_0$) WECs for (**a**) $\beta = 0^o$; (**b**) $\beta = 22.5^o$; (**c**) $\beta = 45^o$.

4.4. Power Absorption Ability of the Most Efficient Array Configuration by Utilizing an "Active" PTO Mechanism

In all previous sections, the assessment of the performance of the array for different WEC geometries, d_{net} and β values, along with the determination of the most efficient array configuration (WECs' geometry and net radial distance) were implemented by deploying a linear PTO mechanism of constant damping coefficient, b_{PTO}, appropriately tuned in terms of maximizing energy absorption at the natural frequency of the corresponding single, isolated WEC (i.e., $b_{PTO} = B_{33}^{ISO}(\omega = \omega_{n3}^{ISO})$). It is evident, however, that the power absorption ability of an array configuration could also be enhanced by adequately exploiting the characteristics of the PTO mechanism itself. Motivated by this, in the present section, we investigate and assess the power absorption ability of the most efficient array configuration defined in Section 4.3 (i.e., array with oblate spheroidal

WECs and $d_{net} = 0.125r_0$) by utilizing a PTO mechanism (hereafter called "active" PTO mechanism) of variable with frequency damping coefficient, $b_{PTO}(\omega)$, which enables the appropriate adjustment of its damping coefficient along the examined frequency range. It is noted that the technology of an "active" damper is widely deployed in the automotive industry, where the relevant contemporary technological advances facilitate the adjustment of the damping coefficient within fractions of a second (e.g., [37–39]).

In order to determine the $b_{PTO}(\omega)$ values of the "active" PTO mechanism along the examined frequency range, we utilize the envelope of a family of power absorption curves obtained for the case of the isolated oblate spheroidal WEC for various constant b_{PTO} values. The corresponding results are shown in Figure 16a, where $p(\omega)$ for the isolated WEC is plotted for nine constant b_{PTO} values equal to $qB_{33}^{ISO}(\omega = \omega_{n3}^{ISO})$, with q varying between 1.0 and 5.0 with a step of 0.5 (Table 3). As expected, the consideration of $b_{PTO} = B_{33}^{ISO}(\omega = \omega_{n3}^{ISO})$ results in the maximization of $p(\omega)$ at $\omega = \omega_{n3}^{ISO} = 2.282$ rad/s. However, the gradual increase of the aforementioned b_{PTO} value successively enhances the power absorption ability of the isolated WEC at lower wave frequencies, since for $b_{PTO} = qB_{33}^{ISO}(\omega = \omega_{n3}^{ISO})$ and $q \neq 1.0$, larger $p(\omega)$ peaks compared to the case of $b_{PTO} = B_{33}^{ISO}(\omega = \omega_{n3}^{ISO})$ occur at $\omega < 2.282$ rad/s. For example, by deploying a b_{PTO} value equal to $5B_{33}^{ISO}(\omega = \omega_{n3}^{ISO})$, the $p(\omega)$ peak is increased by approximately 40% and it occurs at $\omega = 1.2$ rad/s. It is noted that the b_{PTO} values of Table 3 are within the limits of commercial dampers existing nowadays in the market (e.g., [40]), while, moreover, values larger than $5B_{33}^{ISO}(\omega = \omega_{n3}^{ISO})$ were not taken into account, since they did not lead to any significant increase of the WEC's power absorption ability.

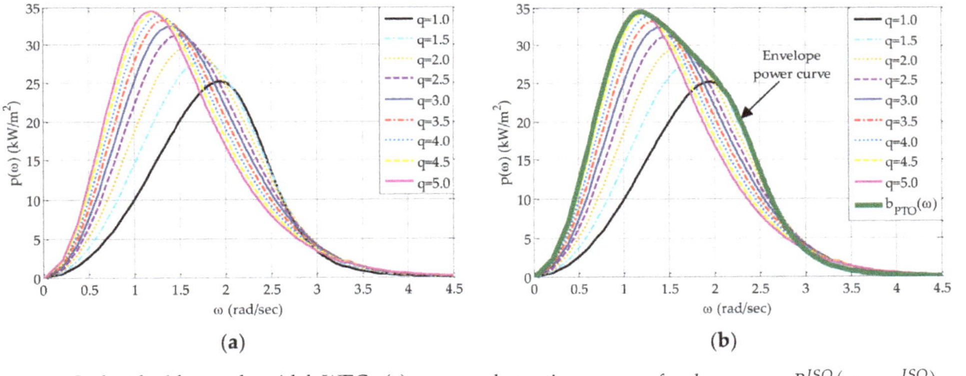

Figure 16. Isolated oblate spheroidal WEC: (**a**) power absorption curves for $b_{PTO} = qB_{33}^{ISO}(\omega = \omega_{n3}^{ISO})$, q = 1.0, 1.5, 2.0, 2.5, 3.0, 3.5, 4.0, 4.5, and 5.0; (**b**) envelope power curve representing the utilization of an "active" PTO mechanism with $b_{PTO}(\omega)$.

Table 3. Values of $b_{PTO} = qB_{33}^{ISO}(\omega = \omega_{n3}^{ISO})$ used for determining $b_{PTO}(\omega)$ for the "active" PTO mechanism.

q	b_{PTO} **(kNs/m)**
1.0	20.615
1.5	30.923
2.0	41.230
2.5	51.538
3.0	61.845
3.5	72.153
4.0	82.460
4.5	92.768
5.0	103.075

Having obtained the family of the nine power curves, we determined their envelope (green curve in Figure 16b). This envelope curve represents the power absorbed by the isolated oblate spheroidal WEC in the case of an "active" PTO mechanism, where the variable with frequency b_{PTO} (i.e., $b_{PTO}(\omega)$) is appropriately tuned in terms of maximizing energy absorption at a given incident wave frequency or at a given frequency range. The advantage of this "active" PTO mechanism is twofold. Firstly, it gives the potential to absorb adequate amount of power at a very wide frequency range. For example, by utilizing the "active" PTO mechanism, the power absorption ability of the oblate spheroidal WEC becomes larger than 25 $\mathrm{kW/m^2}$ at 0.75 rad/s $< \omega <$ 2.1 rad/s, contrary to any of the nine cases, where a constant b_{PTO} was considered. Secondly, it leads to the occurrence of the $p(\omega)$ peak at lower wave frequencies, facilitating the efficient employment of this WEC and, thus, of the proposed hybrid offshore wind and wave energy exploitation system at marine areas characterized by low-frequency sea waves.

Having determined the $b_{PTO}(\omega)$ values of the "active" PTO mechanism along the examined frequency range, we applied this mechanism for the most efficient array configuration consisting of oblate spheroidal WECs situated at $d_{net} = 0.125r_0$ around the monopile in order to assess the array's power absorption ability under both regular and irregular waves of incident wave direction $\beta = 45^o$. For regular waves, the corresponding results are shown in Figure 17, where the power absorbed by the whole array in the case of the "active" PTO mechanism (i.e., utilization of $b_{PTO}(\omega)$) is compared with the power absorbed for a PTO mechanism of constant b_{PTO} equal to $B_{33}^{ISO}(\omega = \omega_{n3}^{ISO})$. It is evident that the utilization of the "active" PTO mechanism leads to a significant improvement of the power absorption ability of the array, since it leads to an $\approx 50\%$ increase of the $p(\omega)$ peak and facilitates adequate power absorption at low wave frequencies.

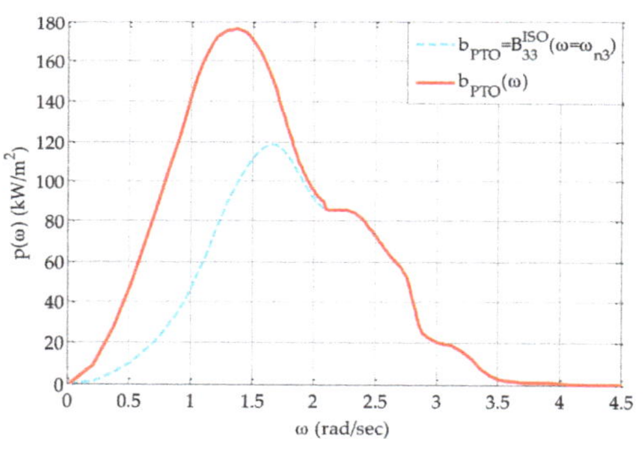

Figure 17. Comparisonof the mean power absorbed by the array with oblate spheroidal WECs and $d_{net} = 0.125r_0$ under the action of regular waves with $\beta = 45^o$ by utilizing (**a**) a PTO with constant b_{PTO} equal to $B_{33}^{ISO}(\omega = \omega_{n3}^{ISO})$; (**b**) an "active" PTO with $b_{PTO}(\omega)$.

Under the action of irregular waves, for a given sea state described by a spectrum with significant wave height, H_s, and peak period, T_p, we obtain the power absorbed by the array, $p(H_s, T_p)$, using the following equation:

$$p(H_s, T_p) = \sum_{j=1}^{4} p_j(H_s, T_p) \tag{14}$$

where $p_j(H_s, T_p)$, $j = 1, \ldots, 4$, is the power absorbed by the jth WEC, calculated as follows:

$$p_j(H_s, T_p) = \int_0^{\infty} S_{JONSWAP}(\omega | H_s, T_p) p_j(\omega) d\omega \tag{15}$$

In Equation (14), $S_{JONSWAP}(\omega|H_s, T_p)$ is the spectral density of the JONSWAP spectrum, which is deployed in the present paper for describing the incident waves, while the symbol " | " is used to denote given values of H_s and T_p. The spectral density $S_{JONSWAP}(\omega|H_s, T_p)$ is obtained by applying the following equation [41]:

$$S_{JONSWAP}(\omega|H_s,\ T_p) = [1 - 0.287\ln(\gamma)]\frac{5}{16}H_s^2\omega_p^4\omega^{-5}\exp\left(-\frac{5}{4}\left(\frac{\omega}{\omega_p}\right)^{-4}\right)\gamma^{\exp\left(-0.5\left(\frac{\omega-\omega_p}{\sigma\omega_p}\right)^2\right)} \quad (16)$$

where $\omega_p = 2\pi/T_p$, γ is the non-dimensional peak shape parameter equal to 3.3 and σ is the spectral width parameter equal to 0.07 for $\omega \leq \omega_p$ and 0.09 for $\omega > \omega_p$. It is noted that in Equation (15), $p_j(\omega)$ for each frequency component of the spectrum is calculated using Equation (13), which corresponds to incident waves of unit amplitude.

The power absorption ability of the most efficient array configuration is assessed for 225 sea states (i.e., H_s and T_p combinations), with H_s varying between 0.5 m and 7.5 m with a step of 0.5 m and T_p between 3.0 s and 17 s with a step equal to 1.0 s. The corresponding results are shown in Figure 18, where the contours of p for the aforementioned sea states are plotted for the array with and without the utilization of the "active" PTO mechanism. In the latter case (Figure 18a), the maximum value of p corresponds to 340 kW (H_s = 7.5 m and T_p = 4.0 s), while by utilizing the "active" PTO mechanism (Figure 18b), the maximum power absorbed by the array becomes equal to 535 kW (H_s = 7.5 m and T_p = 4.0 s), representing an $\approx$ 60% increase. One more important aspect to note is that the power iso-contours for the array with the "active" PTO mechanism are much wider along the T_p-axis compared to the array, where the PTO mechanism has a constant b_{PTO} value. This in turn illustrates that the array with the "active" PTO mechanism shows a better power absorption ability for a larger number of sea states. Even for small H_s values, the power absorption ability of the array is still maintained at adequate levels along the examined T_p range.

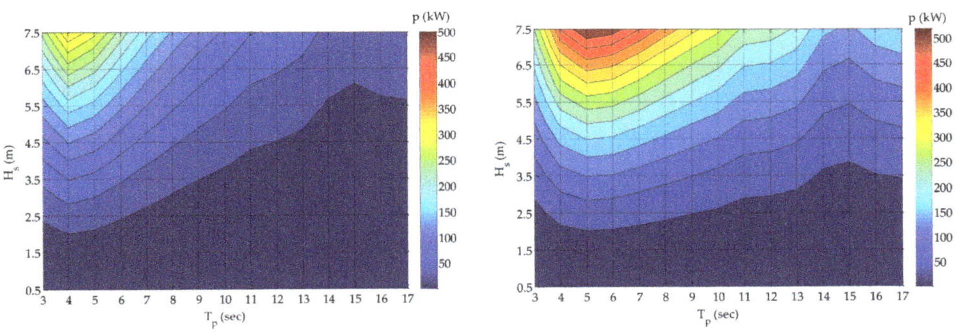

Figure 18. Comparisonof power absorbed by the array the oblate spheroidal WECs and $d_{net} = 0.125r_0$ under the action of irregular wave with $\beta = 45°$ by utilizing (**a**) a PTO with constant b_{PTO} equal to $B_{33}^{ISO}(\omega = \omega_{n3}^{ISO})$; (**b**) an "active" PTO with $b_{PTO}(\omega)$.

5. Conclusions

In this paper, a frequency-based numerical analysis was implemented to investigate the performance (hydrodynamic behavior and power absorption) of a circular array of four semi-immersed heaving WECs distributed uniformly around a hybrid wind–wave monopile support structure. Arrays consisting of oblate spheroidal and hemispherical-shaped WECs were examined and compared, while focus was given to the effect of characteristic design parameters (WECs' net radial distance from the monopile, incident wave direction, and PTO characteristics) on the power absorption ability of the array.

The main conclusions of the present investigation for the specific WECs' geometrical characteristics and water depth conditions considered can be summarized as follows:

- With regard to the effect of the net radial distance from the monopile on the array's power absorption ability, this effect depends strongly upon the examined frequency range for the array consisting of oblate spheroidal WECs. Specifically, at low wave frequencies, the aforementioned array shows the best power absorption ability by placing the WECs at the smallest examined net radial distance ($d_{net} = 0.125r_0$), while at higher wave frequencies, the exact opposite trend is observed, namely, the best power absorption ability occurs for the array placed at the largest examined net radial distance ($d_{net} = 1.5r_0$). This trend advocates the potential efficient utilization of the oblate spheroidal WECs at marine areas characterized by either low-frequency or high-frequency sea waves. It is noted, however, that for $d_{net} = 0.125r_0$, the frequency range where adequate amount of power is absorbed becomes quite wide. This feature combined with constructability issues as well as with the fact that most real sea states are characterized by low-frequency sea waves leads to the selection of $d_{net} = 0.125r_0$ as the most appropriate net radial distance for placing the array around the monopile. On the other hand, the array consisting of hemispherical-shaped WECs shows the best power absorption ability when the WECs are situated at small net radial distances from the monopile (i.e., for $d_{net} = 0.125r_0, 0.25r_0$, and $0.5r_0$). The value of $d_{net} = 0.25r_0$ is considered the most appropriate net radial distance for placing the array, since it facilitates adequate power absorption ability for both the seaward and the leeward WECs.
- The power absorption ability of the array with oblate spheroidal WECs and $d_{net} = 0.125r_0$ is greatly improved under the action of oblique waves with $\beta = 45^o$, while the opposite trend is observed for the array with hemispherical-shaped WECs and $d_{net} = 0.25r_0$, where the best power absorption ability occurs for head waves (i.e., $\beta = 0^o$).
- The deployment of oblate spheroidal WECs around the monopile enhances the power absorption ability of the array, since compared to the hemispherical-shaped WECs, adequate power is absorbed at much wider frequency ranges, while, moreover, larger power peak values may be observed depending upon the incident wave direction. Hence, the array with the oblate spheroidal WECs situated at $d_{net} = 0.125r_0$ around the monopile presents the most efficient array configuration (WECs' geometry and net radial distance) in terms of power absorption ability.
- The utilization of an "active" PTO mechanism, facilitating the consideration of a variable with frequency PTO damping coefficient, $b_{PTO}(\omega)$, that maximizes energy absorption at a given incident wave frequency or at a given frequency range, significantly enhances the power absorption ability of the array compared to the deployment of a PTO mechanism with a constant b_{PTO}. For the most efficient array configuration (i.e., oblate spheroidal WECs with $d_{net} = 0.125r_0$), the deployment of this "active" PTO mechanism under the action of oblique regular waves results in an $\approx 50\%$ increase of the power peak, as well as to a wider frequency range, extending mainly towards lower frequencies, where adequate power absorption is achieved. Positive conclusions are also derived in the case of irregular waves, since by utilizing the "active" PTO mechanism, an $\approx 60\%$ increase of the maximum power absorbed by the array occurs while, moreover, the array configuration shows a better power absorption ability for a larger number of sea states.

The present investigation can be extended by appropriately coupling the hydrodynamic model with an optimization algorithm, aiming at defining the optimum in terms of power absorption maximization and array configuration (WECs' geometry and net radial distance) while utilizing an "active" PTO mechanism. Recalling that the WECs along with the monopile present components of a bottom-mounted hybrid offshore wind and wave energy exploitation system, it would also be interesting to analyze in time domain the performance of the whole system and assess the effect of the existence of the WECs on the structural integrity of the monopile.

Author Contributions: Conceptualization, E.L., S.G., and P.C.; methodology, S.G., P.C., and E.L.; software, S.G. and P.C.; formal analysis, S.G. and P.C.; investigation, S.G., P.C., and E.L.; writing—original draft preparation, S.G. and P.C.; writing—review and editing, E.L.; supervision, E.L. All authors have read and agreed to the published version of the manuscript.

Funding: This research received no external funding.

Data Availability Statement: The data presented in this study are available on request from the corresponding author.

Conflicts of Interest: The authors declare no conflict of interest.

References

1. Bahaj, A.S. Generating electricity from the ocean. *Renew. Sustain. Energy Rev.* **2011**, *15*, 3399–3416. [CrossRef]
2. Iglesias, G.; Carballo, R. Wave energy potential along the Death Coast (Spain). *Energy* **2009**, *34*, 1963–1975. [CrossRef]
3. Oceans of Energy—European Ocean Energy Roadmap 2010–2050. Available online: https://www.icoe-conference.com/publication/oceans_of_energy_european_ocean_energy_roadmap_2010_2050/ (accessed on 2 November 2020).
4. Perez-Collazo, C.; Astariz, S.; Abanades, J.; Greaves, D.; Iglesias, G. Co-located Wave and Offshore Wind Farms: A Preliminary Case Study of a Hybrid Array. In Proceedings of the 34th International Conference in Coastal Engineering, Seoul, Korea, 16–20 June 2014; Lynett, P., Ed.; Curran Associates Inc.: New York, NY, USA, 2016; p. 34. [CrossRef]
5. Astariz, S.; Perez-Collazo, C.; Abanades, J.; Iglesias, G. Towards the optimal design of a co-located wind-wave farm. *Energy* **2015**, *84*, 15–25. [CrossRef]
6. Astariz, S.; Iglesias, G. Enhancing wave energy competitiveness through co-located wind and wave energy farms. A review on the shadow effect. *Energies* **2015**, *8*, 7344–7366. [CrossRef]
7. Pérez-Collazo, C.; Jakobsen, M.M.; Buckland, H.; Fernández-Chozas, J. Synergies for a Wave-Wind Energy Concept. In Proceedings of the European Offshore Wind Energy Conference, Frankfurt, Germany, 19–21 November 2013; The European Wind Energy Association: Brussels, Belgium, 2013.
8. Jeffrey, H.; Sedgwick, J. ORECCA European Offshore Renewable Energy Roadmap. Available online: http://orecca.rse-web.it/doc_info/ORECCA_EU_offshore_Renewable_Energy_Roadmap.pdf (accessed on 2 November 2020).
9. Karimirad, M. *Offshore Energy Structures*; Springer International Publishing: Cham, Switzerland, 2014.
10. Colin, W. (Ed.) Offshore Wind in Europe—Key Trends and Statistics 2019. 2020. Available online: https://windeurope.org/wp-content/uploads/files/about-wind/statistics/WindEurope-Annual-Offshore-Statistics-2019.pdf (accessed on 2 November 2020).
11. Lynch, K.; Murphy, J. *Overview of Offshore Wind and Ocean Energy Technologies*; Public Report of Marine Renewable Integrated Application (MARINA) Platform Project: Cork, Ireland, 2012.
12. Michailides, C.; Gao, Z.; Moan, T. Experimental and numerical study of the response of the offshore combined wind/wave energy concept SFC in extreme environmental conditions. *Mar. Struct.* **2016**, *50*, 35–54. [CrossRef]
13. Michailides, C.; Gao, Z.; Moan, T. Experimental study of the functionality of a semisubmersible wind turbine combined with flap-type wave energy converters. *Renew. Energy* **2016**, *93*, 675–690. [CrossRef]
14. Michailides, C.; Luan, C.; Gao, Z.; Moan, T. Effect of Flap Type Wave Energy Converters on the Response of a Semi-Submersible Wind Turbine in Operational Conditions. In Proceedings of the 33rd International Conference on Ocean, Offshore and Arctic Engineering, San Francisco, CA, USA, 8–16 June 2014; p. OMAE2014-24065. [CrossRef]
15. Muliawan, M.J.; Karimirad, M.; Moan, T. Dynamic response and power performance of a combined Spar-type floating wind turbine and coaxial floating wave energy converter. *Renew. Energy* **2013**, *50*, 47–57. [CrossRef]
16. Muliawan, M.J.; Karimirad, M.; Gao, Z.; Moan, T. Extreme responses of a combined spar-type floating wind turbine and floating wave energy converter (STC) system with survival modes. *Ocean Eng.* **2013**, *65*, 71–82. [CrossRef]
17. Gao, Z.; Moan, T.; Wan, L.; Michailides, C. Comparative numerical and experimental study of two combined wind and wave energy concepts. *J. Ocean Eng. Sci.* **2016**, *1*, 36–51. [CrossRef]
18. Wang, Y.; Zhang, L.; Michailides, C.; Wan, L.; Shi, W. Hydrodynamic response of a combined wind-wave marine energy structure. *J. Mar. Sci. Eng.* **2020**, *8*, 253. [CrossRef]
19. Bachynski, E.E.; Moan, T. Point Absorber Design for a Combined Wind and Wave Energy Converter on a Tension-Leg Support Structure. In Proceedings of the 32nd International Conference on Ocean, Offshore and Arctic Engineering, Nantes, France, 9–14 June 2013; p. OMAE2013-10429. [CrossRef]
20. Karimirad, M.; Koushan, K. WindWEC: Combining Wind and Wave Energy Inspired by Hywind and Wavestar. In Proceedings of the 2016 IEEE International Conference on Renewable Energy Research and Application, Birmingham, UK, 20–23 November 2016; pp. 96–101. [CrossRef]

21. Karimirad, M.; Michailides, C. Effects of Structural Design Parameters on the Hydrodynamic Interaction and Response of the Combined WindWEC Concept. In Proceedings of the 12th European Wave and Tidal Energy Conference, Cork, Ireland, 27 August–1 September 2017.
22. Karimirad, M.; Michailides, C. Effects of Misaligned Wave and Wind Action on the Response of the Combined Concept WindWEC. In Proceedings of the 37th International Conference on Ocean, Offshore and Arctic Engineering, Madrid, Spain, 17–22 June 2018; p. OMAE2018-77078. [CrossRef]
23. Mazarakos, T.; Konispoliatis, D.; Katsaounis, G.; Polyzos, S.; Manolas, D.; Voutsinas, S.; Soukissian, T.; Mavrakos, S.A. Numerical and experimental studies of a multi—purpose floating TLP structure for combined wind and wave energy exploitation. *J. Mediterr. Mar. Sci.* **2019**, *20*, 745–763. [CrossRef]
24. Lee, H.; Poguluri, S.K.; Bae, Y.H. Performance analysis of multiple wave energy converters placed on a floating platform in the frequency domain. *Energies* **2018**, *11*, 406. [CrossRef]
25. Perez-Collazo, C.; Greaves, D.; Iglesias, G. A novel hybrid wind-wave energy converter for jacket-frame substructures. *Energies* **2018**, *11*, 637. [CrossRef]
26. Perez-Collazo, C.; Pemberton, R.; Greaves, D.; Iglesias, G. Monopile-mounted wave energy converter for a hybrid wind-wave system. *Energy Convers. Manag.* **2019**, *199*, 111971. [CrossRef]
27. Ren, N.; Ma, Z.; Fan, T.; Zhai, G.; Ou, J. Experimental and numerical study of hydrodynamic responses of a new combined monopile wind turbine and a heave type wave energy converter under typical operational conditions. *Ocean Eng.* **2018**, *159*, 1–8. [CrossRef]
28. Power Technology. Available online: https://www.power-technology.com/projects/greenoceanenergywav/ (accessed on 2 November 2020).
29. Wave Star. Available online: http://wavestarenergy.com/ (accessed on 2 November 2020).
30. Miquel, A.M.; Antonini, A.; Archetti, R.; Bozzi, S.; Lamberti, A. Non-linear modelling of a heaving point absorber: The surge effect. *Int. J. Mar. Energy* **2017**, *19*, 95–109. [CrossRef]
31. Bozzi, S.; Miquel, A.M.; Scarpa, F.; Antonini, A.; Archetti, R.; Passoni, G.; Gruosso, G. Wave energy production in Italian offshore: Preliminary design of a point absorber with tubular linear generator. In Proceedings of the 4th International Conference on Clean Electrical Power: Renewable Energy Resources Impact, Alghero, Italy, 11–13 June 2013; pp. 203–208. [CrossRef]
32. Gaeta, M.G.; Segurini, G.; Moreno, A.M.; Archetti, R. Implementation and validation of a potential model for a moored floating cylinder under waves. *J. Mar. Sci. Eng.* **2020**, *8*, 131. [CrossRef]
33. WAMIT Theory Manual. Available online: https://www.wamit.com/Publications/tmanual.pdf (accessed on 2 November 2020).
34. Lee, C.H.; Newman, J.N. Computation of wave effects using the panel method. In *Numerical Models in Fluid-Structure Interaction*; Chakrabarti, S., Ed.; WIT Press: Southampton, UK, 2005; pp. 211–251.
35. Falnes, J.; Kurniawan, A. *Ocean Waves and Oscillating Systems: Linear Interactions Including Wave-Energy Extraction*, 2nd ed.; Cambridge University Press: Cambridge, UK, 2020; pp. 20–23.
36. Beels, C.; Henriques, J.C.C.; De Rouck, J.; Pontes, M.T.; De Backer, G.; Verhaeghe, H. Wave Energy Resource in the North Sea. In Proceedings of the 7th European Wave and Tidal Energy Conference, Porto, Portugal, 11–13 September 2007.
37. Acura, Active Damper System. Available online: https://www.acura.com/mdx/modals/active-damper-system#:~{}:text=The%20Active%20Damper%20System%20can,road%20requires%20high%2Dperformance%20handling (accessed on 2 November 2020).
38. KONI, KONI ACTIVE Technology. Available online: https://www.koni.com/en-US/Cars/Technology/ACTIVE_Technology/ (accessed on 2 November 2020).
39. General Kinetics. Active Shock Brand Semi Active Dampers. Available online: http://www.generalkinetics.com/products/military/active-shock-semi-active-dampers.php (accessed on 2 November 2020).
40. ENDINE. Available online: https://www.itt-infrastructure.com/en-US/Home (accessed on 2 November 2020).
41. Det Norske Veritas–Germanischer Lloyds (DNV–GL). *Environmental Conditions and Environmental Loads, Recommended Practice DNVGL-RP-C205*; DNV-GL: Oslo, Norway, 2017.

MDPI AG
Grosspeteranlage 5
4052 Basel
Switzerland
Tel.: +41 61 683 77 34

Journal of Marine Science and Engineering Editorial Office
E-mail: jmse@mdpi.com
www.mdpi.com/journal/jmse

www.ingramcontent.com/pod-product-compliance
Lightning Source LLC
Chambersburg PA
CBHW061828210926
51850CB00006B/833
9783725865161